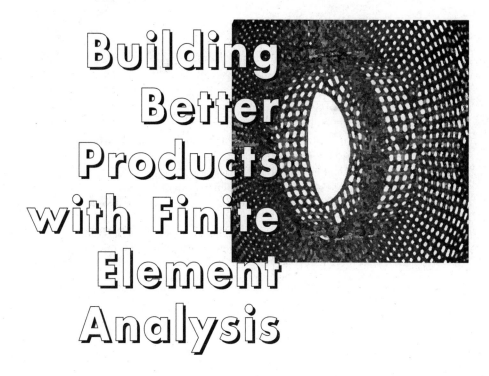

Building Better Products with Finite Element Analysis

Vince Adams and Abraham Askenazi

ONWORD® PRESS

Building Better Products with Finite Element Analysis

Vince Adams and Abraham Askenazi

Published by:

OnWord Press
2530 Camino Entrada
Santa Fe, NM 87505-4835 USA

Carol Leyba, Publisher

David Talbot, Acquisitions Director

Barbara Kohl, Associate Editor

Daril Bentley, Senior Editor

Andy Lowenthal, Director of Production and Manufacturing

Cynthia Welch, Production Manager

Liz Bennie, Director of Marketing

Lynne Egensteiner, Cover Illustration and Design

Michael Kline, Indexer

Copyright © OnWord Press

First edition, 1999

SAN 694-0269

10 9 8 7 6 5

Printed in the United States of America

Library of Congress Cataloging-in-Publication Data

Adams, Vince, 1963-
 Building Better Products with finite element analysis / Vince Adams and Abraham Askenazi.
   p.   cm.
  Includes index.
  ISBN 1-56690-160X
  1. Finite element method. I. Askenazi, Abraham, 1969-   . II. Title.
TA347.F5A25   1998
620'.001'51535–dc21

98-28578
CIP

Trademarks

OnWord Press is a registered trademark of High Mountain Press, Inc. All other terms mentioned in this book that are known to be trademarks or service marks have been appropriately capitalized. OnWord Press cannot attest to the accuracy of this information. Use of a term in this book should not be regarded as affecting the validity of any trademark or service mark.

Warning and Disclaimer

This book provides information on the current state of finite element analysis technology, underlying theory, and its use in engineering design. Every effort has been made to make the book as complete, accurate, and up to date as possible; however, no warranty or fitness is implied.

The information is provided on an "as is" basis. The authors and OnWord Press shall have neither liability nor responsibility to any person or entity with respect to any loss or damages in connection with or arising from the information contained in this book.

Acknowledgments

I would like to thank Catherine Campbell for her tireless and caring aid, and her willingness to do anything to help keep the book on track; Mike Koch, Chris Vasiliotis, Randy Andracki, and Rick Fischer for their technical contributions and support throughout; WyzeTek, Inc. and my clients, whose patience provided a little breathing room in an already full schedule; and finally, Bill Krasson, who was responsible for pushing me to learn more about the technology and who suggested some of the original material that, after several overhauls, formed the basis for this book.

Vince Adams

I am grateful to my beautiful wife Sofi for all her love, care, and support; my mother Paz and father Eduardo for always being a phone call away; my brother Pepe for much needed comic relief; and my cats Anselmo and Sesame for their solidarity in staring at the computer screen for too many hours. I would also like to thank my close friends for their *ooohs!* and *aaahs!* regarding this book, and last, but not least, the Buell Motorcycle Co. and its people for their investment in this technology, trust in my work, and willingness to listen, discuss, and provide helpful advice.

Abraham Askenazi

About the Authors

Abraham Askenazi serves as senior analyst for the Buell Motorcycle Company. He has both B.S. and M.S. degrees in mechanical engineering from the University of California at Berkeley, with a thesis on "Dynamics of Single Track Vehicles." Abraham utilizes finite element analysis as a powerful tool in the analysis-driven design process of the complete motorcycle system. By incorporating this and other simulation technologies, Abraham assists the Buell development team in enhancing its state-of-the-art vehicle design reputation.

Committed to helping others grow in FEA technology usage, Abraham has helped elevate the Chicago/Milwaukee MECHANICA User Group (MUG) program to an educational format. Abraham shares his Pro/MECHANICA expertise with Wisconsin, Illinois, Indiana, and Iowa peers as a regular "fill-in" presenter at MUG meetings. Recognized for his Pro/MECHANICA knowledge and contributions to this organization, he has been an elected officer for the past two years.

Over the past 16 years, *Vince Adams* has established a reputation for excellence. He began his tenure in product development as a product design engineer and project manager, accumulating six U.S. and European patents through the launch of successful and innovative programs such as the Life Fitness Lifestride Treadmill and Zebra Technologies' Stripe Bar Printer.

As regional technical support representative for The MacNeal-Schwendler Corporation (MSC), followed by serving as North American product manager for Pro/MECHANICA and other analysis software products at Rand Technologies, Vince established several unprecedented training and support programs. Earning the Top Customer Support Award in his first year at MSC and developing a strong customer base for Rand's analysis software product suite, Vince has been recognized by his peer group of engineering professionals as a knowledgeable mentor. He was elected to the MECHANICA User Group vice-chair position, followed by chairperson in 1996 and re-elected to the chair position for the 1997-98 and 1998-1999 periods.

Currently, as founder and president of WyzeTek, Inc., Vince chartered a mission to help companies *"build better products through 'wyze' application of advanced technologies."* Specializing in simulation tools, such as finite element analysis, providing project support, customized training, and industry leading software, WyzeTek is involved in all facets of product development from conceptual design to failure verification. Known for his FEA coaching style and programs, Vince facilitates the WyzeTek sponsored FEA advocacy and educational program entitled the "MESH" series.

Contents

Chapter 10: Convergence   313

Chapter 11: Displaying and Interpreting Results   325

List of Figures

List of Tables

Introduction

Few tools have shown such great power and promise for the future of product design as finite element analysis (FEA). The ability to simulate the performance of a part or system prior to building a physical proto-type is only beginning to filter into the world of design engineering. This book was conceived and written with the goal of clearing the air around the technology and bringing some of the knowledge and techniques of expert analysts to design analysts. It is intended to be as much about building better products as using FEA, although the focus is, of course, on applying FEA to the task. The book is focused on getting the job done as efficiently and as accurately as possible. The format is light-hearted, yet the compilation of general theory, background information, and techniques for the design analyst is unprecedented.

The structure of the book provides access to different levels of the technology to readers with varying interest levels. The first and last sections provide valuable insight to all members of the design team, including managers and project leaders who might have only a sideline role in the actual analysis process. The remainder of the text builds on the more general knowledge of these sections with detailed descriptions of the "do's" and "don'ts" of FEA. Taken as a whole, the book should prove to be an excellent primer for the new user and a trusted reference for even the most experienced analysis specialist. While the primary focus of the book is on linear static analysis, which should be mastered before venturing into the more complex areas of FEA, a solid review of more advanced techniques in FEA is provided for the user who is ready to apply this powerful tool to even greater challenges.

Who Will Benefit From This Book

The thought occurred to the authors that this book would be beneficial to all of humanity. If such claim proves to be a bit extravagant, the following people will surely benefit from it.

New FEA Users

If you are new to FEA, consider this as the book that the authors wish they had access to when they embarked on career paths in simulation technology. Take the time to understand the historical and technical background behind FEA in addition to its various techniques and modeling methods. Growth in FEA requires a patient approach to problem solving and results interpretation. The techniques provided in this book have been compiled from numerous experiences and lessons learned by new users like you.

Experienced FEA Users

This all-encompassing category is intentionally broad due to differences in the definition of "experienced." Many newer users are considered experts in their companies and forget, quite inadvertently, that they have only just begun. This book contains concepts, techniques, and methodologies ranging from introductory to advanced. They are presented in such a manner that you can easily focus on the level of detail appropriate to your needs without wading through unnecessary explanation. As you grow into the position of experienced user, the concepts in this book should become second nature to you. Moreover, these concepts should suggest additional techniques that you can develop on your own to further improve your modeling and analysis skills.

All Engineers

If you are an engineer, you cannot afford to be blind to FEA technology. FEA is driving the engineering design path at a much faster pace than ever before, and you must become familiar with it if you would like to remain competitive in your field. This book is quick to point out the hazards of trusting the technology blindly, but it also points out its awesome power when used with discretion and in conjunction with other engineering tools.

Managers and Supervisors

Whether implementing simulation into the design process is a strategic goal for you or your company or it is simply a technology used on an as needed basis in your project work, you also cannot be blind to the capabilities and limitations of FEA. One important factor behind the success of a simulation program is its proper positioning and expectations at the upper levels of management. Chapters 1 through 4 and 19 through

21 provide a large amount of overview information that will allow you to discuss the technology intelligently and judge the quality of the data being generated for your design challenges.

How This Book Is Organized

This book consists of 21 chapters organized into four parts. It starts with an overview of the historical and technical basis behind FEA, presented in a clear and comfortable format that walks you all the way up to the most sophisticated techniques in FEA methodology. Toward the end of the book, the state-of-the-art technology within the FEA industry is reviewed with recommendations for implementation based on the practices of companies who have been successful with FEA in the design process. Some repetition has been intentionally built into the format of the book to provide a complete reference for all intended readers. This format separates the detailed technical information from the overview information so that you can more readily locate what is most relevant to you.

Part 1, Introduction to FEA and the Analytical Method. Chapters 1 through 4 provide a historical overview of the technology and its role in the product design process. A detailed refresher on engineering mechanics and theory, as it relates to the use of FEA, is presented. This review of basic engineering principles contains a summary of "must know" concepts that every user of FEA should be familiar with. However, if you are new to the technology, it might be best to first skim this chapter so that you are familiar with its contents. Then, as you encounter references to it later in the book flip back to the appropriate section to clarify the concepts. The importance of these concepts should not be downplayed, however. Accurate modeling and results interpretation will, in the end, depend on your familiarity with fundamental engineering concepts. Finally, the first section provides an overview of the key concepts in FEA and the assumptive process. The importance of the assumption set and modeling choices to effective use of the technology are stressed via examples and discussion.

Part 2, The Finite Element Method. Chapters 5 through 12 enter into the meat of FEA. Detailed information for developing finite element models will be presented from automeshing to boundary conditions to CAD geometry. Detailed descriptions of element types, properties, meshing techniques, convergence topics, and results interpretation are included to shore up your FEA knowledge base.

Part 3, Advanced Modeling Techniques and Applications. Chapters 13 through 18 build on topics covered in Part 2 by presenting more advanced topics. Assembly modeling and thermal expansion accompany nonlinear and dynamic analysis. As mentioned previously, grounding in the basics presented in the earlier chapters is crucial to success in the more advanced areas of the technology.

Part 4, Integrating Simulation into Product Design Strategy. Finally, Chapters 19 through 21 take a step back to provide an overview of the tasks and topics related to implementation of the technology in a product design environment. The key concepts for understanding the current FEA marketplace are provided with a list of software configurations and key vendors. Tables in Chapter 19 should serve as guidelines for capabilities in the codes reviewed as well as a cross check for your current implementation to ensure that you are getting the most out of the technology.

The Importance of "Classical Engineering"

As will be pointed out throughout the book, FEA is not a panacea for design. It is an extremely powerful tool that every engineer should feel comfortable using. However, it does not replace other "classical" tools. With the advent of affordable and user-friendly FEA codes, "push-button simulation" is being provided to design engineers, often within the confines of their CAD program, in order to produce beautiful stress-contour pictures of their "just finished" designs. Unfortunately, without proper training in analytical methods, no matter how sophisticated the program, garbage in will always equal garbage out. That is why the type of work that the engineering community conducts and that this book is based on is so important. This book reinforces the value of working from first principles and using technology as a tool and not a substitute for good, solid engineering.

It is hoped that this book will fill a void in your personal library if you are trying to or planning to utilize FEA on the job. Written by design engineers for design engineers, the lessons learned from years of practical applications in the product design environment are summarized to provide a reference as you grow into this exciting technology.

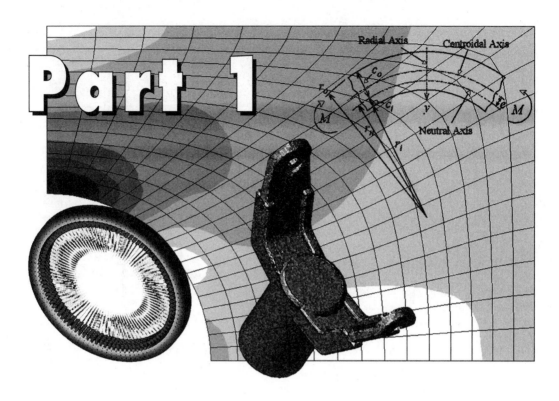

Part 1

Introduction to FEA
and the Analytical Method

1

Introduction to FEA in the Product Design Process

The history of engineering is, in essence, the history of human culture. The organization and quality of life have been categorized as much by the tools and technologies of the age as any other factor. The Bronze Age, Iron Age, and Industrial Revolution all attest to the importance of technology, or engineering, to the story of who we are today. This chapter will focus on events as they relate to the growth of FEA. However, Samuel C. Florman provides insight into the psychological and sociological background of engineering in his book, *The Civilized Engineer* (St. Martin's Press, 1987).

To fully appreciate the complexity and capabilities of the tools available to design engineers today, one must understand this story and acknowledge its impact on their daily work.

A Brief History of Computer-aided Engineering

Fig. 1.1. Tensile testing was performed by Galileo.

The classical period of engineering relied on extensive testing and the development and use of fundamental principles. Galileo, Newton, Da Vinci, Hooke, and Michelangelo all contributed to the body of knowledge on mechanics and materials. Design engineers were often asked to put their money where their mouth, or pencils, were. Early railroad bridge engineers often took the first ride across a new structure to show confidence in their calculations. When the designer's life was on the line, the importance of a sound understanding of the tools used was clear.

Fig. 1.2. Railroad engineering utilized truss calculations and drove the development of fatigue analysis methods.

In the late 1800s Lord John William Strutt Rayleigh, better known as Lord Rayleigh, developed a method for predicting the first natural frequency of simple structures. It assumed a deformed shape for a structure and then quantified this shape by minimizing the distributed energy in the structure. Walter Ritz then expanded this into a method, now known as the Rayleigh-Ritz method, for predicting the stress and displacement behavior of structures. The choice of assumed shape was critical to the accuracy of the results and boundary or interface conditions had to be satisfied as well. Unfortunately, the method proved to be too difficult for complex shapes because the number of possible shapes increased exponentially as complexity increased. However, because the number of possible shapes increased exponentially as complexity increased, this predictive method was critical in the development of FEA algorithms in later years.

By the 1940s, numerical methods had been developed to predict behavior of more general structures. Most were frame and truss based and utilized energy methods from Alberto Castigliano and William Rowan Hamilton. In 1943, Richard Courant proposed breaking a continuous system into triangular segments. The 1940s also saw the "birth" of digital computing with the unveiling of ENIAC at the University of Pennsylvania. This room-sized computer was commissioned by the U.S. Army to calculate ballistic trajectories during WWII but provided, as its first historic project, the preliminary calculations for the original hydrogen bomb. It used electromechanical devices for data storage and vacuum tubes for calculations, and required a team of operators to keep it running at a maximum processor speed of 46 operations per second. While primitive by today's standards, it marked a turning point in the potential of computing and paved the way for today's engineering tools. However, there was a long way to go.

In the 1950s, analog computers were developed to process more complex structural problems. With the promise of more powerful computers, analytical methods were advanced to include matrix based solutions of frame and truss structures. A team from Boeing demonstrated that complex surfaces could be analyzed with a matrix of triangular shapes. The benefit to the growing aerospace business was clear and most major manufacturers in this industry were developing in-house programs for structural analysis on computers. The basic concepts of finite elements were born, although the process was still time consuming and limited.

Dr. Ray Clough coined the term "finite element" in 1960. A finite element differed from previous 2D methods for computer simulation by replacing combinations of 1D elements with a single entity which could model 3D strain. The 1960s saw the true beginning of commercial FEA as digital computers replaced analog ones with the capability of thousands of operations per second. While many aerospace companies had their own in-house codes, few codes were available for purchase or lease by other industries. STARDYNE was a notable exception at this time. However, most codes were still industry, company, and even product specific.

In the early 1960s, a small analog computer manufacturer and consultant for the aerospace industry in southern California was awarded a contract from NASA to develop a general purpose FEA code. This company, The MacNeal-Schwendler Corporation (MSC), ensured the growth of commercial FEA by developing what is now known as NAS-TRAN. This original code had a limit of 68,000 degrees of freedom, which was believed to be larger than anyone would ever need. When the NASA contract was complete, MSC continued development of its own version called MSC/NASTRAN, while the original NASTRAN became available to the public and formed the basis of dozens of the FEA packages available today. Around the time MSC/NASTRAN was released, ANSYS, MARC, and SAP were introduced. FEA was still considered a specialist's tool and engineers who wished to utilize this technology typically bought time from a computing center running IBM 7094 or UNIVAC 1107 mainframes. Linear static and limited dynamic analyses were available to engineers who could justify the expense.

By the 1970s, minicomputers became more readily available and were more powerful than earlier mainframes. In fact, the HP35 calculator introduced in the 1970s was more powerful than ENIAC. The power and availability of FEA software matched the growth of the computer industry. While shared computing centers were still the norm, frequent users were migrating to in-house software, either internally developed or leased from the commercial FEA vendors. Although most analysis was linear, nonlinear solvers were developed and made available. These solutions were still too resource intensive for more casual users despite the enhancements in hardware. The speed of computers was improved to 10,000 and even 100,000 operations per second. Computer-aided design, or CAD, was introduced later in the decade.

If the 1960s marked the birth of commercial FEA, the 1980s represented its coming of age. Computing centers were becoming a thing of the past as workstations and PCs on the engineer's desktop began to dominate the market. Another key advancement was the use of FEA and CAD on the same workstation with developing geometry standards such as IGES and DXF. Standards permitted limited geometry transfer between the systems. These workstations were capable of over 1 million operations per second and took advantage of optimized algorithms for working with the math behind FEA. In the 1980s, CAD progressed from a 2D drafting tool to a 3D surfacing tool, and then to a 3D solid modeling system. The developments in graphics processing left their mark on FEA as graphical 'pre' and 'post' processors became available and engineers could examine colored stress contours instead of poring over tabular output on green-bar fanfold. Thanks to this advancement, design engineers began to seriously consider incorporating FEA into the general product design process. The link to CAD was the catalyst for this natural next step.

As the 1990s draw to a close, the PC platform has become a major force in high end analysis. The technology has become so accessible that it is actually being "hidden" inside CAD packages. It is not uncommon for a product engineering company to have nonspecialists performing nonlinear, vibration analysis, computational fluid dynamics (CFD), and multiphysics simulation. Models with 1 million degrees of freedom are being run on "deskside" supercomputers capable of running 1 trillion operations per second. This means more computations in a second than ENIAC could have completed in over 650 years!

Enter the Design Analyst

This literal explosion of speed and capabilities has been a mixed blessing for design engineers. From the near-seamless, near-painless integration with CAD, a growing number of design analyst nonspecialists with a part-time interest in FEA has emerged. The growth of this segment of the industry is accelerating rapidly as technology providers scramble to fill the need for easy-to-use FEA software tools. Some of these tools are so easy to use, little thought is required to develop stress contours on parts with complex geometry. With little thought comes little chance of accuracy. The early analysts sweated every node and element and paid dearly for lengthy run times. Good practices were developed to ensure boundary conditions and properties were well thought out before a model was submitted. Hand calculations and results correlation increased users' awareness of the capabilities and limitations of their tools. Today, new users tend to believe that any result that looks right probably is right. Material properties are assigned carelessly and boundary conditions are applied more out of convenience than based on actual environmental interactions.

Is the role of the analysis specialist doomed? Hardly! Most design analysts would agree that their usage of the technology is, by the nature of their responsibilities, only surface level. Many do not want nor do they have time to pursue more complex analysis types or assembly simulation with ambiguous interactions. Engineers have come to expect the instant gratification provided by photorealistic CAD models and rapid prototyping systems. Consequently, most design analysts who learn to appreciate the complexity of FEA hesitate to jump head first into a problem that requires lengthy setup, and run times which often yield less than spectacular data. These data, despite sometimes ordinary appearances, are a key element of the rapid product development process.

Rapid Product Development Process

The effectiveness of the process or the concept, depending on its implementation, of rapid product development (RPD) has a direct relationship to the cost and speed of computers. As computers have become faster and cheaper, new and more powerful uses have been developed. The implementation of a rapid product development strategy maximizes the use of computer tools to provide three fundamental activities: (1) communication, (2) visualization, and (3) simulation.

Improvements in communication are reflected in the popularity of concurrent and collaborative engineering. Design groups can share detailed and accurate data across phone lines with suppliers, customers, internal support groups, and peer design teams.

Visualization allows engineers to better understand geometric and component interactions. While a spreadsheet or well-organized set of equations can provide valuable design data, the ability to review concepts in virtual environments produces invaluable insight into the cause and effect relations of various design decisions. An engineer working in the 3D solid modeling world will be less tempted to put the note "Blend

Here" on a difficult-to-represent drawing view. Tool makers can see a shaded image of the difficult area and better plan their approach.

Simulation tools have dramatically reduced the product design cycle in many companies. Form, fit, and function can be checked in three-dimensional (3D) space. Photorealistic rendering provides marketing with samples before a single physical part has been produced. The stress levels or displacement behavior of a part or system can be examined under operating conditions as well as extreme situations which can be difficult and costly to test. Potential failures and valuable cost reductions can be identified early in the process to minimize the cost of change requests or field repairs.

Three enabling technologies have emerged to provide the communication, visualization, and simulation capabilities required by RPD. These technologies are 3D solid modeling, finite element analysis, and rapid prototyping.

Solid models of parts and assemblies have allowed designers to quickly represent their ideas in an unambiguous manner (visualization and communication). Team members can qualify assembly techniques, manufacturability, and "look and feel" (simulation). This solid model is often considered the master database for the part or system, and is used for all downstream applications such as detailing, documentation, prototyping, analysis, manufacturing, and marketing.

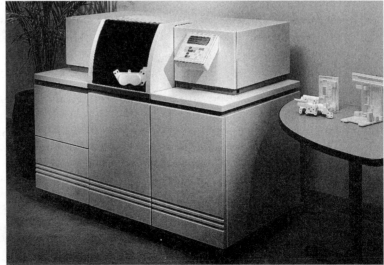

Fig. 1.3. Actua rapid prototyping equipment from 3D Systems.

Rapid prototyping (RP) has bridged the virtual and physical worlds. The technology has progressed to the point where an engineer can literally make a 3D "print" of a part in a matter of hours. Engineering and marketing can test subtle variations of a concept and incorporate suggestions into prototypes in near "real time" (simulation). While the solid model can convey much data about a part using shaded, dynamic, on-screen manipulation, it is very difficult to pass a monitor and mouse around to meeting participants. In contrast, a physical part can answer many questions.

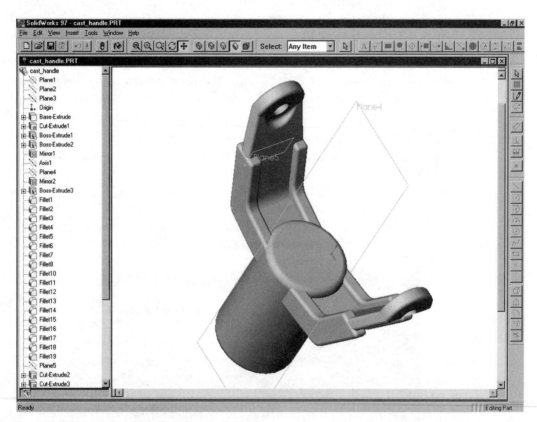

Fig. 1.4. Solid model built in SolidWorks.

Although the resultant deformations calculated by an analysis can often shed light on form and fit, finite element analysis (FEA) focuses on the function part of the form-fit-function equation. Understanding stress levels, deformation, temperatures, and response to vibration or fluid flow characteristics up front in the design process is clearly the embodiment of simulation. Multiple design iterations on conceptual geometry can easily lead to a radical new design or a significant reduction in the number of prototypes. Today's analysis tools also greatly enhance an engineer's ability to visualize and communicate with rendered and animated color contour plots of parts under test. While a page of equations might not get the attention of management or manufacturing when a change is requested, a 3D color plot showing large areas of red reflecting critical locations or twisted components in an exaggerated deformation will make people sit up and take notice—often regardless of the actual result magnitudes and accuracy of the underlying model.

Most engineers have recognized the value of RPD and the importance of all the enabling technologies contributing to that process. However, corporations have accepted these technologies to varying degrees. It is interesting to note that the acceptance of these technologies is not proportional to their maturity. The next illustration shows the relative age and acceptance level in product design of the core tools.

Fig. 1.5. Relative acceptance and age of the three enabling technologies.

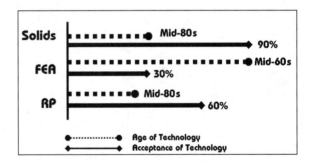

The different level of acceptance of these tools sheds light on the reluctance of traditional design organizations to break away from familiar processes. Drawings and prototypes have been the mainstay of design groups throughout history while actual engineering has often been a small part of the overall process. In the traditional product design process used at most companies, increasingly shorter time lines preclude lengthy engineering evaluations of the performance of chosen configurations. Even today, many engineering companies would prefer to take the chance that a "best guess" design will work, thereby risking possible redesign. This seems more attractive due to that instant gratification alluded to earlier than delaying the creation of initial prototypes through expanding up-front engineering and analysis in order to reduce testing and prototyping. Fig. 1.6 depicts the traditional product design process used at most companies.

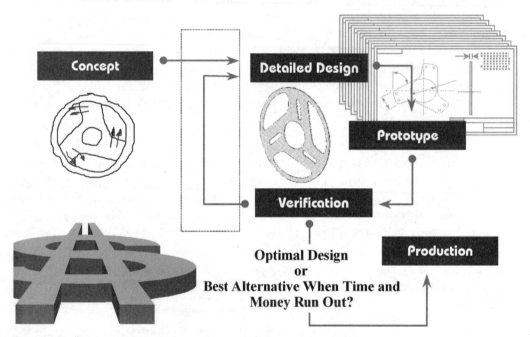

Fig. 1.6. Traditional product development process.

Conceptual designs are rushed through the system. Here, the primary task of engineering is to develop layouts and drawings. When structural functionality is clearly an issue up front, it may be addressed by calculations, analysis, or overdesign during the drawing creation stage. The true performance of a part or system is typically not understood until the prototyping and testing phase. If no issues arise during these phases, the design is often deemed acceptable and considered finished. If time and budget permit, some degree of optimization is performed on a trial and error basis until schedules and costs dictate the end. If the part fails the testing phase, companies typically scramble to resolve the problem with as little change as possible. FEA is often brought in at this stage. However, the engineer's hands are somewhat tied by the design envelope dictated by other released and in-process components. True optimization is very difficult at this stage and the changes required may not be the most cost effective due to these restrictions. Fig. 1.7 indicates the relative cost in time, dollars, and complexity of changes at various stages of the design process.

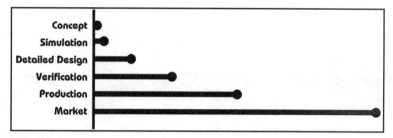

Fig. 1.7. Relative cost of product change at the different stages of the design process.

One of the reasons for the relative cost of change later in the process can be better understood by the following diagram which indicates the disparity of a company's investment in a design compared to their level of understanding. The "build and break" cycle controls the area between the curves in Fig. 1.8. The more cycles required, the greater the gap in knowledge versus cost.

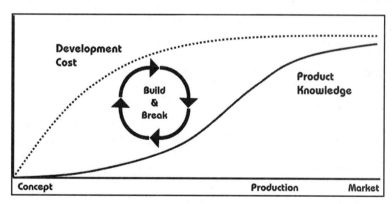

Fig. 1.8. Cost versus knowledge dilemma.

It should be clear that the value of the up-front predictive analysis or simulation lies in the reduction of the knowledge gap and in the potential for savings in the later stages of the product development cycle. The next figure shows the ideal predictive engineering enabled design process where the test/redesign cycle is performed on software prototypes. Not only can planned testing then be done more quickly and with less cost, simulation can allow the engineer to explore design options or loading scenarios which would be too costly to address in the physical world. One has to simply examine the cost of prototype tooling for a complex plastic part to see the value in qualifying design options digitally.

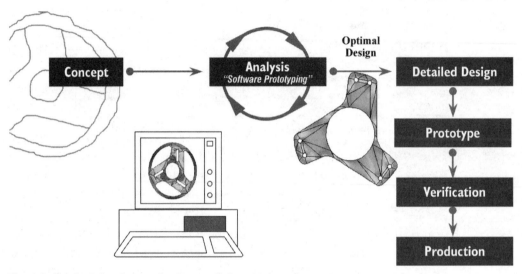

Fig. 1.9. Product development using predictive engineering.

This revised design process results in rapid closure of the knowledge gap or the disparity between investment and understanding.

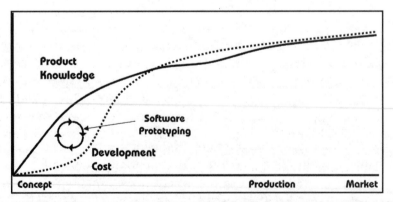

Fig. 1.10. Improved tracking of cost versus product knowledge with simulation.

Who Should Use FEA?

The role of the design analyst is here to stay. As engineers clamor for more user-friendly CAD integrated tools, the software industry responds with more creative ways to fill the void. But who are the design analysts? Their background, experience level, and enthusiasm level runs the gamut. A design analyst, as opposed to an analysis specialist, may be executing CAD design in the morning, be on the shop floor in the afternoon, and running analyses during intermittent breaks at his or her desk.

The typical design analyst is the only FEA user in his/her company. They are usually thought of as experts upon return from software training but the value of their contribution is usually unappreciated. As long as most companies still consider FEA as a "fire extinguisher," the design analyst will always have to fight for hardware upgrades and additional training.

The design analyst will typically fall into one of two categories: the champion or the designate. The champion was integral in the acquisition of the technology. The champion would like to make FEA become a key part of the design process and will lobby for this constantly. Champions tend to be good engineers with sound grounding in engineering fundamentals and basic analytical procedures, and would like to see their time on the system increase but do not want to become specialists.

The designate, on the other hand, was selected to be the FEA guy or gal once management made the decision to bring in FEA. Designates are often chosen because of their CAD prowess or general computer knowledge. Nearly as often, they are chosen because of a temporary lull in work load. Designates may immerse themselves into the task with enthusiasm but will never match the zeal of champions. Most designates are genuinely interested in the technology but will not, for various reasons, push to increase usage. While happy to analyze when asked, these individuals might be just as happy to let the software tool go unused. Designates may think of themselves as champions because they are the only persons suggesting upgrades or expressing interest. However, the designate can easily be identified by comments such as "I'll never be given time to…," "Accuracy isn't as important as…," or "Seamless CAD integration is critical…" Essentially, designates will have a handy excuse for not pushing themselves to build better models. Champions, on the other hand, take a certain thrill in the challenge of a tough model.

Notable among both types of design analyst is a great disparity in appreciation for the complexity of FEA. Many champions lose sight of their limitations in their enthusiasm to validate their tool while some designates proceed with methodical caution to ensure that results are accurate. Also common to both types is isolation from peer interaction to talk about modeling techniques and results interpretation. While some do not know where to look for this support, others do not know they should.

Which design analysts are successful in the implementation of the technology into the design process? Successful design analysts are typically champions. The enthusiasm for the potential of FEA may be the most important characteristic of a successful design analyst. Running a close second is a healthy respect for the complexity of the subject. With this respect comes a desire to learn and excel at the finer points practiced by specialists, almost to the extent of becoming part-time specialists. Unfortunately, the resources available to these enthusiasts are limited. Software user groups are a good place to start, although code independent forums such as the MESH FEA Advocacy group in Chicago are better.

Pointing FEA in the Right Direction

Success with FEA in product design is very dependent on the way it is utilized and the expectations of the organization. Few companies succeed by simply purchasing a tool, picking a user, and waiting for a need to arise. All engineers and managers who come in contact with the technology should have an awareness of the capabilities and limitations of the tool so that misconceptions are clarified and expectations are set correctly. In addition, those using the technology should approach all problems methodically following the guidelines below. Good practices and the right frame of mind will allow the user to focus on the problem at hand.

Analytical Problem Solving Process

The merits of planning have been stressed in many sources for many types of problem solving techniques. Planning is also important for solving problems in FEA. The thought process for solving a problem using FEA is similar to that for solving a problem using lab tests, free body diagrams, classical equations, or spreadsheets. Understanding the "big picture" as well as specific data objectives will bring the selection of tools, the problem solving approach, and the interpretation of the results into clear perspective. Moreover, setting proper expectations and being technically prepared for the meaning of any and all data are critical to the difference between "pretty pictures" and solid engineering with advanced tools. Many would-be analysts plunge right into FEA without first becoming comfortable building free body diagrams, navigating simple beam calculations, and investigating the meaning of von Mises stress and why it is a poor indicator of failure in many engineering problems. These issues are addressed in later chapters.

The ideas presented in Chapters 1 and 2 should be considered prerequisites for further pursuit of finite element based design. Be prepared to refer back to these ideas periodically and to challenge your analysis results based on engineering first principles.

Process

Most engineering problems are solved using the following four steps. They will not always be explicitly documented but they will be considered at some point in the process. Shortchanging any of them is the quickest path to failure.

1.   Establish a clearly defined goal.

2.   Compile and qualify the inputs.

3.   Solve the problem with the most appropriate means.

4.   Verify and document the results.

What is the goal of the analysis?

The goal of the analysis should include two important decisions. First, is it important to predict an exact solution to a problem or is it sufficient to get in the ballpark or show trends? The steps taken to solve the problem and the precision of the inputs ride on the answer to this question. Second, what specific pieces of data will help you make the engineering decision required? When can you stop analyzing? Which shortcuts are allowed to focus on the problem at hand? An excellent illustration of this decision involves developing an analysis for overall displacements versus localized stress data. If stresses are not deemed important, many details can be left off the model and your options for idealizations to improve efficiency are expanded.

Predictive engineering versus failure verification

The most efficient and cost-effective point to begin the structural analysis of a component or a system is at the conception stage of design. Letting the results of the analysis drive the choice of materials, features, wall thickness, and so forth is called predictive engineering. Using this methodology, data are generated when they can have the greatest impact on the cost and quality of the design at the lowest cost of change and schedule.

Failure verification represents the most rigid usage of structural analysis. Correlating FEA data to an existing condition requires a careful examination of the boundary conditions, material properties, geometry, operating environment, and the actual failed part. Assumptions and approximations must be minimized. Prior to undertaking an analysis, evaluate whether failure is consistent across several parts or is isolated to one sample. An isolated instance might suggest that a defect in geometry or material, and/or an unexpected loading scenario was responsible. An analysis that does not take these factors into account may not provide useful data.

If analysis is to be used early in the design process, start on simplified geometry. Complex solutions can be more easily approximated as simple ones at this stage. Refine your analyses as the part definition and behavior become more defined.

Trend analysis versus absolute data

Geometry or boundary conditions are frequently impossible to model, unknown, or beyond the analyst's control. In these cases, trend analysis is an effective technique which allows the engineer to develop useful data that can quickly drive design decisions. It must be understood from the start that a trend analysis will not necessarily yield actual performance data, but rather will show the effect of geometric changes and the sensitivity to parameters such as material properties and applied load.

For instance, assume that the true stress-strain data for a highly deforming plastic component are not available. However, a linear trend analysis shows that decreasing the nominal wall thickness reduces cost by 10% with less than a 1% increase in stress and deformation. Does this change make sense?

The actual values for stress and deformation are not required to evaluate this modification. While nonlinear behavior most likely renders the linear results inaccurate, the trend shown by this study clearly indicates that the strength of the part will not be compromised by the cost savings.

Selecting required output data

The need for specific data is often intuitive and obvious. If strength is a consideration, solve for stress and deflection. If cooling is an issue, solve for temperatures. However, is stress in a localized area the only concern? Is the cooling solution steady state or are transients to be considered? Will natural convection or radiation effects significantly impact the behavior? Such specific questions will drive the choice of tools and dictate the level of detail or skill required to use these tools. This book provides information on preparation to ask such questions as well as on answering the same.

What input is required for the solution and what level of uncertainty does it introduce?

The next step in solving an engineering problem is compiling all inputs required by the chosen solution technique or tool. A hand calculation might only require cross-sectional areas and moments of inertia whereas a CFD analysis might require temperature dependent viscosity, density, and specific heat properties of several interacting fluids. In addition to filling in the blanks on a data form or in a closed form equation, care must be taken to qualify the data being used. A material or geometric parameter is rarely consistent for all produced components, and loading measured in a test apparatus may differ dramatically from actual field usage. An engineer's ability to evaluate results with such an incomplete data set separates an analyst from a skilled button pusher.

What is the most efficient means to solve the problem?

Some complex problems can be simplified with little or no loss in solution quality in order to be solvable with hand calculations and more classical methods. Some seemingly simple problems undergo behaviors that require advanced computer based simulation to sufficiently model the true behavior. It is assumed that as a reader of this book, which is focused primarily on computer based, finite element analysis, that you have identified the need for more powerful tools to solve engineering challenges. As the techniques for approaching these solutions will be described in greater detail in later chapters, it is sufficient to state at this point that one should not default to FEA as the tool of choice. Relatively manual methods may be more efficient and cost effective, as well as provide greater insight into a new challenge or problem. Even when FEA as a requirement has been identified, try to explore simplifications which can be solved manually to provide a starting point for the selection of inputs, and to qualify the general accuracy of the results.

Introduction to the Assumptive Approach

One of the most useful realizations a design analyst can come to is the extent of the assumptions going into even the most basic analysis. An engineer from Caterpillar Corporation recently commented, "If you make enough assumptions, you can analyze anything." Reversing cause and effect, it might be said, "Since you can make enough assumptions, you can analyze—period." In a friendly discussion, a design engineer and a material scientist debated the wisdom of picking a linear Young's modulus for cast iron. The material scientist asserted that a "linear" portion of the cast iron stress-strain curve did not exist, and that the variation in processing could result in a property value of ±50% from lot to lot. The design engineer's position could be summarized as follows: "I understand all that, but which Young's modulus should I use?"

Engineers are charged with making decisions based on incomplete data sets every day. Assumptions regarding material uniformity, assembly variability, user inconsistency, and general unpredictability need to be weighted, qualified, and documented. In the end, however, a decision must be made. The design analyst must take this process a step further. In addition to the uncertainty mentioned previously, s/he must use an idealized computer model with clearly unrealistic boundary conditions to represent a physical phenomenon which could rarely be repeated in testing. The key to success in light of these seemingly uncontrollable circumstances is a scientific qualification of the uncertainties involved. Qualification of uncertainty is critical to the success of design analysis. While a specialist may have the time or resources to test every conceivable option or analyze every possible combination, the design analyst must quickly assess his/her uncertainty and make a decision based on limited results. Occasionally, the decision may be to analyze further. Typically, a production decision is needed to keep schedules on track. Careful specification and understanding of required assumptions will make basing decisions on incomplete data more palatable.

Common Misconceptions About FEA

Those who are not intimate with the finite element method will certainly harbor some misconceptions about the ease of use or the degree of accuracy entailed. It is interesting to note that nonusers or casual users with similar levels of exposure can have diametrically opposed viewpoints on these issues. Typically, most preconceptions are based on some defining experience with the technology. Some common incidents which have a profound impact on one's opinion might be unsupported use of a difficult preprocessor, inconsistent or even disastrous correlation to test or field data, a bad experience with a consultant who was not much more qualified than the nonuser, or a successful project that seemed effortless and nearly "pushbutton." However they came about, misconceptions about the capabilities and limitations of the technology at any level of the organization can slow or stunt the growth of simulation in the product design process. The following descriptions clarify some of the more common misconceptions about finite element analysis.

Meshing is Everything

This attitude is a driving misconception behind the recent proliferation of CAD embedded analysis tools, or tetrahedral meshers working directly within the CAD environment. It is often believed that if a part can be meshed, the battle is over. The exact opposite is often closer to the truth. With the efficiency and quality of today's automeshers, developing a solid automesh of clean CAD geometry is probably the easiest step in the process. Ensuring that each mesh is clean with good quality elements and that the final mesh has been converged on the desired behaviors is a more difficult proposition. No commercial FEA tool is insensitive to poorly shaped elements. The actual effects of these elements will be discussed in Chapters 7, 10, and 11. Accuracy local to a poorly shaped element will be affected by that element. The design analyst must qualify the mesh once completed. Local mesh refinement tools are critical to ensuring a good mesh with gradual transitions between densities. Choosing the right type of mesh or element for the problem is equally important to well-shaped elements. A linear tetrahedral mesh on a thin-walled plastic part is probably not going to provide reliable results, regardless of the uniformity of the elements. Analysts must learn how to build models best suited for the geometry and problem to be solved.

It is also important to note that meshing is only one input to the problem. Accuracy is also controlled by boundary conditions, material properties, and adherence of the model to the physical part geometry. Improperly assigned data in any of these areas will result in significant inaccuracies. While a good mesh can be obtained with patience in most preprocessors, well-conceived boundary conditions and representative material properties will never be automated because the engineer's judgment must direct and qualify the necessary assumptions. Simply building a mesh in no way ensures that the results of an analysis will be meaningful. New users must resist the temptation to utilize mindless automeshing.

FEA Replaces Testing

A rigorous analysis program may actually *increase* the amount of testing required on a design at early stages of the implementation. There are a couple of reasons for this. First, confidence must be developed in the analytical methods. It is unwise to assume that all answers kicked out of an FEA program are accurate, given all the assumptions and uncertainties which must accompany any study. Only through correlation of test models and actual prototypes can the methods and assumptions used in FEA be qualified. After qualification, it may be decided that the analysis results for similar studies in the future are reliable and some developmental testing may be eliminated. A second reason for increased testing is the fact that more is learned about the structure which might suggest unforeseen problems. A physical test can only examine a small number of quantities and typically ends when a catastrophic failure occurs. In most cases, the observer has no way to evaluate the nearness to failure of other portions of the system. A well-developed analytical model will show the relative quality of all parts of the structure. Previously undetected displacements or stress risers may suggest a test revision to ensure that these potential problem areas are monitored.

It is best to say that FEA augments testing and vice versa. Analytical results can suggest strain gage placement and orientation. A test can provide valuable data about the validity of boundary conditions. However, when applied correctly, it can be said that a solid predictive engineering program can reduce testing in the design stage as the confidence in simulation results grows. The levels of uncertainty inherent in the process will always require that final products, at a minimum, be tested. These tests should be even more efficient due to the existence of FEA data.

Finite Element Analysis is Easy

The surest route to failure in FEA is to underestimate the complexity of the technology. Without a healthy respect for the variability of the physical systems being modeled and the sensitivity of output to multiple interrelated input quantities, a would-be analyst may jump to conclusions too quickly and make design decisions based on poorly qualified data. Modern tools have made getting a mesh on a CAD part extremely easy. Consequently, obtaining answers is literally a few button pushes away from the completed geometry. Knowing what those answers mean and how to adjust the model to improve, or even evaluate the accuracy, requires a deeper understanding of the workings of the technology and the meaning of the various parameters involved.

Very few problems have obvious boundary conditions. Test models and additional research are almost always required to determine loads and constraints. The same can be said about material properties. Many problems must be bracketed (run at extremes of their inputs) to fully understand the sensitivity of the results to these inputs. Like it or not, building the occasional mesh portion by hand, element by element, is unavoidable in a general program of simulation. The proverbial "brass ring" of seamless, painless, effortless FEA has been offered by many of the leading technology providers with some of the CAD embedded technology. The wise analyst must look beyond the promise of instant gratification. With simplicity of use comes the removal of many of the options required for accurate analyses. Results developed too quickly without the due diligence required by the complexity of the technology will always be questionable.

Getting results is only half the battle. Interpreting results requires engineering knowledge of materials and their failure modes. Results interpretation also requires that the analyst know the impact of each assumption made in the model. For example, entering any value for Young's modulus in the material definition makes a significant assumption about the uniformity and repeatability of the material and the predictability of the environment in which the part is being used. Assumptions escalate from there. Evaluating the needs of each specific model requires some work and it is best done with the input from multiple advisory sources. Few will doubt the benefit of sharing ideas and having to defend assumptions to a constructive audience. Growth in the technology often depends on the interaction of users at various levels. Very few self-taught analysts ("islands of technology"), left to learn and evaluate their work alone, will produce consistent results that will hold up to scrutiny by experts. While these users are often the only experts in their companies, they may never know or have to admit the limits of their skills. When such users move on, their unfortunate successors quickly appreciate the complexity of the analysis problem when called upon to understand what has been left behind.

Finite Element Analysis Is Hard

With all of the above taken into consideration, sound work in FEA is 20% creativity and 80% hard work and patience. Any would-be analyst with the desire to learn the technology and a solid understanding of the fundamental principles of engineering and material mechanics can be taught to be an adequate analyst. The basics of FEA can be learned and reinforced through practice and interaction with a mentor or coach. Geometry based preprocessors have taken much of the drudgery out of building models and have reduced the need to "know the code." However, the need to understand the problem being approached and the limitation of the tools being used still rests with the analyst. While exceptions exist, analysts become successful through the exercise of patience and thoroughness, not necessarily exceptional intelligence or ability. It cannot be stressed enough that the technology, with reliable software, will provide correct answers to questions defined by mesh, properties, and boundary conditions. It has been said that there are no wrong answers in FEA, only wrong questions. If users can learn how to pose the questions correctly, they will get good answers. What could be simpler?

Learning the Interface Equals Learning FEA

As discussed previously, the usual candidates for the designate type of design analysts are those most proficient at CAD. Such speedsters on the keyboard are expected to learn the interface quickly and make the most out of the FEA investment. In fact, the misconception that CAD power users are the avenue through which the FEA investment will pay for itself often blinds users and management to the limitations of the chosen software and the user. As long as instant gratification is still rewarded, the speedy CAD jock will appear to have a good handle on the technology by producing plot after plot of stresses and displacements.

Most experienced analysts would agree, however, that the first models should be performed slowly and repeated with minor variations of inputs to better understand the sensitivities and anomalies of the technology. The sooner all interested parties up and down the management chain understand that getting some kind of result is easy, while getting the right result requires the analytical problem solving of an experienced engineer, the sooner the qualifications for choosing the best users will be adjusted. An analyst from Harley-Davidson Motor Company once commented that experienced analysts will acknowledge they have much to learn while newer users who think they know what they are doing are probably in trouble. Learning the interface or the preprocessor is relatively simple compared to the subtle nuances of assembly interaction or the ability to ascertain the meaning of local stress concentrations.

Summary

Learning design analysis can best be characterized by the phrase, "Check your egos at the door." Harnessing the power of simulation in the product design process requires hard work, patience, a certain zeal for the technology, and countless hours of review with peers and mentors. As the book progresses, "must know" techniques and information will be reviewed, but you must remember to put these lessons into practice with the caution of the early railroad bridge engineers. Test each assumption and question all results. Your reward will be a skill that is, and will continue to be, increasingly valuable in the product design marketplace. The penalty for proceeding hastily without a firm understanding of the tools and technologies involved may mean lost revenue to the company and lost employment for the overconfident ex-analyst. The penalty for a flawed design will not likely be a plunge into an icy river for the responsible engineer, but there may be similarly hazardous conditions for end users of the product designed.

2

Fundamentals

The importance of achieving a clear understanding of engineering fundamentals should not be underestimated. As a design analyst who cannot wait to get started with FEA, you might find this chapter to be a little heavy on theory. If so, do not get bogged down: you can proceed to later chapters and refer to this one as necessary. In fact, you will find many references to specific sections in this chapter throughout the rest of the book. You should always take the time to understand the concepts being discussed. In the end, this chapter will likely become a very powerful engineering companion in your analysis challenges.

First Principles

Body Under External Loading

When performing engineering analysis, you are virtually always concerned with how a body will behave under external loading. Newton's laws, or the laws that will most generally govern this behavior, are listed below.

- *First Law:* A body will remain at rest or will continue its straight line motion with constant velocity if there is no unbalanced force acting on it.

- *Second Law:* The acceleration of a body will be proportional to the resultant of all forces acting on it and in the direction of the resultant.

- *Third Law:* Action and reaction forces between interacting bodies will be equal in magnitude, collinear, and opposite in direction.

The most important engineering equation arising from these laws follows:

$$Eq.\ 2.1 \quad F = ma$$

where F is the resultant force vector, m is the mass of the body under consideration, and $\mathbf{a}$ is its acceleration vector.

Because acceleration is the time derivative of velocity (dv/dt), and $G = mv$ constitutes the *linear momentum* vector of a body, the above equation can also be written as follows.

$$Eq.\ 2.2 \quad F = m\frac{dv}{dt} = \dot{G}$$

In other words, Newton's second law may also be interpreted as stating that the time rate of a body's change of momentum will be proportional to the resultant force acting on it and in the same direction.

Fig. 2.1. General free body diagram (a). Resultant forces and moments (b). Second law equivalent (c).

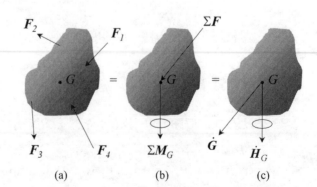

The most useful tool for understanding and implementing the loads and constraints, or *boundary conditions* that govern a body's behavior, is the *free body diagram.* The general free body diagram above (a) represents the body in space removed from its operating system. All externally applied loads and reaction forces are represented with vectors on the body. If the body is in *equilibrium,* all these force vectors must add up to zero, both in magnitude and direction.

In the most general sense, externally applied loading on a three-dimensional rigid body cannot only alter its translation, but its rotation as well. Referring to resultant forces and moments (b) and the second law equivalent (c) in Fig. 2.1, the corresponding spatial equations of motion for a rigid body follow:

Eqs. 2.3
$$\sum F = \dot{G}$$
$$\sum M = \dot{H}$$

where $\Sigma\mathbf{F}$ and $\Sigma\mathbf{M}$ are the force and moment vector sums, respectively, of all externally applied loading, including reactions, and $\mathbf{H}$ is the *angular momentum* vector of the body. Both $\Sigma\mathbf{M}$ and $\mathbf{H}$ must be calculated about the same point on the body.

Fig. 2.1 shows free body motion where this point corresponds to G, the center of gravity of the body. For constrained motion, it corresponds to O, the fixed point about which the body rotates. In Eqs. 2.3, the time derivative of $\mathbf{H}$ is a complex quantity to deal with mathematically, but suffice it to say that $\mathbf{H}$ is a function of both the angular velocity and angular acceleration of the body. Its inertia component is not the mass of the body but its mass *moment of inertia tensor* ($\mathbf{I}$), which is a 3 x 3 matrix comprised of *mass moments of inertia* (I_{ii}), and *mass products of inertia* (I_{ij}), derived with respect to the body coordinate axes. These quantities describe how the mass of a rigid body is distributed with respect to the chosen axes. The general equations for these quantities follow:

Eqs. 2.4
$$I_{ii} = \int (j^2 + k^2)\, dm$$
$$I_{ij} = \int ij\, dm$$

where i, j, and k are any combination of the three coordinate axes chosen.

Fig. 2.2. Uniaxial spring and
damper system (a). Planar
body motion (b).

(a)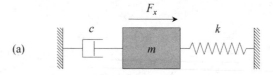

Fig. 2.2. Uniaxial spring and
damper system (a). Planar
body motion (b).

(b)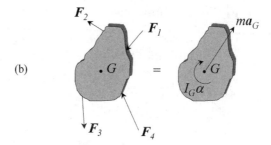

Constraining the body to uniaxial motion and allowing for an external spring and damper in the system, as shown in Fig. 2.2(a), Eq. 2.1 expands to the following:

Eq. 2.5 $F_x = m\ddot{x} + c\dot{x} + kx$

with k denoting the spring stiffness, c the damper coefficient, and F_x, x, $\dot{x}$ and $\ddot{x}$ the body's resultant applied force, position, velocity, and acceleration along the x axis.

Constraining the body to planar motion [see Fig. 2.2(b)], Eqs. 2.3 simplify to the following expression.

Eq. 2.6 $$\sum F = ma_G$$
$$\sum M_G = I_G \alpha$$

In the above equation, a_G is the vectorial acceleration of the center of gravity (c.g.) of the body. ΣM_G is the sum of all moments about the same point. I_G is the mass moment of inertia of the body about an axis normal to the plane of motion through the c.g., and α is the body's angular acceleration.

Barring dynamic analyses, FEA will always deal with bodies in equilibrium. By definition, a body in such a state must have zero acceleration, so that the result of all externally applied forces must be zero. This type of analysis is called *static*. Although this condition sounds very limiting, consider that many times a well-understood dynamic system may be effectively reduced to a quasi-static state at an instance of interest by applying a "freeze-frame" acceleration force, *ma*, to the body as an external force.

Rigid body motion implies movement of a body in space with little or no deflection, bending, or, more generally, no mechanical strain. Any condition which creates a nonstatic equilibrium regardless of the strain levels is a case of rigid body motion as well. Pivoting freely about an axis is considered rigid body motion. Boundary conditions in an FEA model must remove all possibility of rigid body motion unless a specific dynamic solution is requested which can resolve nonstatic equilibrium. Rigid body motion is represented in a modal analysis by a natural frequency of zero.

Area Moments of Inertia

There are many types of applied loads that cause a continuously distributed force over an area. In these cases, it is often necessary to calculate the resultant moment caused by this force about an axis either on or normal to the plane of the area. This is known as the *area moment of inertia* and is the solution to the integral $\int(distance)^2 d(area)$.

For example, Fig. 2.3(a) shows a submerged vertical wall subject to a distributed pressure (p) that is proportional to the depth (y) below the horizontal surface line. The total moment experienced by the wall about the surface line is $k\int y^2 dA$, where k is the constant of proportionality. In the same manner, as will be discussed later in this chapter, an elastic beam under pure bending [Fig. 2.3(b)] will develop in its cross section a linear distribution of normal force intensity (or stress, σ) that is proportional to the vertical distance (y), from a neutral axis. Hence, the total moment on this cross section will once again be $k\int y^2 dA$. A final example concerns an elastic bar under torsion [see Fig. 2.3(c)]. This torsional moment will cause a distribution of tangential shear stress τ, that is proportional to the radial distance (r), from the shaft center. In this case, the total moment is $k\int r^2 dA$.

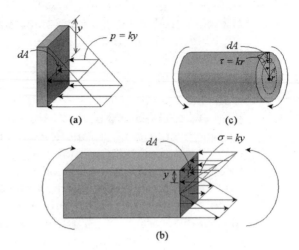

Fig. 2.3. Submerged wall (a).
Beam in pure bending (b).
Bar in pure torsion (c).

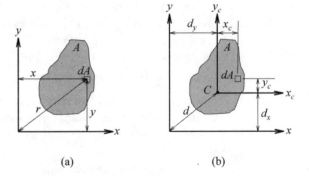

Definitions

Fig. 2.4. Rectangular and
polar moments of inertia (a).
Parallel axis theorem (b).

The first two examples in Fig. 2.3 utilize what is known as the *rectangular*
moments of inertia of a section, illustrated in Fig. 2.4(a). These are area
moments of inertia about a rectangular or *Cartesian* coordinate system
on the plane of the section of interest, and are given by the following
equation.

$$Eq.\ 2.7 \quad I_x = \int y^2 dA$$
$$I_y = \int x^2 dA$$

On the other hand, the last example makes use of a *polar* moment of
inertia about an axis normal to the plane of the section. Taken about
the origin of the same rectangular coordinate system,

Eq. 2.8 $J_z = \int r^2 dA$

It is useful to note that, because $x^2 + y^2 = r^2$,

Eq. 2.9 $J_z = I_x + I_y$

Expressions for the rectangular and polar moments of inertia for a section will often be known about a set of axes x_c, y_c with the origin at C, the centroid of the section. To express these inertia terms about any set of parallel axes x, y, as in Fig. 2.4(b), the *parallel axis theorem* gives rise to the following equations:

Eq. 2.10 $I_x = I_{cx} + A d_x^2$
$I_y = I_{cy} + A d_y^2$

Eq. 2.11 $J_z = J_{cz} + A d^2$

where the subscript c denotes the terms about the centroidal axes.

When a geometrically complex section can be divided into a number of simple ones, it is possible to obtain a resulting *composite* moment of inertia. This is done by adding the individual centroidal moments of inertia together with a parallel axis theorem correction for each of the simple sections as follows:

Eq. 2.12 $I_x = \sum I_{cx} + \sum A d_x^2$
$I_y = \sum I_{cy} + \sum A d_y^2$

Eq. 2.13 $J_z = \sum J_{cz} + \sum A d^2$

If the body of interest has a section that is completely asymmetric, or the coordinate axes chosen are placed in such a way that the body is asymmetric with respect to this coordinate system, a product of inertia results. This is expressed by the next equation.

Eq. 2.14 $I_{xy} = \int xy dA$

Note that as soon as one of the axes becomes a symmetry axis, this inertia term becomes zero.

Using the product of inertia, it is possible to mathematically rotate a set of axes about a point and compute new moment of inertia terms as functions of both the initial inertia terms and the angle of rotation. The angle is the only variable in the equations, so it is possible to solve for the critical angle (α) that gives the axes an orientation of maximum and minimum inertia, as described in the following expression.

$$\textit{Eq. 2.15} \quad \tan 2\alpha = \frac{2I_{xy}}{I_y - I_x}$$

Eq. 2.15 yields two angles that differ by $\pi/2$. One angle defines the axis of the maximum moment of inertia, and the other is the minimum. These two rectangular axes are called the *principal axes of inertia*. Note that if a chosen axis is a symmetry axis, the right side of Eq. 2.15 is zero, and the corresponding angle of rotation must be zero or $\pi/2$. *This means that if a set of axes is chosen, at least one of which is a symmetry axis, these must be principal axes.*

For an arbitrary set of axes, the magnitudes of the principal moments of inertia are as follows.

$$\textit{Eq. 2.16} \quad I_{max,\,min} = \frac{I_x + I_y}{2} \pm \frac{1}{2}\sqrt{(I_x - I_y)^2 + 4I_{xy}^2}$$

It is useful to note that many CAD and FEA packages can simply utilize a sketched section to calculate cross-sectional properties of complex geometry about a user-defined coordinate system. In addition, these packages usually have an extensive library of properties for standard shapes.

Stress and Strain

It is surprising how many engineers engage in stress analysis without being able to define structural stress. The primary failure quantity sought in an FEA solution has many facets and formulations. Choosing the correct stress quantity for the problem at hand requires advance knowledge of the available options. Many texts, including most college mechanics books, will cover this material in more detail if you wish to pursue it. You can never know too much about these topics if you are to utilize finite element analysis in your career.

What Is Stress?

When a body is subjected to an applied load, a stress state is caused inside the body. The stress can be described as the internal force exerted by either of any two adjacent sections of the body upon the other, across an imaginary plane of separation. When the forces are parallel to this plane, the stress is called *shear stress*, (τ). When the forces are normal to it, the stress is called *normal stress* (σ).

Subdividing the body into many imaginary stress elements is useful at this point (see Fig. 2.5). For the body to be in static equilibrium, both shear and normal stresses must act on each one of these elements in such a way as to place it in static equilibrium. If the normal stress is directed toward the element on which it acts, it is called *compressive* stress and, by convention, is negative in value. If it is directed away from that element, it is called *tensile* stress and is positive. All of these stresses result from the cohesive nature of the body's material; if the body came apart with no resistance under applied loading, it would experience no stress.

Fig. 2.5. General stress elements in equilibrium.

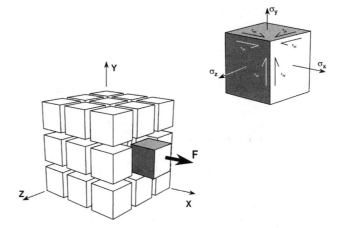

As seen in Fig. 2.5, various subscripts are employed to denote special characteristics of these stress quantities. For example, τ_{xy} denotes a shear stress parallel to the y axis, on an element face normal to the x axis, and σ_y is a normal stress acting along the y axis on a face normal to it. Note that for static equilibrium, $\tau_{xy} = \tau_{yx}$, $\tau_{yz} = \tau_{zy}$, and $\tau_{zx} = \tau_{xz}$.

Principal Stresses

In a given loaded structure, a particular element orientation exists for which all the shear stress components are zero. The normals to the faces of an element in this orientation are called the *principal directions* and the stresses along these normals are the *principal stresses*. They are referred to as the maximum (σ_1), middle (σ_2), and minimum (σ_3) principal stresses and are ordered from the most positive to the most negative. It must be pointed out that the principal stress orientation is a function of loading, not geometry only.

When one of the principal stresses is zero, the stress state is considered to be *biaxial* or *plane* stress. These problems can be deconstructed into planar approximations in which the loading and boundary conditions are in that plane and identical on any parallel plane. No stress is normal to that plane, although strain usually is.

A case of plane stress in which there is no strain in the direction normal to the plane of loading is called *plane strain*. This condition assumes that a cross section cuts an infinite depth and any member end effects can be ignored.

When two of the principal stresses are zero, the condition is known as *uniaxial* stress.

Fig. 2.6. General plane stress element (a). Orientation of principal stress (b). Orientation of maximum shear stress (c).

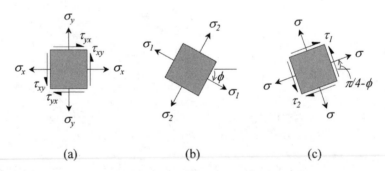

(a)   (b)   (c)

In a plane stress situation [see Fig. 2.6(a) above], the maximum and minimum principal stresses are easily determined from the solution of the following quadratic equation.

$$Eq.\ 2.17 \quad \sigma_{1,2} = \frac{\sigma_x + \sigma_{xy}}{2} \pm \sqrt{\left(\frac{\sigma_x - \sigma_y}{2}\right)^2 + \tau_{xy}^2}$$

The solution of this equation [see Fig. 2.6(b)] assumes that you have access to the local normal (σ_x, σ_y) and shear (τ_{xy}) stress data. Note that the angle between the principal stress orientation and the measured stress state is denoted by ϕ.

For the same biaxial condition, there is an additional element orientation of interest for which the shear stresses are a maximum, although the corresponding normal stresses are not zero (see c in previous illustration). This orientation is 45° away from the orientation of principal stress, and its stress state is given by the next two equations.

Eq. 2.18 $\quad \tau_{1,2} = \pm\sqrt{\left(\dfrac{\sigma_x - \sigma_y}{2}\right)^2 + \tau_{xy}^2}$

Eq. 2.19 $\quad \sigma = \dfrac{\sigma_x + \sigma_y}{2}$

In a general *triaxial* case, the calculation of the three principal stresses entails finding the roots of the following third order equation:

Eq. 2.20 $\quad \sigma^3 + I_1\sigma^2 + I_2\sigma - I_3 = 0$

where

Eq. 2.21 $\quad I_1 = \sigma_x + \sigma_y + \sigma_z$

$\qquad\qquad I_2 = \sigma_x\sigma_y + \sigma_y\sigma_z + \sigma_z\sigma_x - \tau_{xy}^2 - \tau_{yz}^2 - \tau_{zx}^2$

$\qquad\qquad I_3 = \sigma_x\sigma_y\sigma_z + 2\tau_{xy}\tau_{yz}\tau_{zx} - \sigma_x\tau_{yz}^2 - \sigma_y\tau_{zx}^2 - \sigma_z\tau_{xy}^2$

I_1, I_2, and I_3 are called *stress invariants*. I_1, or the first invariant, is internal hydrostatic pressure. This value is important in crack and fracture analysis.

In this general triaxial case, the maximum shear stress is given in terms of the maximum and minimum principal stresses as follows.

Eq. 2.22 $\quad \tau_{max} = \dfrac{\sigma_1 - \sigma_3}{2}$

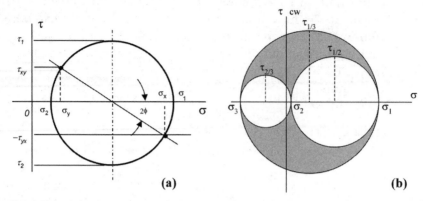

Fig. 2.7. Mohr's circle diagram for plane stress (a) and triaxial stress (b).

All elemental stress state expressions introduced above may be readily derived and represented with the use of *Mohr's circle diagram*. This very efficient graphical method is shown in Fig. 2.7 for the plane stress condition (a), and the general triaxial case (b). Mohr's circle diagrams can provide much useful information once you are comfortable with the format. Refer to a mechanics of materials textbook for a more complete description.

Strain

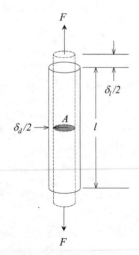

As depicted in Fig. 2.8, the change in size of an element in a body with respect to its original size is known as *unit strain* or simply *strain* (ε). This quantity is related to the total elongation or *total strain* (δ_l) in a bar of length l under uniaxial loading.

Eq. 2.23 $\varepsilon = \dfrac{\delta_l}{l}$

For a stress element subjected to pure shear, the change in an edge angle from 90° is known as *shear strain* (γ).

Fig. 2.8. Deformation of a uniform bar under uniaxial loading.

There are two intrinsic elastic properties that enable a material to regain its original dimensional shape after an applied load has been removed. The properties are known as the material's *modulus of elasticity* or *Young's modulus (E), and modulus of rigidity (G)*. According to Hooke's law, stress in a material will be linearly proportional to strain within certain limits. Although not all materials that regain their original shape obey Hooke's law, the materials that do are considered elastic and are governed by the following relations.

Eq. 2.24 $\sigma = E\varepsilon$
$\tau = G\gamma$

A uniformly distributed stress takes place in the normal, cross-sectional area A away from the ends of a bar under a uniaxial loading F, and it is calculated based on the following equation.

Eq. 2.25 $\sigma = \dfrac{F}{A}$

In this case, making use of Eq. 2.23, the total elongation of the bar will reduce to the next equation.

Eq. 2.26 $\delta_1 = \dfrac{Fl}{AE}$

In the same bar, under similar uniaxial loading, within the deformation region governed by Hooke's law, there will exist an additional lateral deformation (δ_d) which will cause a lateral strain proportional to the axial strain and given by *Poisson's ratio*.

Eq. 2.27 $v = \dfrac{-\varepsilon_d}{\varepsilon}$

Note the negative sign, which indicates "narrowing" of the bar under tension and "bulging" of the bar under compression.

It turns out that the three elastic constants mentioned above are related by the following equation.

Eq. 2.28 $E = 2G(1 + v)$

Principal Strain

The strains that occur in the direction of principal stresses are known as *principal strains*. Note that all shear strains will be zero for this element orientation. Hence, if you were to experimentally obtain the principal strains in the location of a body, they could be related to the principal stresses at that point. Table 2.1 shows these relations for the three types of stress states.

Table 2.1. Converting principal strain values to principal stress

Uniaxial stress	Biaxial stress	Triaxial stress
$\sigma_1 = E\varepsilon_1$	$\sigma_1 = \dfrac{E(\varepsilon_1 + v\varepsilon_2)}{1 - v^2}$	$\sigma_1 = \dfrac{E\varepsilon_1(1 - v) + vE(\varepsilon_2 + \varepsilon_3)}{1 - v - 2v^2}$
$\sigma_2 = 0$	$\sigma_2 = \dfrac{E(\varepsilon_2 + v\varepsilon_1)}{1 - v^2}$	$\sigma_2 = \dfrac{E\varepsilon_2(1 - v) + vE(\varepsilon_1 + \varepsilon_3)}{1 - v - 2v^2}$
$\sigma_3 = 0$	$\sigma_3 = 0$	$\sigma_3 = \dfrac{E\varepsilon_3(1 - v) + vE(\varepsilon_1 + \varepsilon_2)}{1 - v - 2v^2}$

Fundamental Stress States

Most of the time in real world scenarios, the boundary conditions to which a part is subject place it in a complex state of stress. This stress state is extremely difficult to quantify with simple analytical means, unless it can be broken into a linear sum of more basic states. To this end, this section continues by presenting the most significant of these fundamental stress states which arise in mechanical systems. For completeness and reference, each of the following topics will be introduced with a "screen capture" of its corresponding representative FEA solution.

Stress in Flexure

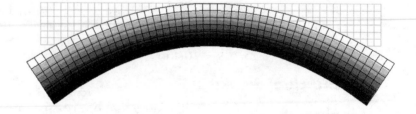

Fig. 2.9. FEA of a beam in flexure.

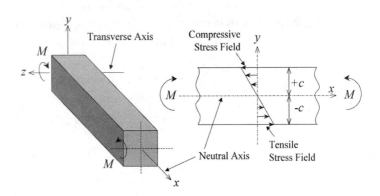

Fig. 2.10. Straight beam in flexure.

Assume a material that is both isotropic and homogeneous, such that its properties are direction independent and do not change throughout its expanse (see Figs. 2.9 and 2.10). Assume as well that this material obeys Hooke's law. Now, consider an initially straight beam of constant cross-section, which is made up of this material and is subjected to a pure bending moment (M). If its cross sections remain planar and either one of the section's principal axes coincides with the plane of bending, the normal stresses developed follow.

$$Eq.\ 2.29 \quad \sigma = -\frac{My}{I}$$

In Eq. 2.29, I denotes the area moment of inertia of the beam's cross section about its transverse axis, and y is the normal distance away from the *neutral axis* in the plane of bending. The neutral axis is defined by the intersection of the plane of bending with the longitudinal surface of zero fiber stress, or *neutral surface*. The location of the neutral surface may be calculated by finding the horizontal cross-sectional line about which the area moments are equal.

Note that the sign notation of Eq. 2.29 follows the use of a right-hand coordinate system. It follows that this flexural stress will be a maximum at $y_{max} = c$, where the sign of c will govern the direction of the stress per this equation.

For certain materials, it is useful to measure a *flexural*, or *bending modulus* (E_{BR}) using the following equation:

$$Eq.\ 2.30 \quad E_{BR} = \frac{MR}{I}$$

where R is the radius of curvature of the deformed neutral axis. For materials that meet all of the assumptions stated in the development of Eq. 2.29, the flexural and elastic modulii are equal.

Fig. 2.11. Curved beam in flexure.

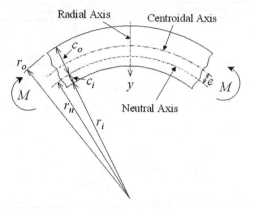

If the beam under consideration is originally curved as seen in Fig. 2.11, the neutral axis is no longer located by the centroidal axis of the section. These two are separated by a distance (e), and the former can be located by a neutral axis radius of curvature (r_n) as calculated by the following equation:

Eq. 2.31 $$r_n = \frac{A}{\int \frac{dA}{r}}$$

where A is the usual cross-sectional area of the beam. Eq. 2.29 then becomes the following equation.

Eq. 2.32 $$\sigma = \frac{My}{Ae(r_n - y)}$$

The critical stresses are found at the inner- and outermost fibers and are given by

Eq. 2.33 $$\sigma_i = \frac{Mc_i}{Aer_i}$$

$$\sigma_o = -\frac{Mc_o}{Aer_o}$$

where c_i and c_o are the distances from the neutral axis to the inner and outer fibers, respectively, and r_i and r_o are their respective fiber curvature radii.

Stress in Shear

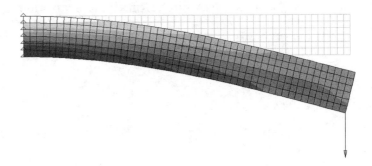

Fig. 2.12. FEA of a cantilever beam in shear.

Fig. 2.13. Straight rectangular section beam in shear.

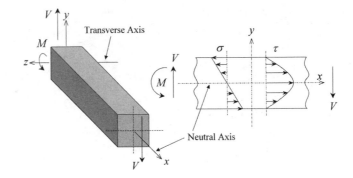

In the real world, subjecting a beam to pure bending is very rare. A beam will typically be subjected to both shearing forces and bending moments. Although more difficult to derive mathematically, the presence of a shearing force will not invalidate the results from the previous section. Yet, in addition to the normal stresses due to the bending moment, there will now be additional shear stresses due to the shearing force. These shear stresses are governed by the following equation:

Eq. 2.34 $\tau = \dfrac{VQ}{Ib}$

where V is the shear force, b is the width of the stress section, and Q is the *first moment of area* of the section about a transverse axis with origin on the neutral axis. Fig. 2.13 shows the resulting general shear stress distribution for a rectangular section beam. Note that it is a maximum at the neutral axis and zero at the outer surfaces. Maximum magnitude solutions for Eq. 2.34 for a number of standard cross sections are provided in Table 2.2.

Table 2.2.
Maximum shear stress formulas for selected standard cross sections

X-Section	Formula
Rectangular, solid	$\tau_{max} = \dfrac{3V}{2A}$
Circular, solid	$\tau_{max} = \dfrac{4V}{3A}$
Circular, hollow	$\tau_{max} = \dfrac{2V}{A}$

Stress in Torsion

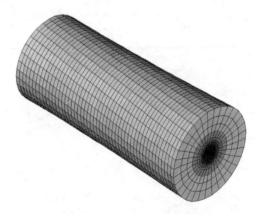

Fig. 2.14. FEA of a round
bar in torsion.

Fig. 2.15. Solid round bar in
torsion.

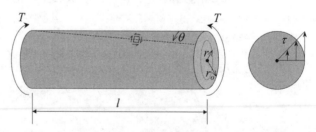

When a torque (T) is applied to an elastic beam about its longitudinal
axis, a state of shear stress develops away from the torsional axis. For a
round section (hollow or solid in Fig. 2.15), if plane and parallel cross
sections remain plane and parallel, and radial lines continue to be
straight, this stress is given by the following equation.

$$Eq.\ 2.35 \quad \tau = \frac{Tr}{J}$$

Here, r is the radius from the torsional axis and J is the section's polar area moment of inertia. Of course, the maximum torsional shear stress will occur at $r_{max} = r_o$, the outer radius of the bar. The angular deflection (θ) at the end of a solid round bar of length l follows:

$$Eq.\ 2.36 \quad \theta = \frac{Tl}{GJ}$$

where G is the material's modulus of rigidity.

A note must be made here regarding Eq. 2.36. Although this equation as presented may not be used for beams with cross sections that are not perfectly circular (hollow or solid), by substituting the variable K for J the equation may be generalized for all other sections. K represents the section's *torsional stiffness factor*. This factor is equal to J for round sections, yet it is less than J for all other sections. When assigning line element properties in FEA, you must input K. Beware of the fact that many FEA preprocessors confuse the nomenclature by referring to K as J. However, by erroneously inputting the section's polar area for its torsional stiffness factor, you will effectively understiffen the line elements in torsion. K values for a variety of beam sections are tabulated in many engineering textbooks, notably *Roark's Formulas for Stress and Strain* by Warren C. Young (McGraw Hill, 1989). Refer to Chapter 7 for more information on line element properties.

It is very difficult to obtain formulas for both stress distribution and maximum stress in beams with noncircular sections. Hence, most are done experimentally. The following is an approximate formula for a rectangular section beam of width (w) and thickness (t).

$$Eq.\ 2.37 \quad \tau_{max} = \frac{T}{wt^2}\left(3 + 1.8\frac{t}{w}\right)$$

Stress in Pressure

Fig. 2.16. FEA of a thick cylinder under pressure.

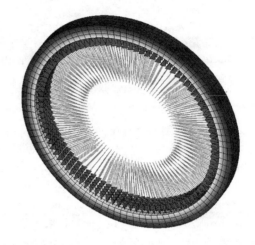

Fig. 2.17. Cylinder under pressure (a). Press fit cylinders (b).

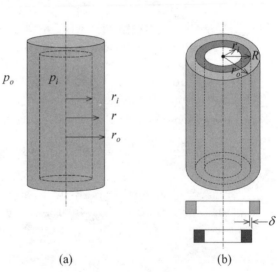

(a)   (b)

A capped cylinder subject to internal and/or external pressure, as shown in Fig. 2.17(a), will develop tangential, radial, and longitudinal normal stresses. For a cylinder with an internal radius and pressure (r_i and p_i) and external radius and pressure (r_o and p_o), the tangential and radial stress magnitudes at a given radius (r), a good distance away from the end caps are calculated as follows:

$$Eqs.\ 2.38 \quad \sigma_t = \frac{p_i r_i^2 - p_o r_o^2 - r_i^2 r_o^2 (p_o - p_i)/r^2}{r_o^2 - r_i^2}$$

$$\sigma_r = \frac{p_i r_i^2 - p_o r_o^2 + r_i^2 r_o^2 (p_o - p_i)/r^2}{r_o^2 - r_i^2}$$

where σ_t is also known as the *hoop stress* in the cylinder.

The longitudinal stress due to pressure on the end caps is constant throughout the cylinder and is given by the next equation.

$$Eq.\ 2.39 \quad \sigma_l = \frac{p_i r_i^2 - p_o r_o^2}{r_o^2 - r_i^2}$$

For a thin-walled cylinder with thickness (t) less than about one-twentieth of its radius, the maximum tangential and constant longitudinal stresses under internal pressure (p) and negligible outside pressure are calculated via the next equations.

$$Eqs.\ 2.40 \quad \sigma_{t,\,max} = \frac{p(d_i + t)}{2t}$$

$$\sigma_l = \frac{pd_i}{4t}$$

Of course, the radial stress in this case would simply be the internal pressure on the inside of the cylinder and zero on the outside.

It is useful to note that if the pressure created due to an equal-length cylindrical press-fit was known, Eqs. 2.38 could be used for obtaining the stress state on both the outside and inside cylinders. Hence, this pressure is given here as a function of the radial interference (δ):

$$Eq.\ 2.41 \quad p = \frac{\delta}{R}\left[\frac{1}{E_o}\left(\frac{r_o^2 + R^2}{r_o^2 - R^2} + v_o\right) + \frac{1}{E_i}\left(\frac{R^2 + r_i^2}{R^2 - r_i^2} - v_i\right)\right]^{-1}$$

where now [see Fig. 2.17(b)], r_i is the inside cylinder's internal radius, r_o is the outside cylinder's external radius, R is the transition radius, and E_i, E_o, v_i, and v_o are the inside and outside material Young's moduli and Poisson's ratios, respectively.

Stress in Contact

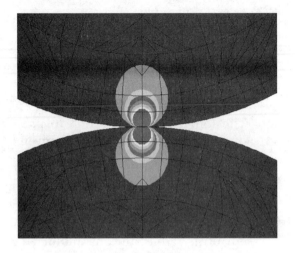

Fig. 2.18. FEA of two spheres in contact.

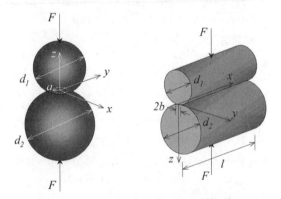

Fig. 2.19. Forced contact of two spheres (a) and two cylinders (b).

The resulting stress state developed within each of two bodies being pressed together is a complex nonlinear phenomenon that can only be resolved analytically in closed form for a handful of special cases. In this section, *Hertzian stresses* resulting from the forced contact of two spheres and two equal-length cylinders will be presented (see Fig. 2.19).

When two solid spheres of diameters d_1 and d_2 are pressed together with a force F, their contact point turns into a contact circular area of radius a given by

$$Eq.\ 2.42 \qquad a = \sqrt[3]{\frac{3F}{8}\frac{(1-v_1^{\,2})/E_1 + (1-v_2^{\,2})/E_2}{1/d_1 + 1/d_2}}$$

where E_1, E_2, v_1, and v_2 are the respective elastic constants of the two spheres' materials. At the center of this area, a maximum pressure p_{max} will occur of the following magnitude.

$$Eq.\ 2.43 \quad p_{max} = \frac{3F}{2\pi a^2}$$

Placing a coordinate system at this center point, with the contact circle on the xy plane and the z axis normal to either one of the spheres, the principal stresses coincide with these axes and are a function of the z coordinate according to the following equation:

$$Eq.\ 2.44 \quad \sigma_x = \sigma_y = -p_{max}\left[\left(1 - \frac{z}{a}\tan^{-1}\frac{1}{z/a}\right)(1 + v) - \frac{1}{2(1 + z^2/a^2)}\right]$$

$$\sigma_z = \frac{-p_{max}}{1 + z^2/a^2}$$

where v is the Poisson's ratio of the sphere under consideration. The maximum principal shear stress is then given by the next equation.

$$Eq.\ 2.45 \quad \tau_{xz} = \tau_{yz} = \frac{\sigma_x - \sigma_z}{2} = \frac{\sigma_y - \sigma_z}{2}$$

Note that the three normal stresses are compressive and highest at the contact surface, yet the shear stress reaches a maximum slightly below this surface.

For two cylinders of equal length l and diameters d_1 and d_2, the resulting contact surface is a rectangle of length l and width $2b$, where

$$Eq.\ 2.46 \quad b = \sqrt{\frac{2F}{\pi l}\frac{(1 - v_1^2)/E_1 + (1 - v_2^2)/E_2}{1/d_1 + 1/d_2}}$$

The maximum pressure occurs along the long center line of the rectangle.

$$Eq.\ 2.47 \quad p_{max} = \frac{2F}{\pi b l}$$

Placing a similar coordinate system on this contact region, but this time orienting the x axis parallel to the axes of the cylinders, the resulting normal principal stresses are given by the following equations.

Eqs. 2.48 $\sigma_x = -2vp_{max}\left(\sqrt{1 + \dfrac{z^2}{b^2}} - \dfrac{z}{b}\right)$

$$\sigma_y = -p_{max}\left[\left(2 - \dfrac{1}{1 + z^2/b^2}\right)\sqrt{1 + \dfrac{z^2}{b^2}} - 2\dfrac{z}{b}\right]$$

$$\sigma_z = \dfrac{-p_{max}}{\sqrt{1 + z^2/b^2}}$$

The resulting principal shear stresses are provided in Eqs. 2.49.

Eqs. 2.49 $\tau_{xy} = \dfrac{\sigma_x - \sigma_y}{2}$

$$\tau_{xz} = \dfrac{\sigma_x - \sigma_z}{2}$$

$$\tau_{yz} = \dfrac{\sigma_y - \sigma_z}{2}$$

Again, note that the normal stresses are compressive and maximum at the contact surface. The maximum shear stress is τ_{zy} at about $z = 0.75b$.

The equations in this section apply to a sphere or cylinder in contact with a planar surface by letting d for this surface equal infinity, and a sphere or cylinder in contact with an internal spherical or cylindrical surface, respectively, by letting d be negative.

Stress in Thermal Expansion

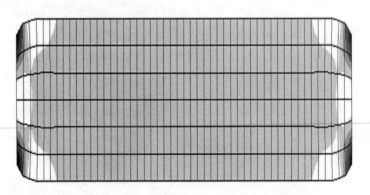

Fig. 2.20. FEA of a shaft subject to a temperature change with the ends fully constrained.

If an unconstrained body is subjected to a temperature change (ΔT), it will expand proportionally in all directions according to the next equation.

$$\text{Eq. 2.50} \quad \varepsilon_x = \varepsilon_y = \varepsilon_z = \alpha(\Delta T)$$

The constant of proportionality α is known as the material's coefficient of thermal expansion. This strain state will simply result in a volumetric expansion and the stress state will be zero.

If, instead, the body is constrained, the resulting stress state is nonzero and its complexity will depend on both the geometry of the body and the configuration of the constraint. For a straight beam constrained at both ends, the resulting compressive stress at a distance from the ends is given by the next equation.

$$\text{Eq. 2.51} \quad \sigma = \alpha(\Delta T)E$$

Stress Concentration Factors

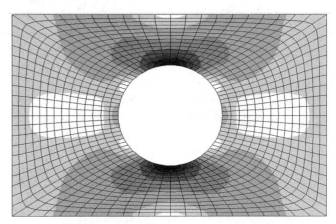

Fig. 2.21. FEA of a drilled plate under tension.

Thus far, this section has concerned itself with very idealized geometries. Yet, real parts will usually have design features, such as holes, notches, fillets, steps, grooves, and so forth, that will cause the idealized stress distribution to develop highly localized regions of concentrated stress in their vicinities. These *stress concentrations* will also appear close to unplanned irregularities in the part, such as cracks and pits. To account for this, the following *theoretical stress concentration factor K_t or K_{ts}* is introduced that depends on the type of stress involved.

$$\text{Eqs. 2.52} \quad K_t = \frac{\sigma_{max}}{\sigma_o}$$

$$K_{ts} = \frac{\tau_{max}}{\tau_o}$$

Here, σ_0 and τ_0 are the nominal stresses found in the part without the feature. Sometimes, depending on the relative size of the feature with respect to the size of the part, these nominal stresses will use the net instead of the gross cross-sectional area of the part in their calculation.

Stress concentration factors are mostly found through testing. Hence, with the aid of controlled experimentation, many of these factors have been tabulated in the literature for many different feature types existing in many different part types. These can often be used accurately in analysis with Eqs. 2.52.

Material Properties

Material properties can be found in various references with varying degrees of accuracy. It is not so much that the publisher of such data was neglectful or intentionally misleading. The fact of the matter is that most materials behave differently under different conditions (i.e., temperature, strain rate, processing conditions, etc.). Even steel, one of the more predictable engineering materials, has different failure properties depending on alloying, heat treatment, cold working, or manufacturing method. When the source or testing method for material properties is unknown or questionable, consider results based on these properties to be questionable. When accurate stress data are required for a failure or a fatigue calculation, independent testing of the material should be performed under anticipated loading conditions on a part with a shape similar to the part being studied.

Types of Materials

- *Isotropic.* Properties are the same in any direction or at any cross section.

- *Anisotropic.* Properties differ in two or more directions.

- *Orthotropic.* Specific type of anisotropic in which planes of extreme values are orthogonal (i.e., perpendicular to one another).

Most analyses are performed under the assumption of isotropic and homogeneous material properties. Homogeneous materials have consistent properties throughout the volume. This is an important approximation to understand in predictive engineering. It is critical when performing failure verification. Many parts fail due to inconsistencies in processing, heat treating, or internal voids which locally reduce the inherent stiffness of the material.

Most FEA systems allow for the specification of orthotropic properties, such as those exhibited by composite materials. The definition of composite material properties can be extremely complex and should not be undertaken the first time without the advice of an expert.

Common Material Properties

In the previous section, three material dependent properties were presented: the modulus of elasticity (E), modulus of rigidity (G), and Poisson's ratio (v). These properties remain virtually constant for all materials of the same type. The properties introduced in this section are more manufacturing process dependent and can vary greatly between two bodies of the same material type which were manufactured differently.

Arguably, the most useful of all material properties to have access to is its *strength*. Note that the units of strength are the same as those of stress. Yet, whereas stress in a body is always a function of the applied loading and cross section, strength is an inherent property of the body's material/manufacturing process and governs the overall performance of its design.

Strength and other typical material properties are often obtained from a standard tensile test which subjects a sample bar to uniaxial stress, extracting deformation versus applied load data. This information is then plotted on a stress-strain diagram to illustrate the relationship between the various parameters of interest. If a high degree of accuracy is required, material properties used in a structural analysis should be obtained from testing under conditions similar to the actual operating conditions of the part or system. These operating conditions should be known, even if such a level of accuracy is not necessary, in order to adjust the data to compensate. Conditions to consider include operating temperature, strain rate, material grain or flow direction with respect to loading, and torsion versus tension versus bending.

*Fig. 2.22. Typical
engineering stress-strain
diagram of a ductile
material specimen subject
to uniaxial tensile loading.*

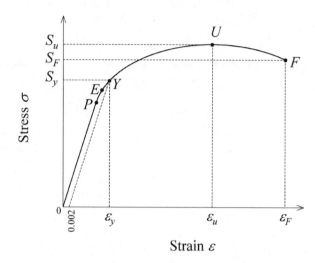

A typical stress-strain plot reflects *engineering stress* (see Fig. 2.22). This is a somewhat inaccurate quantity because it is calculated by dividing the applied load (F) by the original cross-sectional area (A). Typically, a localized, irreversible decrease in cross-sectional area (or *necking*) occurs in a tensile test. This occurrence is shown in an engineering stress-strain curve when the stress appears to decrease with increasing strain. In actuality, the reduced cross-sectional area causes the resulting stress state to keep rising as a function of strain. The *true stress-strain curve*, which accounts for the specimen's necking, typically shows greater stress values than the engineering stress. However, because it is difficult to track the change in the cross-sectional area of the specimen, the engineering stress-strain curve is used almost exclusively.

As seen in Fig. 2.22, the stress-strain relation will initially be linear up to point P, known as the *proportional limit*. This portion of the curve is governed by Hooke's law and Eq. 2.24. Then, although not proportional anymore, the material will continue to behave elastically up to point E, the *elastic limit.* Beyond this point, permanent deformation will be observed upon removal of the load.

Point Y is the *yield point* of the material, corresponding to its *yield strength* (S_y). Some materials, such as ferrous materials, have a distinct yield point or "knee" on the stress-strain curve. For these materials, points E and Y will coincide. Yet, for many other materials, the yield point is less clear and is often determined by an *offset method.* Using this method, Y is considered to be the intersection of an offset line, parallel to the linear portion of the stress strain curve typically at .002 axial strain, and the plastic portion of the curve.

Farther up the stress-strain curve, point U indicates the maximum stress that can be achieved by the material. This corresponds to its *ultimate* or *tensile strength* (S_u or S_{ut}). Beyond this, as already described, necking may occur up to the *fracture point* (F), which marks the *fracture strength* (S_F) of the material.

An additional material property that is usually reported and proves useful to the engineer is its *percent elongation*. This gives, as a percentage, the strain of the tensile specimen prior to failure and is indicative of the *ductility* of the material as will be described below.

Although all terms presented here are derived from a tensile test, analogous compressive strength numbers exist that, although harder to obtain, are sometimes useful. Torsional tests can also be performed on bars to come up with a *torque-twist diagram* that provides values for *torsional yield strength* (S_{sy}) and the *modulus of rupture* (S_{su}). For the latter, there is actually an equation:

$$Eq.\ 2.53 \quad S_{su} = \frac{T_u r}{J}$$

where T_u is the maximum point U on the torque-twist diagram, r is the radius of the bar, and J its second moment of area.

Ductile versus Brittle Material Behavior

According to the previous discussion, a body is said to have yielded or to have undergone plastic deformation if it does not regain its original shape when a load is removed. The resulting deformation is called *permanent set*. If permanent set is obtainable, the material is said to exhibit *ductility*. A measure of ductility comes from the percent elongation, or strain at failure, of a tensile test specimen. Brittle materials will have a much lower elongation and area reduction than ductile ones. Hence, the amount of necking, and the corresponding dip in the engineering stress-strain curve, is indicative of the ductility of the material.

For ductile materials, the ultimate tensile and compressive strengths have approximately the same absolute value. Brittle materials on the other hand are stronger in compression than in tension.

Brittle materials exhibit the behavior described below.

- A graph of stress versus strain is a smooth, elastic curve until failure which manifests as fracture. Materials behaving in this manner do not have a "yield strength."

- Compressive strength is usually many times greater than tensile strength.

- Modulus of rupture is approximately the same as tensile strength.

- Rapid crack propagation along cleavage planes occurs with no noticeable plastic deformation.

The structural analyst should have a feel for whether the material being studied will behave as ductile or brittle at the temperatures and strain rates expected. Most materials become more brittle as strain rate increases and as temperature decreases. The method of results evaluation and failure quantities used are dependent on this property.

Rules of thumb used to determine if brittle or ductile behavior should be expected are summarized below.

- If the percent elongation is at or below 5%, assume brittle behavior.

- If the published ultimate *compressive* strength is greater than the ultimate tensile strength, assume brittle behavior.

- If no yield strength is published, suspect brittle behavior.

Safety Factor

Having introduced both stress and strength, now is a good time to present a quantity that relates the two in the design process: the *safety factor* (n). This number is defined as the quotient of the strength divided by the stress in a part, and it provides an indication of the level of confidence in not only the accuracy of the inputs used and their representation in the analysis, but also in the accuracy of the analysis tool itself.

$$Eq.\ 2.54 \quad n = \frac{strength}{stress_{max}}$$

For example, if (a) you were to obtain material properties, geometry, and boundary conditions that were known to be 100% indicative of every operating state of a part, (b) representation of these in the analysis model were known to be 100% accurate, and (c) the tool used for the analysis, FEA or otherwise, were known to be 100% accurate, you could design the part in question so that the maximum stress output of the analysis is equal to the material strength, $n = 1$. Questions regarding the accuracy of any of the above require you to assign a safety factor greater than unity to account for both foreseen and unforeseen variances. Of course, the larger the safety factor, the less efficient the design becomes. Hence, the safety factor chosen for each design must be a solid compromise utilizing the engineering judgment of the entire design team.

The discussion above assumes, in the strictest sense, that the design parameter of interest in the analysis is maximum stress. If this quantity is of a different nature, such as maximum displacement or maximum temperature, the same concepts apply. To obtain a parameter-specific safety factor, you need only modify Eq. 2.54 by placing the limiting quantity in the numerator and the measured quantity in the denominator.

Failure Modes

Results Interpretation

The first step in results interpretation is to review the goals set forth at the beginning of the study. These should tell you where to look and what to look for. In most cases, you will be looking for some evidence of failure or assurance that failure is unlikely. With this in mind, a few words on the nature of failure in engineering design are warranted before specific failure predictors are discussed.

Typical Failure Modes

Appearing below are summary descriptions of the more common types of mechanical failure.

- *Fracture.* Fracture is said to occur when new cracks appear or existing cracks are extended. A brittle fracture is one that exhibits little or no permanent (plastic) deformation.

- *Yielding.* A body which experiences stresses in excess of the yield strength is said to have failed only when this yielding compromises the integrity or function of the part. Yielding near stress concentrations is not considered a failure if it produces localized strains which merely redistribute the stress, whereupon yielding ceases.

- *Insufficient stiffness.* Parts must be stiff enough to hold tolerances and support required loads. Moving parts may have undesirable resonant frequencies if they are too flexible.

- *Buckling.* The sudden loss of stability or stiffness under applied load. Stress levels need not be high for buckling to occur.

- *Fatigue.* Parts that are subject to variable loading will lose strength with time and may fail after a certain number of cycles.

- *Creep.* Bodies under load gradually deform over time. The *apparent modulus* property is derived from empirical creep data for various materials and may be used to compensate for the effects of creep.

In most engineering problems, two or more of these failure modes may be possible given the operating conditions of the system. It is important to review the data for all occurrences of any potential failure.

Classic Failure Theories

In cases where failure due to yielding or fracture is of interest, choosing the correct stress quantities and applying the appropriate failure predictor or theory is important. Several of the more general and widely used failure theories are discussed below. These classic failure theories exclude existing macroscopic cracks, buckling, creep, or excessive elastic failure and are primarily concerned with material failure.

These theories are not derivable laws but tend to provide unifying accounts of experimental data. Published material data are frequently determined by testing in uniaxial stress states, whereas actual engineering problems are frequently biaxial or complex stress states. Moreoever, a material may perform in a ductile manner at one temperature or loading condition, yet fail in a brittle mode at another. It is extremely important to understand the load path and material behavior for the particular test condition regardless of the means of analysis.

Ductile Failure Theory

A ductile material under static load can redistribute stress by yielding without fracture. This is illustrated by the forming of a stamped metal part. A load that produces yielding sets up residual stresses that extend the elastic range under future loads in the same direction but decrease the elastic range under future loads in the opposite direction. This is called the *Bauschinger effect*. Ductile failure is characterized by slow crack or void propagation after significant plastic deformation.

- *Maximum normal stress theory.* Failure occurs whenever σ_1 or σ_3 equals the failure strength of the material in tension or compression, respectively. Failure by yielding occurs when the yield strength is reached. Failure by fracture occurs when the ultimate strength is reached.

- *Maximum shear stress theory (Tresca criterion).* Yielding begins when the maximum shear stress becomes equal to one-half the yield strength. Failure in tension of ductile materials occurs on one of the 45° maximum shear planes. Annealed ductile materials tend to fail according to this theory. This theory only predicts yield failure, hence it is only good for ductile materials. The maximum shear stress τ_{max} is calculated using Eq. 2.22 and is oriented on planes at 45° from the σ_1 and σ_3 planes. This theory is suggested by the fact that yielding is related to shear slip at the atomic level of materials.

- *Distortion energy (Von Mises-Hencky) theory.* Probably the most widely used, this theory predicts that failure by yielding will occur whenever the von Mises, or effective stress (σ'), equals the yield strength of the material. This stress quantity is derived using a strain energy hypothesis and is given by the following equation.

$$Eq.\ 2.55 \quad \sigma_{vm} = \left[\frac{(\sigma_1 - \sigma_2)^2 + (\sigma_2 - \sigma_3)^2 + (\sigma_1 - \sigma_3)^2}{2} \right]^{1/2}$$

The beauty of the above equation is that it represents the entire stress state, no matter how complex it is.

Fig. 2.23 depicts the three ductile failure theories for a plane stress situation. The stress states defined by the locus of points enclosed by the ellipse, the square, and the truncated polygon describe the safe stress combinations as predicted by the distortion energy, maximum normal, and maximum shear stress theories, respectively.

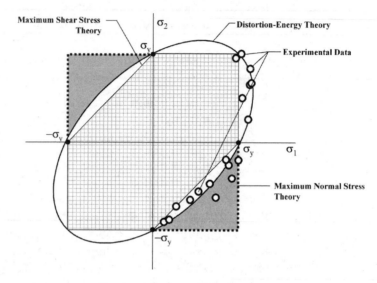

Fig. 2.23. Comparison of three ductile failure theories.

If σ_1 and σ_2 are similar in sign and magnitude, the maximum normal stress theory reasonably predicts behavior, but knowing all circumstances in which this theory applies is difficult. The maximum shear stress theory will be conservative but acceptable from a design standpoint. Meanwhile, the best match with experimental data is provided by the distortion energy theory.

Brittle Failure Theory

A brittle material cannot be considered to have failed until it has broken. This can occur either through a tensile fracture, when the maximum tensile stress reaches the ultimate tensile strength, or through what appears to be a shear fracture, when the maximum compressive stress reaches the ultimate compressive strength. The latter fracture occurs on a plane oblique to the maximum compressive stress but not, as a rule, on the plane of maximum shear stress. Therefore, it cannot be considered to be purely a shear failure.

- *Maximum normal stress.* Similar to that defined for ductile materials. Failure occurs when the ultimate strength, not yield, is reached.

- *Coulomb-Mohr theory.* Fracture occurs when the maximum and minimum principal stresses combine for a condition which satisfies the following:

Eq. 2.56 $\quad \dfrac{\sigma_1}{S_{ut}} - \dfrac{\sigma_3}{S_{uc}} \geq 1$

where S_{ut} and S_{uc} represent the ultimate tensile and compressive strengths, and both σ_3 and S_{uc} are always negative, or in compression.

Although this theory is applicable to both ductile and brittle materials, it is applied more frequently to brittle materials because they are stronger in compression. When compression is dominant ($\sigma_c \gg \sigma_t$), the Mohr criterion is the most reliable predictor.

- *Modified Mohr theory.* Fracture occurs as defined in the Coulomb-Mohr theory except in the fourth quadrant condition where σ_1 is in tension and σ_2 is in compression. In this situation, the material is somewhat stronger than a Coulomb-Mohr plot would suggest. The impact of the modified Mohr theory is shown in Fig. 2.24.

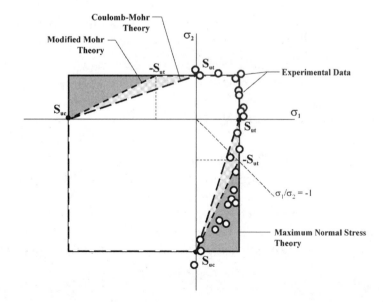

Fig. 2.24. Comparison of three brittle failure theories.

Fig. 2.24 shows the stress states defined by the locus of points enclosed by the outer rectangle, the outer truncated polygon, and the inner truncated polygon. These stress states illustrate the safe stress combinations in a plane stress condition as predicted by the maximum normal stress, Coulomb-Mohr, and modified Mohr theories, respectively.

Highest match rate in experimental results for tests with brittle materials occur with the modified Mohr theory, yet the Coulomb-Mohr theory yields a more conservative prediction and is acceptable for design.

Other Failure Theories

Failure theories excluded in the previous discussion are presented below.

Buckling

In certain situations, the maximum load a member will sustain is determined not by the strength of the material but by the stiffness of the member. The *critical load* (P_{cr}) is defined as the compressive force for which the resulting deformation state of a body is considered to be in unstable elastic equilibrium. Any increase in load over this critical load will cause the body to elastically collapse. Typical conditions where buckling is a concern include a slender (*Euler*) column under axial loading, a thin-walled cylinder under external pressure, a thin plate under edge pressure, and a deep, thin, cantilevered beam under a transverse end load applied at the top surface.

The Euler equation for obtaining the critical central load on a straight column of constant cross-section follows:

$$Eq.\ 2.57 \quad P_{cr} = \frac{\pi^2 EI}{L_e^2}$$

where E is the modulus of elasticity of the column's material, I is the smallest or least moment of inertia of its cross-sectional area, and L_e is its *effective length*. The last term, $L_e = KL$, depends on the actual length L of the column and an effective length factor K, which is assigned according to the constraint conditions of the column ends. Common end conditions and their corresponding K values are shown in Fig. 2.25.

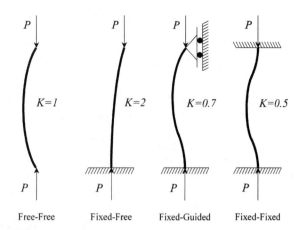

Fig. 2.25. Effective length factors for common end conditions of centrally loaded columns.

Free-Free   Fixed-Free   Fixed-Guided   Fixed-Fixed

It is useful at this point to introduce the quantity r as the smallest radius of gyration of the column's cross-sectional area (A).

$$Eq.\ 2.58 \quad r = \sqrt{\frac{I}{A}}$$

With the use of this equation and Eq. 2.57, a corresponding critical stress (σ_{cr}) may be calculated as seen in the next equation.

$$Eq.\ 2.59 \quad \sigma_{cr} = \frac{\pi^2 E}{(L_e/r)^2}$$

It is now possible to define what constitutes an Euler column. This is done by calculating a *slenderness ratio* (L_e/r) against a prescribed criterion. Specifically, if

$$Eq.\ 2.60 \quad \frac{L_e}{r} > \sqrt{\frac{\pi^2 E}{S_y}}$$

the column is considered Euler, and a critical load must be calculated and recorded.

Note from Eq. 2.59 that the critical stress is governed by the elastic proportionality of the material. Hence, this critical stress is valid for Euler columns since, by definition, their stress state must be below the yield point of the material. Yet, for nonEuler columns, the following more general stress equation must be used.

$$Eq.\ 2.61 \qquad \sigma_{cr} = \frac{\pi^2 E_t}{(L_e/r)^2}$$

In Eq. 2.61, a tangent modulus variable (E_t) has taken the place of the elastic modulus. This quantity is defined as the tangential slope of the stress-strain curve, and is a function of the location along this curve. Of course, below the yield point, $E_t = E$ as expected.

The resulting general curve of σ_{cr} versus L_e/r is shown in Fig. 2.26 and has been verified through testing with excellent accuracy. Note that the extension of Eq. 2.59, or *Euler's hyperbola*, into the nonEuler region of the curve provides nonconservative critical stresses. Hence, it must be noted that the use of a linear FEA code for executing buckling problems will be nonconservative for columns that do not satisfy Eq. 2.60. However, as long as the operating stresses on these nonEuler columns remain below the yield point, buckling is not a possible failure mode.

Fig. 2.26. Critical stress in columns as a function of slenderness ratio.

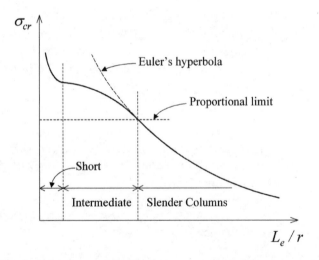

Fig. 2.27 shows FEA results for a complex model in buckling.

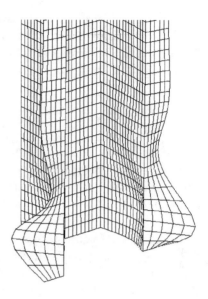

Fig. 2.27. Buckling FEA of a complex shell structure.

Since few columns are perfectly straight, it is useful to present an equation that resolves the maximum axial stress for an eccentrically loaded column shown in Fig. 2.28(a). Although it can be shown that an additional axial load can be supported by such a column before buckling, the development of the exact critical stress is complex and hard to generalize.

Fig. 2.28. Eccentrically loaded column (a). Maximum unit load for different values of load eccentricity and column slenderness ratio (b).

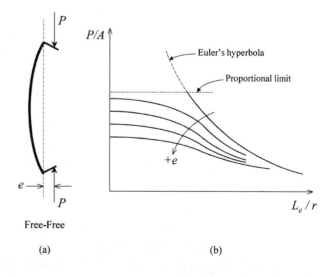

Hence, it is conservatively assumed that failure constitutes material yielding in compression. This assumption yields the following equation for the maximum unit load (P/A) that can be supported by an eccentrically loaded column:

$$Eq.\ 2.62 \quad \frac{P}{A} = S_{yc}\left[1 + \frac{ec}{r^2}\sec\left(\frac{L_e}{r}\sqrt{\frac{P}{4AE}}\right)\right]^{-1}$$

where e is the eccentricity of the load and c is the distance from the neutral axis to the surface fiber in compression. Note that in this case r is not necessarily the minimum radius of gyration—it corresponds to the I associated with the axis around which bending occurs. Note also that this equation cannot be solved explicitly for P/A; hence, a graphical solution is most often utilized. Fig. 2.28(b) shows a typical solution in the shape of patterns of unit load as functions of both eccentricity and slenderness ratio. Note that as either the slenderness ratio increases or the eccentricity decreases, these curves asymptotically approach Euler's hyperbola.

Fatigue

Fig. 2.29. Typical S-N diagram for steel.

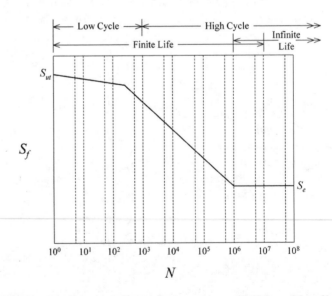

Fatigue data in the form of material strength versus number of loading cycles, or *S-N charts*, are generally derived from a reverse bending, rotating beam test. Fig. 2.29 shows a typical S-N chart for steel. Note how the strength of steel drops at a certain rate up until about 1,000 cycles and then continues to drop at a higher rate. This change in slope separates what are considered *low-* and *high-cycle fatigue* failures. As the cyclic loading of the specimen continues past this point, the strength of the material will stabilize somewhere between 10^6 and 10^7 cycles, the number of cycles that empirically represent "infinite" life. The corresponding *endurance* or *fatigue limit* (S_e) is defined as the maximum cyclic stress which a part can sustain for an "infinite" number of cycles. Note that for nonferrous metals and alloys, the strength of the material never stabilizes but keeps decreasing with time. Hence these materials do not have an endurance limit.

The endurance limit of the actual rotating beam specimen is usually designated as S_e'. For ferrous alloys with an ultimate strength below 200 ksi, S_e' is approximately half of this strength. For ferrous alloys with a strength above 200 ksi, S_e' is approximately equal to 100 ksi. Because nonferrous metals and alloys lack an endurance limit, *a fatigue strength* (S_f') is usually reported for $50(10^7)$ cycles of reversed stress. This strength is often as low as $1/4\ S_{ut}$ for some aluminum alloys.

Of course, one must find a way to correlate the endurance strength of a part to that of the test specimen. This is accomplished via several modifying factors, which are all less than or equal to unity, as seen in the next equation.

$$Eq.\ 2.63 \qquad S_e = k_a k_b k_c k_d k_e S_e'$$

Here, k_a is a surface factor, k_b is a size factor, k_c is a load factor, k_d is a temperature factor, and k_e is an all encompassing, other miscellaneous effects factor. Numbers for some of these factors can be readily obtained in the literature while others are rarely available. All are mentioned here simply to point out the difficulty of executing predictive fatigue analysis.

It is crucial to stress that, prior to the attempt of designing for a fatigue situation, you must have concrete values for three parameters which will govern the analysis. The first is the desired number of cycles the part must withstand, which will dictate the value used for its material strength. The second is the loading history of the part, which will provide values for the mean and amplitude stress states that the part will experience. Third is the parameter which always relates the strength and stress in a part—the desired safety factor in its design. Without having a good handle on the values for these parameters, successful analysis is impossible.

Several methods are available to relate cyclic loading data to fatigue life. All such theories are estimates based on empirical data and should only be used for initial design estimates. As a result of sensitivity to so many factors, nothing can replace actual fatigue testing in realistic environments to guarantee a reliable design for a part experiencing cyclic loading.

To obtain the fatigue strength at N cycles for a part experiencing *alternating* or completely reversed stress, you can curve-fit the *S-N* curve using the following equation:

$$Eq.\ 2.64 \quad S_f = aN^b$$

where a and b are provided by

$$Eqs.\ 2.65 \quad a = \frac{(0.9S_{ut})^2}{S_e}$$

$$b = -\frac{1}{3}\log\frac{0.9S_{ut}}{S_e}$$

Note that S_e' may be substituted for S_e in Eqs. 2.65 to predict S_f'.

If the completely reversed stress has an amplitude (σ_a), the corresponding number of cycles of life is calculated via the next equation.

$$Eq.\ 2.66 \quad N = \left(\frac{\sigma_a}{a}\right)^{1/b}$$

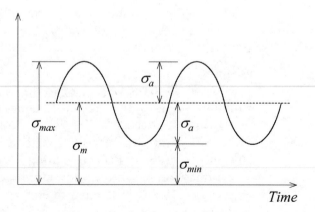

Fig. 2.30. Sinusoidal fluctuating stress, amplitude versus time.

When the *mean stress* (σ_m) is at a level other than zero (see Fig. 2.30), the cyclic loading is classified as a *fluctuating* stress case. One of the most accepted equations that provides a solution to this scenario is the *modified Goodman relation*:

$$Eq.\ 2.67 \quad \frac{\sigma_a}{S_e} + \frac{\sigma_m}{S_{ut}} = \frac{1}{n}$$

where S_{ut} is the ultimate tensile strength of the material and n is the safety factor used in the design. This relation is good for predicting fracture life. If you are more interested in yield as a failure criterion, the *Soderberg relation* should be used, which replaces S_{ut} with the tensile yield strength, S_{yt}. Either of these relations allows an engineer to evaluate the changes brought about by varying either the alternating stress amplitude or the mean stress in a fluctuating load situation. In both of these relations, S_f can replace S_e if one is more interested in finite life analysis. Of course, for materials with no endurance limit, you must use the former according to the desired life (N).

Both *Miner's rule* of cumulative damage and *Manson's method* relate the effects of cyclic loading at different stresses for different durations. Neither of these methods constitutes a closed form solution; they are mentioned here simply because they currently constitute the most accepted approximations to the problem. Both methods require accurate *S-N* plots and adjust the apparent endurance limit based on damage from finite cycle overstressing that does not cause part failure. Mathematically, Miner's rule is expressed as follows:

$$Eq.\ 2.68 \quad \sum \frac{n_i}{N_i} = 1$$

where n_i is the number of cycles of stress σ_i applied to the part, and N_i is the fatigue life corresponding to σ_i. Manson's method is readily found in the literature. It uses a graphical approach and is typically preferred because it correlates to empirical data more consistently.

When a ductile material is subjected to a fatigue-type loading, there are basic structural changes that occur. In chronological order, the changes are summarized below.

1. *Crack initiation.* A crack begins to form within the material.

2. *Localized crack growth.* Local extrusions and intrusions occur at the surface of the part because plastic deformations are not completely reversible.

3. *Crack growth on planes of high tensile stress.* The crack proceeds across the section at those points of greatest tensile stress.

4. *Ultimate ductile failure.* When the crack reduces the effective cross section to a size that cannot sustain the applied loads, the sample ruptures by ductile failure.

Although this section did not specifically expand on all the modifying factors that affect the endurance life of a part (Eq. 2.63), some of the more significant factors and their effects are worth mentioning, at least in a general sense. These factors are summarized below.

- *Stress concentrators.* General part features as described in the "Stress and Strain" section, which cause high local stresses and thus decrease fatigue life.

- *Surface roughness.* Smooth surfaces are more crack resistant because roughness creates stress concentrators.

- *Surface conditioning.* Hardening processes tend to increase fatigue strength while plating and corrosion protection tend to diminish fatigue strength.

- *Environment.* A corrosive environment greatly reduces fatigue strength. A combination of corrosive attack and cyclic stresses is called *corrosion fatigue.*

Creep

Creep, or viscoelasticity, is a time dependent, plastic deformation under a sustained load. The amount of creep experienced is a function of time, temperature, and applied load. For most plastics, significant creep can be seen at room temperatures under certain loading conditions. Creep is not a significant concern for most metals until operating temperatures reach 35 to 70 percent of respective melting points.

The stress state for a viscoelastic material can be expressed as

Eq. 2.69 $\sigma = E\varepsilon + \eta\dfrac{d\varepsilon}{dt}$

where η is the material's coefficient of viscosity and $d\varepsilon/dt$ its strain rate. Because the latter term is a function of the applied load and operating environment, it must be obtained experimentally.

Viscoelastic analysis solutions are available with some FEA codes but require detailed material studies to provide property information.

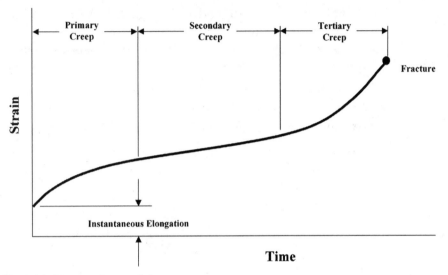

Fig. 2.31. Stages of creep failure.

The four stages of creep failure, as shown in Fig. 2.31, are described below.

- *Instantaneous elongation.* Normal deformation under applied load.

- *Primary creep.* Material strain hardens under load to decrease creep rate.

- *Secondary creep.* Material elongates at a steady rate, called *minimum creep rate.*

- *Tertiary creep.* Due to necking and formation of voids, elongation proceeds at an increasing rate until fracture.

The secondary phase is of significant interest to engineers because it dominates the actual creep process from a time standpoint. *Creep strength* is defined as the stress which produces a minimum creep rate of $10^{-5}\%$ per hour.

Apparent (or *effective*) *modulus* is a value somewhat lower than the actual modulus of elasticity. It is chosen from empirical tables based on the anticipated stress level of the system to compensate for creep effects.

Dynamic Analysis

Dynamic analyses as applied to FEA involve loads and corresponding response states that vary with time. Strictly speaking, such analyses should be referred to as vibration and time response analyses, because large displacement, completely rigid body motion is not in the realm of FEA.

Vibration and time response analyses can be subdivided into the following three related categories.

- Modal or natural frequency analysis
- Frequency response analysis
- Transient response analysis

The first of the categories involves the *free vibration* of the dynamic system. This analysis characterizes the system in the absence of external loading and serves to define its dynamic properties. Conversely, the last two are known as *forced response* analyses. These involve systems under externally applied loading functions, which can be either frequency or time dependent. The analysis type required to solve the problem at hand will depend on the type of information that is needed to reach a design decision.

Modal Analysis

The building block of all dynamic analyses is the *modal analysis*, which reports the *natural frequencies* and corresponding *principal mode shapes* of the system under evaluation. In other words, when performing a modal analysis, you solve for the distinct deformation shapes that the vibrating system will assume at each of its preferred oscillating frequencies. These concepts are better presented with the aid of a simple example.

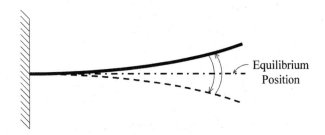

Fig. 2.32. Free vibration of a cantilever beam, first mode.

Equilibrium Position

Referring to Fig. 2.32, it is intuitive that a thin beam fixed at one end will vibrate or fluctuate most easily about its fixed point with no additional "nodes" or bends (inflection points) in its deformed shape. The natural frequency (ω_n) corresponding to this mode shape is essentially the oscillatory speed with which the beam moves from one extreme to the other and back. This speed is defined by two fundamental physical parameters of the beam: *mass* (m) and *rigidity* (k) or "spring-back."

Eq. 2.70 $\qquad \omega_n \propto \sqrt{\dfrac{k}{m}}$

The mass contribution to this equation is understood by considering inertia. The more mass (inertia) that the beam has, the harder it is for the beam to change directions when fluctuating, and consequently, the slower the motion. Spring-back is the force that resists the displacement of the beam from its equilibrium position. When the beam is bent past this position and then released, its material elasticity tries to snap it back into place. Inertial effects prevent the beam from immediately returning to its equilibrium point. Consequently, the beam overshoots its mark, returning to a spring-back condition on the other side, and the cycle begins again. The more rigidity, the faster this happens.

The interaction of these two parameters balance out to provide a constant oscillation speed, which is the first natural frequency of the system. This first natural frequency is the lowest speed at which the beam will vibrate after all external excitations are removed, a state known as *free vibration* and governed by the following equation:

Eq. 2.71 $\ddot{\theta} + \omega_n^2\theta = 0$

where $\ddot{\theta}$ is the angular acceleration of the beam and θ is its angular position away from equilibrium. The solution to this equation will give the mode shape corresponding to the natural frequency. Because oscillatory motion is expected, the solution type can be assumed to be of the form

Eq. 2.72 $\theta = C\sin(\omega_n t + \psi)$

where C and ψ are constants determined by the initial conditions of the system. Letting θ_o and $\dot{\theta}_0$ be the initial position and velocity of the beam, Eq. 2.72 becomes

Eq. 2.73 $\theta = \sqrt{\theta_o^2 + \left(\dfrac{\theta_o}{\omega_n}\right)^2} \sin\left[\omega_n t + \tan^{-1}\left(\dfrac{x_o\omega_n}{\dot{x}_o}\right)\right]$

and it describes the first oscillatory mode of the beam.

Note that the ω_n units are radians per unit time. It is often more convenient to describe this natural frequency in terms of cycles per unit time (cycles per second is common) using the following new variable.

Eq. 2.74 $f_n = \dfrac{\omega_n}{2\pi}$

Either of these two descriptions of the natural frequency can be used to calculate the time required for the system to complete one full cycle of oscillation at this frequency. This is known as the system's *natural period*.

Eq. 2.75 $\tau_n = \dfrac{1}{f_n} = \dfrac{2\pi}{\omega_n}$

In reality, no system is free to vibrate indefinitely. *Damping* represents inefficiencies of the material due to energy loss at a molecular level or of the system due to component interaction. In general, damping decays the vibration of the system and returns it to its equilibrium position in a time period that depends on *its damping coefficient* (c). This coefficient is proportional to the velocity $\dot{\theta}$ of the system and modifies Eq. 2.71 as follows:

Eq. 2.76 $\ddot{\theta} + 2\zeta\omega_n\dot{\theta} + \omega_n^2\theta = 0$

where the *damping ratio* (ζ) has been conveniently introduced as a measure of the severity of the damping. This ratio is calculated by

$$Eq.\ 2.77 \quad \zeta = \frac{c}{2m\omega_n}$$

where the denominator $2m\omega_n$ is defined as the *critical damping* of the system. Hence, the damping ratio may be used to specify the amount of damping present in a system as a percentage of its critical damping.

By assuming harmonic solutions of the form

$$Eq.\ 2.78 \quad x = A^{\lambda t}$$

a general solution to Eq. 2.76 is found as follows:

$$Eq.\ 2.79 \quad \theta = A_1 e^{(-\zeta + \sqrt{\zeta^2 - 1})\omega_n t} + A_2 e^{(-\zeta - \sqrt{\zeta^2 - 1})\omega_n t}$$

where A_1 and A_2 are constants determined by the initial conditions of the system.

Fig. 2.33. Free vibration of systems which are overdamped, critically damped (a), or underdamped in cases where $\theta_0 = A_0$ and $\dot{\theta}_0 = 0$ (b).

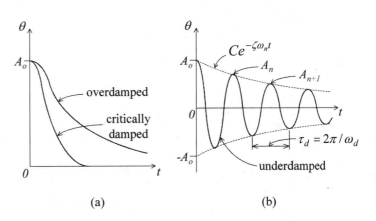

(a)   (b)

The three distinct ranges of damping ratio values that characterize system behavior can be categorized as follows.

- $\zeta > 1$. *Overdamped* [see Fig. 2.33(a)]. The system is so well damped that it will return to its equilibrium point without a single oscillation, following Eq. 2.79.

- $\zeta = 1$. *Critically damped* [see Fig. 2.33(a)]. This system is on the verge of oscillating. It will return to its equilibrium position the fastest. System motion is governed by the next equation.

Eq. 2.80 $\theta = (A_1 + A_2 t)e^{-\omega_n t}$

- $\zeta < 1$. *Underdamped* [see Fig. 2.33(b)]. This system exhibits a decaying oscillatory motion,

Eq. 2.81 $\theta = Ce^{-\zeta \omega_n t}\sin(\omega_d t + \psi)$

at a damped natural frequency ω_d given by

Eq. 2.82 $\omega_d = \omega_n \sqrt{1 - \zeta^2}$

Note in Eq. 2.81 that two new constants, C and ψ, have been mathematically introduced to obtain an equation similar in form to Eq. 2.72. The value of the constants is still determined from initial conditions.

The rate of decay of an underdamped system is represented by the ratio of consecutive amplitudes as follows:

Eq. 2.83 $\dfrac{A_n}{A_{n+1}} = e^{(2\pi \zeta / \sqrt{(1 - \zeta^2)})}$

where A_{n+1} is the peak amplitude occurring in the cycle immediately following A_n.

Most mechanical structures are underdamped. In fact, their damping ratio is usually well below 10%. For a damping ratio of 10%, Eq. 2.82 indicates a difference between ω_n and ω_d of about 0.5%. Hence, for a modal analysis, because the increased complexity of the solution has virtually no effect on its numerical value, damping is generally not taken into account.

Note that by replacing θ and its derivatives with their x counterparts, Eq. 2.76 is the *damped free vibration* or zero force solution to Eq. 2.5.

In reality, the beam structure is much better described as a series of masses connected to one another by three-dimensional spring and damper units (see Fig. 2.34). For each additional mass used in its description, six *degrees of freedom* are added to the beam system. In other words, each additional mass will need six new variables–three translational and three rotational–to define its position at all times as part of the overall system. Of course, many of these variables can be fixed due to geometric and/or boundary condition constraints. Such set of *generalized coordinates* describes general motion by recognizing constraint. This set will have the minimum number of coordinates required to define the system dynamically; this number must thus be equal to the degrees of freedom of the system.

Fig. 2.34. Discretized beam model with 3D spring and damper systems.

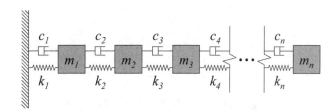

Each possible degree of freedom in a system gives rise to an additional natural frequency, which represents the oscillation of the beam in another deformed shape or principal mode. Fig. 2.35 shows the shape of the beam in its first natural mode, as described at the beginning of this section, plus three higher modes of oscillation.

A continuous structure will actually have infinite degrees of freedom. An infinite number of modes for the beam can be mathematically described in closed form. Yet, the solution to more complex structures can only be approximated by discretizing the system into a finite number of elements, not unlike the beam of the above illustration. This is the concept used by FEA.

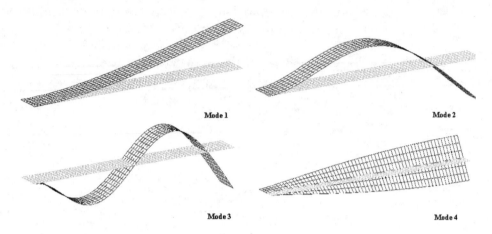

Mode 1   Mode 2

Mode 3   Mode 4

Fig. 2.35. Modal FEA of a cantilever beam, first four modes.

Again, it is useful to study a simple system to gain insight into the concepts at hand. Fig. 2.36 shows an undamped two-degree of freedom system. Its free vibration equations of motion may be written in matrix form as follows:

$$Eq.\ 2.84 \quad \mathbf{M}\ddot{x} + \mathbf{K}x = \begin{bmatrix} m_1 & 0 \\ 0 & m_2 \end{bmatrix} \begin{Bmatrix} \ddot{x}_1 \\ \ddot{x}_2 \end{Bmatrix} + \begin{bmatrix} (k_1 + k_2) & -k_2 \\ -k_2 & k_2 + k_3 \end{bmatrix} \begin{Bmatrix} x_1 \\ x_2 \end{Bmatrix} = \begin{Bmatrix} 0 \\ 0 \end{Bmatrix}$$

where the *scalar mass matrix* (**M**) modifies the *vectorial acceleration matrix* ($\ddot{x}$) and the *scalar stiffness matrix* (**K**) modifies the *vectorial position matrix* (**x**).

Fig. 2.36. Two-degree of freedom system.

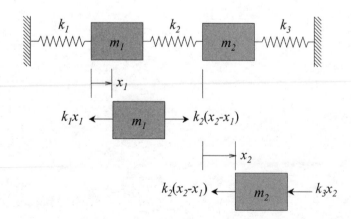

The solution for harmonic motion in the same frequency, or principal mode motion, is provided by the following equation:

$$Eq.\ 2.85 \quad x = \begin{Bmatrix} x_1 \\ x_2 \end{Bmatrix} = \begin{Bmatrix} X_1 \\ X_2 \end{Bmatrix} e^{i\omega t} = X e^{i\omega t}$$

where X_1 and X_2 are the amplitudes of oscillation of m_1 and m_2, respectively. X is known as the system's *eigenvector* or *eigenmode*. Substituting this solution into Eq. 2.84, the resulting matrix form equation appears as follows.

$$Eq.\ 2.86 \quad [\mathbf{K} - \omega^2 \mathbf{M}]X = \begin{bmatrix} (k_1 + k_2) - \omega^2 m_1 & -k_2 \\ -k_2 & (k_2 + k_3) - \omega^2 m_2 \end{bmatrix} \begin{Bmatrix} X_1 \\ X_2 \end{Bmatrix} = \begin{Bmatrix} 0 \\ 0 \end{Bmatrix}$$

Each eigenvector that satisfies Eq. 2.86 is a solution that describes the principal mode shape of the system at its corresponding natural frequency. The number of eigenvector solutions is equal to the number of dynamic degrees of freedom of the system. An additional, trivial solution occurs when this vector is zero, indicating rigid body motion of the system. Nontrivial solutions may only be satisfied when the determinant in Eq. 2.87 equals zero.

$$Eq.\ 2.87 \quad \det[\mathbf{K} - \omega^2 \mathbf{M}] = (k_1 + k_2 - \omega^2 m_1)(k_2 + k_3 - \omega^2 m_2) - k_2^2 = 0$$

For this two degree of freedom system, Eq. 2.87 yields a quadratic in ω^2. The roots of this quadratic are the natural frequencies of the system, also known as its *eigenvalues*. Hence, there are two natural frequencies for this system, each associated with one of its two eigenmodes.

It is useful to note that for this system, and any other linear elastic system for that matter, the deflected shape at any given time will be a linear combination of all its eigenmodes as follows:

$$Eq.\ 2.88 \quad x = \sum X_i \xi_i = \mathbf{X} \xi$$

where $\mathbf{X}$ is the eigenmode matrix, comprised of all individual eigenvectors (X_i) in the system, and ξ is the modal displacement vector, comprised of all individual modal displacements (ξ_i). These ξ_is are also known as the *principal coordinates* of the system, which define the *modal space* (ξ).

It is easy to use the general matrix notation of the equations above to extend their application for solving an undamped n-degree of freedom system. In fact, other derivations of the motion equations for such systems exist that also make use of these matrices, most notably the energy based Rayleigh-Ritz method. However, regardless of the method used, as n grows, these matrix formulations grow by n^2. Hence, performing manual calculations on a system with more than a few degrees of freedom becomes extremely labor intensive. For a complex structure, it is virtually impossible. Yet, with the aid of FEA, these matrix solutions can be carried out numerically in a very time-efficient manner.

The remainder of this section is focused on forced response analyses. It will be shown that when a part is induced to vibrate at a frequency "comfortable" to the part, effects of the input are magnified and may cause premature or catastrophic failure. Hence, sometimes the goal of an analysis is simply to help design the natural frequencies of a system away from its known operating frequencies and to note which modes, if any, are excited as the system ramps up to its operating state. If this is the case, a successful modal solution will be all that is necessary to solve the problem.

Another frequent use of modal analysis is to allow the designer to become acquainted with the induced mode shapes themselves. Deformation patterns at an operating frequency may be deemed acceptable if they do not affect an inherently weak section of the system. In fact, for certain products, such as sonic welding equipment or acoustic components, the design goal is to ensure that a planned deformation is present at an operating frequency.

Frequency Response Analysis

When a system is being excited by known oscillatory frequencies, it may not be feasible to design its natural frequencies out of this operating range. In cases of this nature, evaluating the system in the presence of this enforced vibration proves necessary. When the excitation does not change with time, the solution is a steady state response at the operating frequency of interest. This is known as *frequency response analysis*. The relevant results of this analysis are typically displacements, velocities, and accelerations of the system, which can be used to calculate forces and stresses in the structure.

To illustrate these concepts, a damped single-degree of freedom system will be used (see Fig. 2.37).

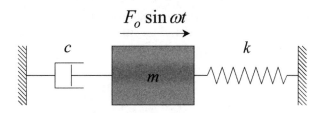

Fig. 2.37. Damped single-degree of freedom system subject to harmonic excitation.

The equation of motion for the above system follows.

Eq. 2.89 $m\ddot{x} + c\dot{x} + kx = F_o \sin\omega t$

Note that this equation is Eq. 2.5 in the presence of a harmonic forcing function $F_x=F_o\sin\omega t$, where F_o is the amplitude of this force, and ω is its driving frequency. By making the same substitutions made in the free vibration derivations, this equation becomes the following.

Eq. 2.90 $\ddot{x} + 2\zeta\omega_n\dot{x} + \omega_n^2 x = \dfrac{F_o}{m}\sin\omega t$

The solution of the above equation is a sum of two terms: the *transient solution*, which is the decaying, free vibration solution (Eq. 2.81), and the *steady state solution*, which is any solution to the complete equation. By assuming solution types similar to those used in the derivation of the free vibration equations, a complete solution to the underdamped system is found as follows:

Eq. 2.91 $x = Ce^{-\zeta\omega_n t}\sin(\omega_d t + \psi) + \dfrac{F_o}{k}A\sin(\omega t - \phi)$

where A is a nondimensional *amplitude ratio* or *magnification factor* of the steady state solution, and ϕ is the *phase angle* of this solution. These terms are provided by the following equations.

Eq. 2.92 $A = \left\{ \left[1 - \left(\dfrac{\omega}{\omega_n}\right)^2\right]^2 + \left[2\zeta\dfrac{\omega}{\omega_n}\right]^2 \right\}^{-1/2}$

Eq. 2.93 $\phi = \tan^{-1}\left[\dfrac{2\zeta\dfrac{\omega}{\omega_n}}{1 - \left(\dfrac{\omega}{\omega_n}\right)^2} \right]$

The following illustration shows patterns of A and ϕ as functions of the *frequency ratio* (ω/ω_n) for different values of the *damping ratio* (ζ). *Resonance* is defined as the point when $\omega/\omega_n=1$, and *resonant frequency* is another term used for natural frequency. Note that for lightly damped systems, as the system approaches resonance, the amplitude ratio becomes divergent. Because excessive response amplitudes are usually not desired, you must either verify that the operating frequency remains away from the resonant frequency of the system, or add damping to the system. The latter is usually accomplished by means of an external viscoelastic mount device. Also note that as the driving frequency goes to zero, the amplitude ratio becomes 1 (static solution), yet as it goes to infinity, the ratio becomes zero. This last observation is an important, albeit surprising, fact: regardless of the driving function's amplitude (F_o), if the system is shaken fast enough, its response will be zero.

Fig. 2.38. Magnification factor and phase angle versus frequency ratio for different damping ratios of single-degree of freedom systems subject to harmonic excitation.

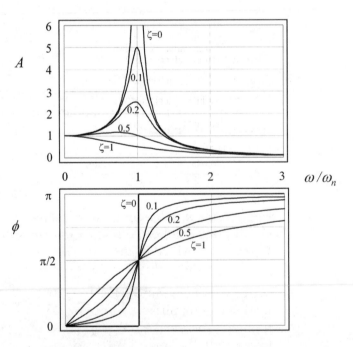

The phase angle of the response may vary from 0 to 180° and represents the portion of the cycle by which the steady state response lags the forcing function (F_x). Note that as the frequency of the input spans the range from zero to infinity, the system response will go from being completely in phase with the input (0° lag) to being completely out of phase (180° lag), always passing through 90° at resonance. Also note that for lightly damped systems, the response will be (approximately) always either in phase or out of phase with the input, with a quick phase shift at resonance.

From the above discussion, it can be gathered that, whereas damping was assumed negligible in free vibration analysis, it plays a big part in forced response and must be quantified. The overall damping of a system is usually the most difficult parameter to obtain. The only accurate means to do so is experimentally, often from the free vibration decay caused by a spike input such as a hit with a rubber hammer, while making use of Eq. 2.83. Selected representative damping ratios are provided in Table 2.3.

Table 2.3. Representative damping ratios

System	Damping ratio (ζ)
Metals (in elastic range)	< 0.01
Metal structures with joints	~ 0.03
Aluminum / steel transmission lines	~ 4×10^{-4}
Auto shock absorbers	~ 0.3
Rubber	~ 0.05
Large buildings during earthquake	0.01 – 0.05

For an n-degree of freedom system, all of the above concepts can be expanded in matrix form, using a derivation similar to that used for the free response. The matrix equation equivalent to Eq. 2.86 follows:

$$Eq.\ 2.94 \quad [-\omega^2 \mathbf{M} + i\omega \mathbf{C} + \mathbf{K}]u(\omega) = F(\omega)$$

where ω is now the driving frequency, C is a system damping matrix, $u(\omega)$ is a generalized coordinate displacement vector, and $F(\omega)$ is the forcing function vector. This matrix equation can then be solved numerically by some FEA systems. Making use of this formulation is known as *direct frequency response analysis*.

An alternate approach for solving an n-degree of freedom system involves transferring Eq. 2.94 into modal space $\xi(\omega)$ with the help of Eq. 2.88 as seen below.

$$Eq.\ 2.95 \quad [-\omega^2\mathbf{M}\mathbf{X} + i\omega\mathbf{C}\mathbf{X} + \mathbf{K}\mathbf{X}]\xi(\omega) = \mathit{F}(\omega)$$

Hence, this approach, known as *modal frequency response analysis,* uses the mode shapes of the structure, instead of its generalized coordinates, to describe motion. Note that for Eq. 2.95 to hold, the solution to every single one of the structure's mode shapes must be obtained prior to the analysis. Of course, this is prohibitive for systems with a large number of degrees of freedom. Yet, Eq. 2.95 turns out to be a good approximation as long as every mode up to a frequency at least two to three times the highest operating frequency of the system is used. Because this number of modes will, in most cases, be less than the number of generalized coordinates that describe the system, using this method is generally much more numerically efficient than direct analysis, although not as accurate.

Depending on the FEA package you use, you might be able to select the method for carrying out your analysis. If the model is small, or subject to only a few excitation frequencies, or subject to high excitation frequencies, the direct method should be used. This method should also be used if a higher level of accuracy is deemed necessary. Yet, for large models or models subject to many excitation frequencies, the modal method will prove to be the better choice.

Transient Response Analysis

When the system under analysis is subject to an excitation that changes with time, the solution is also time-varying. *Transient response analyses* provide solutions to systems of this type, as long as the excitation is explicitly defined in the time domain. The relevant results of these analyses are typically displacements, velocities, and accelerations of the system, which can be used to calculate forces and stresses in the system.

All dynamic concepts and derivations presented thus far in this section come together at this point to provide a solution to the transient response problem. For an *n*-degree of freedom system, the general equation of motion, in matrix form, is the familiar equation below.

$$Eq.\ 2.96 \quad \mathbf{M}\ddot{x} + \mathbf{C}\dot{x} + \mathbf{K}x = \mathit{F}(t)$$

In *direct transient response* analysis, this equation is solved using direct numerical integration.

By utilizing Eq. 2.88 to transfer the formulation of the problem to modal space $\xi(t)$, Eq. 2.96 becomes the following.

$$Eq.\ 2.97 \quad \mathbf{MX}\ddot{\xi}(t) + \mathbf{CX}\dot{\xi}(t) + \mathbf{KX}\xi(t) = F(t)$$

This equation governs *modal transient response* analysis and is solved through numerical integration and summation of the individual modal responses. Note that Eq. 2.97 is only an approximation if the modal content is incomplete. For this approximation to be valid, the frequency content of the transient load must be evaluated to determine a *cutoff frequency*. This will be the frequency above which no modes are noticeably excited. Only modes above this frequency can be left out of the modal transient response analysis.

Once again, depending on your FEA package, you might be able to select the method for carrying out a transient analysis. If the model is small, or subject to only a few time steps, or subject to high excitation frequencies, the direct method should be used. This method should also be used if a higher level of accuracy is deemed necessary. Yet, for large models or models subject to many excitation time steps, the modal method is the preferred choice.

Note that a steady state harmonic forcing function, such as the one used in Eq. 2.89, is explicitly defined for all time. When this function is applied to a structure, there will be a "settling" time period before reaching steady state response. This "settling" is governed by the transient part of Eq. 2.91. Hence, if you are interested in the effects observed at the start of steady state operation, a transient response analysis should be performed. Of course, the results of this analysis will contain all of Eq. 2.91. For this reason, frequency response analysis may be considered to be simply a subset of transient response analysis, which itself can be conceived of as the most generalized form of forced response analysis. Keep in mind though, that if the transient solution is of no interest, frequency response analysis is a much more appropriate and efficient solution. Fig. 2.39 shows FEA transient analysis results for the cantilevered beam problem, using a sinusoidal forcing function applied normally at the free end.

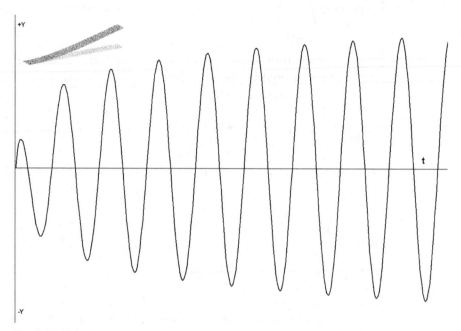

Fig. 2.39. Transient response FEA of a cantilever beam subject to a harmonic excitation applied perpendicularly at its free end.

The excitation in a transient analysis can be a load, acceleration, or displacement of a given duration with a specified amplitude. The ramp of the input, or duration of loading and unloading, can have a significant effect on the resultant behavior. In the structural damping test mentioned earlier, where a system is excited with a hammer strike, it is the transient response which is being studied. These studies can often be used to correlate damping values and model approximations prior to a frequency response analysis.

In designing a system subject to transient loading, it is often helpful to compare the reciprocal of the loading duration to the known natural frequencies. If these values are comparable, it may be worthwhile to evaluate the analytical response to this loading. Transient analyses are typically run for a finite number of cycles. If longer-term response is needed, the system may need to be evaluated using frequency response or other means.

Summary

This chapter should be thought of as a pocket calculator/dictionary gadget—you should not expect it alone to allow you to carry out a complex analysis or enable you to define every engineering condition, yet it should prove to be an extremely helpful and concise companion. Consider making this chapter readily available whenever you are in the process of conducting FEA. The basic engineering fundamentals presented in this chapter should be referenced at all stages of the analysis process.

At the beginning of the FEA process, you should always take the time to understand the nature of the problem at hand because doing so will usually save you time later by helping you avoid unnecessary calculations. What are the fundamental principles that will govern the analysis? How can they be qualified and quantified? Sometimes you might find that a simple manual calculation is all you need to solve the problem. Try not to lose perspective at this initial stage.

When setting up your FEA model, try to envision the fundamental phenomena to be evaluated in the analysis. What kind of stress state will be caused by the geometry and boundary conditions of the model? Will there be bending? Torsion? Can the geometry itself be simplified or broken into a sum of simpler, more manageable shapes? Attempt a prediction of the resulting stress state contour or even some of its numerical magnitudes. This will prove invaluable when evaluating results.

Next, verify that the boundary conditions make sense. Always draw yourself a free body diagram and evaluate the equilibrium state. How confident are you that unexpected and/or undesired reactions are not being introduced by your representation of the constraints? Are the loads themselves adding or subtracting stiffness to the structure by their representation? Remember, any part of the system you decide to leave out of the model must be accurately accounted for in the analysis.

As mentioned above, when you have obtained results in your FEA, attempt to check these by means of a rough manual calculation. See if you are in the ballpark. Of course, you cannot expect yourself to be able to solve every problem accurately; after all, this is the reason you are using FEA. Yet, if your calculation is an order of magnitude off, a closer inspection is warranted. Remember that engineering analysis has been taking place since long before the introduction of the finite element method. FEA is a relatively new tool. When used with care and fore-thought, it can certainly provide excellent results in an extremely effi-cient manner. Yet, as with any powerful tool, if it is used blindly, the magnitude of the error can be just as magnificent.

The material covered in this chapter was intentionally concise and, in addition to serving as a useful reference, it was meant to remind you or introduce you to some basic engineering concepts. It is highly recom-mended that you explore these topics in more depth by using your col-lege textbooks, the FEA software documentation, or references mentioned in the Bibliography. Armed with the appropriate back-ground information on the technology and underlying physics, you will be better equipped to succeed in the exciting world of simulation based design using finite element analysis.

3

FEA Capabilities and Limitations

If you did not recall or were not familiar with the equations and derivations presented in Chapter 2, it is important to reiterate that these solutions represent the most fundamental stress states in engineering design. While a few structures can be accurately represented by one or a combination of these calculations, most real world geometry is much more complex and experiences multiple stress states at any given time. Hence, without physical samples, finite element analysis is often the only way to predict the performance of most structures with any degree of accuracy. In effect, FEA picks up where classical equations end. FEA extends the power of Hooke's Law, $\sigma = \varepsilon * E$, beyond two force members.

As a design engineering tool, the potential of FEA is limited only by the user's time and creativity. However, because the results from even a simple FEA study are much more impressive than a couple of pages of calculations, it is tempting to place more confidence in the former than they deserve. First and foremost, FEA is an approximation. Just as you would make assumptions in a simple beam calculation or in a more complex energy balance on a frame, you must make assumptions in any

FEA study. Regardless of the analysis method used, the accuracy of the solution is always dependent on the validity of all assumptions.

Actual Performance versus FEA Results

One of the most common questions asked when FEA is proposed to solve a problem is "How accurate are the results?" Designers, engineers, and managers alike ask this question. It is also one of the most important questions to ask. A nonanalyst or even a beginning analyst usually interprets that question as "Will I see on the screen what I see in the field?" The more experienced analyst knows that the better meaning is "Will I get the right answer to the question I ask?" Better yet: "Will I be able to ask the right questions or enough questions to fully understand the answers?"

In nearly all cases, there will be noticeable differences between the actual operating performance of the part or system and the FEA simulation. For every obvious difference, there will be many differences that cannot be detected or require special equipment to measure, such as strain gauges and accelerometers.

There are many reasons why a typical analysis will provide different results than a field test. Theoretically, the results do not need to differ. There is excellent and well-documented correlation of FEA results to many of the fundamental stress states defined in Chapter 2. Most codes provide a verification manual, which shows correlation to test or calculated data. So why is it that most analysis results do not match field results? It goes back to the analysts' interpretation of the "accuracy" concern. It is difficult to ask enough questions or the right questions about the performance of a structure. However, once you learn to qualify the assumptions and unknowns, or the questions you cannot ask, extracting useful design information out of FEA becomes easy.

By comparing a simple cantilevered beam calculation to its equivalent FEM, many of the reasons for the differences between FEA results and actual results can be found. Every variable, or piece of data, that you are required to input to the system is an assumption and source of error. The most obvious one is material properties. The assumption in this problem was to use standard steel properties. What are "standard steel properties"? If enough references are consulted, it is easy to find a variance on Young's modulus of 13.8%, or 29 to 33 million psi. Parts made of less predictable materials with less controllable manufacturing methods may have a modulus that varies by twice that percentage. Injection molded parts vary in stiffness due to mold packing, cooling, flow direction, filler direction, operating temperature, and strain rate over and above variations in raw material or any nonlinearity due to the level of strain. Castings have porosity and homogeneity issues. Stampings can have stiffness variation due to cold working and surface treatment.

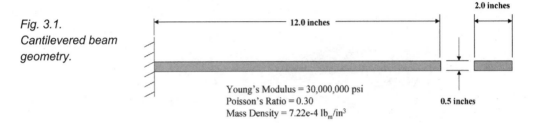

Fig. 3.1. Cantilevered beam geometry.

Young's Modulus = 30,000,000 psi
Poisson's Ratio = 0.30
Mass Density = 7.22e-4 lb$_m$/in^3

Geometry is another source of variance in nearly every analysis. In the simple beam example, the problem requires input for length, depth, and thickness. On a structural steel component of this size, it is not uncommon to expect tolerances of plus or minus 1/32" on the depth and height. Other inherent assumptions are that the beam is square to the load and the mount. Any tolerance in this perpendicularity will cause some component of the load to be oriented axial to the beam, thereby decreasing the transverse load on the part. If the cross section is not aligned orthogonal to the plane of the problem, the effective moment of inertia will increase, making the stiffness in the actual problem greater than expected.

Consider, however, more complex geometry in real world applications, such as that of a pump housing or an engine connecting rod. The number of dimensions required to define the geometry may range in the thousands, each with a tolerance. A CAD solid model is typically constructed at the nominal dimensions and if tooling is programmed directly off of that geometry, there is a good chance that most of the dimensions are within tolerance. However, on a complex part, only the dimensions required on assembly features are checked carefully and it is common to approve a nonconforming dimension if there is no apparent conflict with the rest of the system.

What about dimensions on complex curvature that are difficult to check, even with a coordinate measuring machine (CMM)? These are typically judged visually, again with the understanding that if they look right and do not interfere with the rest of the assembly, they are assumed satisfactory. Laser scanning technology has recently provided more detailed data on geometric variance in complex parts. Fig. 3.2 shows the results of overlaying a 3D scan of a part on the original CAD geometry. Design engineers are often surprised by the error in parts they had assumed correct.

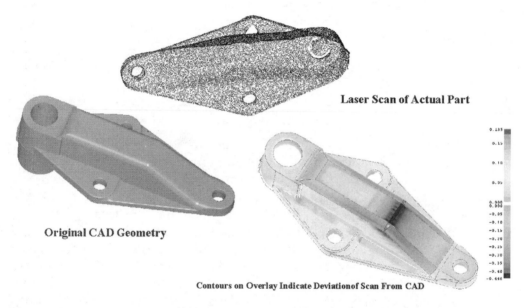

Fig. 3.2. Dimensional laser scan of a part overlayed on the actual CAD geometry.

Bringing this discussion back to FEA accuracy, the effects of geometric variance are nearly impossible to predict. While error on curvature may not affect the mechanics or aesthetics of a part, it will certainly affect stiffness. The effect may be small but nonetheless significant. Minor variations in topology may alter strain gage readings by a sufficient magnitude to call into question the FEA results.

If the manufactured geometry is so complex or ambiguous that conformity to the expected geometry is questionable, the same can be said about the FEM representation of this or any geometry. A finite element model, by definition, breaks up a continuous volume into discrete pieces, or elements. Linear elements–or elements in which the edge connecting two nodes is a line, or the face connecting three of four nodes is planar–will always represent curvature with facets. Theoretically, a second order element can precisely represent a curve such as an arc or other conic. However, most meshers place the midside node at the center of the linear edge. While this edge can deform to represent curved results, the initial geometric representation is chordal.

Beyond the simple element definition, most FEA models are based on some idealization of the actual system. At a minimum, fillets, chamfers, and other small features may be neglected in the FEM. At the other extreme, complex cross sections might be represented mathematically by a beam element or actual part thickness might be represented by a mid-plane surface (shell) element. The use of simplifications and idealizations further increases the gap between the actual and modeled geometry.

The variance between actual behavior and that predicted by an idealized model may be due to tapered walls modeled with a constant thickness shell. Holes, bolts, or other discontinuities will never show up in a beam element. The physical system will typically not be well represented in the shape of joints or transitions between adjoining shells or beams. Calculated mass for either a dynamic study or one where gravity loads are affecting the model may be skewed due to suppressed features or virtual overlap of shells and beams.

This discussion includes only a portion of the uncertainty which can cause FEA results to differ from test results. If this book is your introduction to FEA in the design process, do not despair! As stated earlier, identifying these uncertainties, understanding their impact on results, and making sound engineering decisions based on these results will distinguish you as an analyst from someone who makes pretty pictures on a computer.

How FEA Calculates Data

Some understanding of the inner workings or guts of FEA will enable you to build more accurate models and qualify your assumptions much more accurately. While there are many excellent books on FEA theory and matrix derivation for analysis specialists or ambitious design analysts, a brief overview will be provided here with enough theory to enable you to discuss the technology intelligently.

A Simple Model...

The tension spring assembly shown in Fig. 3.3 represents a simple two-element model. Two end points and an undeformed stiffness define each element. This stiffness in a physical spring is defined by its spring rate, K. The spring rates in this model will be identified as K_1 and K_2, from left to right.

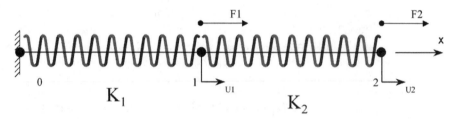

Fig. 3.3. Schematic representation of a two-spring system.

A frame of reference attached to the model can be defined such that element deformations are defined by scalar magnitudes. In the schematic drawing of the springs, the reference direction is called the X direction. Using this frame of reference, or spring coordinate system, each end point of each spring is restricted to displacement or translation in the X direction. The change in position from the undeformed state of each end point can be defined by the variable Ui, where i denotes an end point, which, for convenience, will be numbered *0, 1,* and *2,* left to right. Therefore, the position of each end, with respect to its initial state, will be U_0, U_1, and U_2, respectively. Of course, it is understood that the position in space of this local coordinate system must be known. For this example, the system will be fixed spatially by constraining the leftmost end point, point *0,* completely, so that $U_0 = 0$ for all time.

The final entities required to fully characterize this system are the forces acting on each of the springs' end points that remain unconstrained in *X*. These will simply be called F_1 and F_2, using the same end point nomenclature.

To develop a model that can predict the state of this spring with different combinations of the parameters described previously (K_i, U_i, and F_i), the following linear spring equation will be used:

Eq. 3.1 $F = K x$

where x is commonly understood to mean the change in length of the spring from its undeformed state. Using the spring parameters defined above and assuming a state of equilibrium, equations can be written for the state at each end point as follows.

Eq. 3.2 $F_1 - U_1 K_1 + (U_2 - U_1)K_2 = 0$

Eq. 3.3 $F_2 - (U_2 - U_1)K_2 = 0$

Therefore,

Eq. 3.4 $F_1 = (K_1 + K_2)U_1 + (-K_2)U_2$

Eq. 3.5 $F_2 = (-K_2)U_1 + K_2 U_2$

If this set of equations is written in matrix form, the model then becomes the following statement.

Eq. 3.6 $$\begin{Bmatrix} F_1 \\ F_2 \end{Bmatrix} = \begin{bmatrix} K_1 + K_2 & -K_2 \\ -K_2 & +K_2 \end{bmatrix} \begin{Bmatrix} U_1 \\ U_2 \end{Bmatrix}$$

In this model, if the stiffness due to material properties and geometry were known (K_1 and K_2) and the input were defined (F_1 and F_2), the system could be solved for its resulting deformed shape (U_1 and U_2).

This probably seems like the hard way to describe a spring system but it encompasses most of the key terminology used in FEA. Starting with the matrix representation described in Eq. 3.6, you should be familiar with the following terms.

The *stiffness matrix*, $[K]$, refers to the term appearing below.

Eq. 3.7 $\begin{bmatrix} K_1 + K_2 & -K_2 \\ -K_2 & +K_2 \end{bmatrix}$

The matrix is relatively simple for this two-spring system but turns complex as the pair of in-line springs becomes multiple springs with different orientations, and the one-dimensional elements become to two- or three-dimensional.

Degrees of freedom is the term applied to the ability of a node to translate and transmit load. In this model, we were only concerned with displacements and forces in the X component of the spring reference frame. The three end points or nodes of this model are each allowed one degree of freedom for a total of three degrees of freedom for the entire model. By specifying one end point as grounded or constrained, that degree of freedom is removed from the model and it becomes a two-degree of freedom model. The number of degrees of freedom in a model dictates the number of equations required to define it and is the best indication of model size.

The *boundary conditions* allow the model to be solved. A set of equations that is solvable is meaningless without input for which to derive the solution. The boundary conditions in this model are the constraint, or the specification that $U_0 = 0$, and the input forces, F_1 and F_2. The displacements, U_i, could have been specified in place of F_i as boundary conditions and the model could have been solved for the forces. Either way, the boundary conditions eliminate unknowns in the system.

The mesh in this model consists of two end points or *nodes* per spring for which the connectivity is defined by a point-to-point stiffness called an *element*. The concept of nodes, elements and meshes will be discussed in greater detail throughout the book.

An assumption was made that each spring could only extend or compress and would always remain linear. Based on this assumption of the possible shapes each spring could assume, the resultant behavior could be calculated. The mathematical representation of the assumed shape of an element is called the *shape function*. While the shape function of the spring element is trivial, the shape function for 2D and 3D elements becomes more complex. In addition, second order elements as described previously have even more complex shape functions. In a shape function for even a 1D element that is capable of bending as well as stretching, the shape function must account for slope as well as position. Because slope is a first order derivative of displacement, the defining set of equations becomes a set of differential equations which cannot always be solved using simultaneous methods. Consequently, a finite element solution will assume a shape function form and iteratively calculate the coefficients of that function–much like curve fitting–until the difference between the potential energy of the system and the total work or energy applied is a minimum.

Remember that the shape functions assumed are for each single element. Consequently, for general deformed structures, the minimum energy error can only be as small as the similarity of the desired shape between any two nodes and the ability of an element or edge to approximate that shape. Consequently, as the distance between two nodes decreases, the validity of a linear shape approximation increases and the minimum error decreases to a point where there is little change in error from making the distance smaller.

In FEA, the process of reducing the local error by using smaller and smaller elements or using elements that can approximate more complex point-to-point shapes is called *convergence*. This is most commonly accomplished by increasing or refining the mesh density, either locally to areas of rapid changes in curvature, or to the entire model.

H-elements versus P-elements

Although the most common convergence method is mesh refinement, an alternate element technology allows for increasing the element edge order to better capture desired shapes and reduce error. Where a second order element definition improves the ability of that element to capture load curvature, higher order elements take this a step further. Elements that can assume higher edge orders are called *p-elements*.

The most commonly used technology is *h-element* based, which typically limits the element order to quadratic, and convergence requires mesh refinement. Some p-element tools allow for an edge polynomial order of nine or higher to capture local behavior. However, if this is not sufficient for proper convergence, mesh refinement may still be required. Codes which utilize p-element technology typically provide an internal capability for evaluating convergence and automatically increasing the polynomial order where required. This process is called *automatic adaptivity*; the mesh self-adapts to the needs of the system.

P-elements are allowed to be bigger than h-elements because they can capture more complex behavior over a given distance. Fig. 3.4 shows a p-element mesh of a simple plate with a hole in it.

Fig. 3.4. Simple p-element mesh.

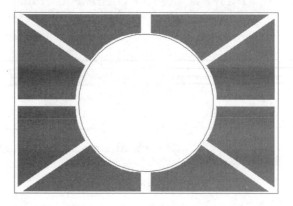

Fig. 3.5 shows the degree of geometric complexity that p-elements with more complex edge definitions can model, even before deformation is induced. By mapping the edges and faces of the elements precisely to the geometry, as in these two Pro/MECHANICA meshes, the resultant deformation will naturally be closer to the accurate results.

Fig. 3.5. Solid p-element mesh.

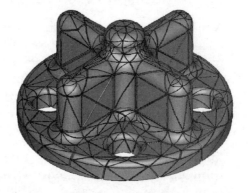

Due to the automatic adaptivity and the complexity of the shape functions, p-element solutions tend to be very resource intensive and computationally time consuming. This is the price you pay for the ease of use. As computers are getting faster, this technology is becoming more popular. Most high end, h-element based solvers include p-elements in their tool boxes. They allow the two elements to be mixed so that local high strain gradients can be captured with resource intensive p-elements, and global behavior and general load transfer can be calculated with faster h-elements. On the other hand, at this time most p-element based tools do not provide a solution for mixing the two element technologies.

Correctness versus Accuracy

In the book, *The Hitchhiker's Guide to the Galaxy* by Douglas Adams (Ballantine Books, 1995), Deep Thought, the greatest supercomputer ever built, was asked to compute the answer to "...Life, the Universe and Everything." Seventy-five thousand generations dedicated themselves to maintaining this computer when suddenly, one day, it awoke from its 7.5 million years of computation. After warning the people that they would not like the answer, the computer pronounced that the answer to life, the universe and everything is "42." When pressed for an explanation, Deep Thought responded that while the answer was right, the people did not really know what the question was.

The similarity of this story to many new users' experience with FEA is remarkable. Regardless of the size and cost of your workstation, if you pose the question poorly or incorrectly, the answer will be meaningless. Although it was fairly easy for the caretakers of Deep Thought to realize that "42" was not the answer they were looking for, identifying the error in an FEA model can be very difficult. It has been said that in FEA, if credible software is used, the answer is always correct but the question may be wrong. It is in this context that the difference between correctness and accuracy must be defined.

The Correct Answer

Referring back to the question posed at the beginning of this chapter, correct results are results expected by those observing or envisioning the parts or systems working in the field. Unfortunately, the concept of correctness does not usually take into account variability as described earlier, which must be considered in FEA.

The Accurate Answer

An accurate answer in FEA is considered the best result obtainable for the properties, geometry, and boundary conditions specified, that is, the best answer to the question posed. Typically, the degree of accuracy refers, in large part, to convergence or the refinement of the mesh necessary to reduce error. It also includes fine tuning of boundary conditions so that fictitious behavior solely related to constraints or loads is eliminated. These fine-tuned inputs are not necessarily the ideal boundary conditions for the problem. However, as the boundary condition scheme approaches the ideal for the system being studied, the accurate FEA results begin to approach the correct system response.

An important concept, which will be reiterated throughout this text, is that a correct solution can be difficult, costly, and even impossible to compute. This is due to the fact that asking the right questions, enough questions, or "phrasing" the questions correctly may never be possible due to part and material tolerances or load variability. However, by understanding the assumptions inherent in your questions, you can compile an impressive stack of data which can enable you to make better, more reliable design decisions than you could without FEA.

Key Assumptions in FEA for Design

Now that all the terms have been introduced, a more pointed discussion of the key assumptions used in FEA is possible. In addition to the four primary assumptions which affect the quality of finite element based solutions, the assumption of linear static, the most commonly used model condition, will be discussed. Finally, a list or glossary of typical project dependent assumptions will be provided.

Four Primary Assumptions

There are, as mentioned above, four assumptions which must be considered in any finite element based solution, be it a structural, electromagnetic, fluid flow, or manufacturing simulation. The assumptions are listed below.

- Geometry

- Properties

- Mesh

- Boundary conditions

Geometry

Geometry must be considered in its proper context. In reality, CAD geometry is only a template for building a mesh. In the current state of commercially available FEA, solvers have no capability to understand geometry as it is thought of in CAD. An FEA solver only understands nodes, and the connectivity of nodes, which are elements. The smaller the element size or the higher the element order, the better the mesh will represent the geometry template it was based on. The inherent assumption when a model is sent to solve is that the mesh represents the geometry adequately for the study goals. In addition, it is common to idealize the geometry "template" by suppressing or removing unnecessary features or by representing solid structures as surfaces (shells) or lines (beams). Any chosen idealization is carried out with the assumption that it will also adequately represent the problem to achieve study goals.

Tips and techniques for preparing CAD geometry to use in FEA are presented in detail in Chapter 5.

Properties

Specifying a single set of material properties for a part in an FEA study makes the significant assumption that all parts in the production run intended to be represented by the analysis have the same properties. It is also typical to assume that most parts will be isotropic and homogeneous. This assumption is only completely true in a few industries. In most companies, the required manufacturing processes for the parts or systems will induce some level of anisotropy such as glass fiber orientation in plastic parts and grain orientation in cold-rolled sheets.

When choosing properties, always remember the limited scope of the assumption. Properties such as material stiffness, shell thickness, or beam properties apply to an idealized understanding of the part under investigation. If variability is not understood, the results of the study cannot be considered representative of the actual problem. When the variability of a property will cause predictable variation in results, it is acceptable to simply interpret the results accordingly. The bending behavior of a beam is linearly proportional to Young's modulus. If the material stiffness can vary plus or minus 15%, the bending behavior will vary plus or minus 15%. However, many responses to changing properties cannot be quantified so easily. In these cases, it is important that you do not assume a single property. Take the time to compute the solution at several combinations of properties or bracket the solution with extremes. While the assumption set becomes more complex, the resulting data set will enable you to make informed design decisions.

The types of properties commonly used in FEA and their meanings are detailed in Chapter 6.

Mesh

As mentioned previously, the mesh is your way of communicating geometry to the FEA solver. The accuracy (not the correctness) of the solution is primarily dependent on the quality of the mesh. The quality of a mesh is best characterized by the convergence of the problem. The global displacements should converge to a stable value and any other results of interest should converge locally. A less tangible, more subjective measure of the quality of a mesh is its appearance and ability to visually convey the geometry it represents. An interesting feature of FEA modeling is that the better a mesh looks, the better it is. Although a good-looking mesh is not necessarily the best mesh, a bad-looking mesh almost always indicates a problem. A good-looking mesh should have well-shaped elements. Equilateral triangles and squares are the ideal. Transitions between densities should be smooth and gradual without skinny, distorted elements. When you determine that a mesh is ready to solve, you are assuming that it will accurately represent the stiffness, or other property of interest, of the intended structure.

Using idealized elements, such as beams and shells, makes assumptions that these elements can adequately represent the geometry *and* that they can capture the structural response of the system. This latter assumption is not always easy to qualify. Take the time to understand the impact of the assumption of element type and mesh quality so that erroneous results can be minimized.

Guidelines for generating meshes and evaluating their quality are presented in Chapter 7.

Boundary Conditions

Choosing boundary conditions requires the greatest leap of faith for new users. Using boundary conditions to represent parts and effects that are not or cannot be explicitly modeled makes the tremendous assumption that the effects of these unmodeled entities can truly be simulated. Complex component interactions can be approximated quite accurately by experienced analysts. However, newer users should take the time to understand the ramifications of simple boundary conditions.

While there are always incorrect modeling practices which should always be avoided, many acceptable ways exist to model the boundary conditions of any given system. These various alternatives can only be considered wrong if the user does not understand the assumptions they represent. Overly constraining a system because it is convenient when in a time crunch is not wrong if the analyst remembers to interpret the results accordingly. Displacements may be understood to be lower than they would have been had the boundary conditions been more appropriate. Stress magnitudes may be higher or lower, depending on the constraint used. If you make the assumption that a coarse boundary condition scheme can be used, you must always interpret your results in light of this assumption.

Types of boundary conditions and techniques for using them are provided in great detail in Chapter 8.

Linear Static Assumption

As mentioned in Chapter 2, there are many types of structural analysis and many other uses for FEA. However, the linear static solution is still the most commonly used in general engineering analysis. The reasons for this are simplicity, speed, and availability. The basic package available from most software developers (read "cheapest") is linear static. Solutions such as nonlinear and vibration are usually modules that require special skill and knowledge beyond a mastery of linear static techniques. However, the popularity of this solution often obscures the fact that it represents another significant assumption. How many events are truly linear and static? To answer this question, you must first be comfortable with what these terms mean in the world of FEA.

Linear

The assumption of linear encompasses several subassumptions, all of which must be valid for the general one to hold true. The three main areas of interest when qualifying the linearity or nonlinearity of a system follow.

- Material properties
- Geometric concerns
- Boundary conditions

These will be introduced briefly here and covered in depth in Chapter 15.

Material Properties

A material is said to be linear if its stress-strain relationship is or can be assumed to be linear. Linear materials must comply with Hooke's Law when stated in terms of stress.

Eq. 3.8 $\sigma = E * \varepsilon$

Here the proportionality constant, E, is Young's modulus, or the modulus of elasticity of the material. Most steels remain linear, for all practical purposes, up to the yield point of the material. Other metals and materials, such as cast iron or plastic, have a small linear range of strains. Their elastic range of behavior is best represented by some curvature of declining stiffness as strain increases. However, it is common to model most of these materials as linear and interpret the results in light of that assumption. In a near-linear condition, the displacement under a given load is underinterpreted and the stress is overpredicted. The displacement error is small and usually more than offset by the lack of geometric stiffening in a linear analysis. The overpredicted shear levels can usually be thought of as a worst case scenario.

Geometry Concerns

As alluded to previously, a phenomenon called *geometric stiffening* can affect FEA results as displacements increase. The primary result of this condition is decreasing displacement under increasing load. This is stiffening over and above the amount resulting from the part's material properties. The primary cause of this stiffening is increased tensile stresses in the areas being deformed. As the axial tensile stress in a beam or the in-plane tension in a shell increases, the physical part will self-stiffen. This is often called *stress stiffening* (not to be confused with strain hardening, a change in material properties). While it is usually straightforward to estimate the nonlinearity of a system due to material properties, it is difficult, if not impossible, to determine if geometric stiffening will be affecting the system.

To account for this effect, a nonlinear solution with a large displacement option must be run. If the difference between the linear solution and the nonlinear run is significant in the area of concern, the nonlinear results should be used.

Boundary Conditions

The last condition of linearity is that the boundary conditions do not change from the point of load application to the final deformed shape. Loading must be constant in magnitude, orientation, and distribution. If any of these are expected to change significantly as the part deforms, a linear assumption is not valid. Contacting surfaces or edges are good examples of this phenomenon. If a pair of surfaces is not in contact in the unloaded state but come in contact as a load is applied, a reaction force is introduced at the contact region, thereby changing the load distribution in the system. The other common example of a nonlinear boundary condition is a load whose orientation is tied to the orientation of the geometry to which it is applied. If the geometry deforms enough, the resultant load vector could significantly alter the overall stress state of the system.

Using the Linear Assumption

In reality, many problems with moderate degrees of nonlinearity are solved satisfactorily with a linear solution. A linear assumption is often valid if the analyst understands the ramifications of any nonlinearity present and adjusts results interpretation. Another common means of handling nonlinearity in a linear solution is by use of safety factors. If your safety factors are high because the uncertainty in one or more of the four primary assumptions is high, the nonlinear error in a near-linear solution may be assumed negligible. It is important to note that you cannot begin to interpret results in light of the linear assumptions unless, first, the possibility of nonlinearity is acknowledged and examined, and second, the effects on the part's behavior due to these nonlinear issues are understood. If your business involves highly deforming, thin-walled parts or materials with a small true linear range, you must take the time to understand the importance of the linear assumption. It is not necessary that you become an expert in nonlinear analysis, but you should have access to such an internal or external expert who can help guide you through these problems.

Static Assumption

Like the linear assumption, it is common to assume that all loads are applied gradually to their full magnitude. Unlike the linear assumption, it is usually clear to any competent engineer, trained in FEA or not, when this assumption is valid. If the load is applied with a vibratory or sinusoidal input, it is probably not static. If the load is generated by impact or collisions with another body, it is probably not static. Many applications can truly be modeled with a static load, even though there is some brief time-dependent, transient response at the actual instant of application; the steady-state results are usually of more interest. In fact, a good way to interpret the static assumption is as that of steady state and constant magnitude. In this interpretation, the structural response of a body spinning with a constant velocity or traveling with a constant acceleration can be modeled as static. The static assumption loses validity when the briefness of the event duration, combined with the stiffness of the structure, causes the load to be removed before the full response can be induced. It is also invalid when the applied impulse or sinusoidal load excites a resonance in the structure and the effects of this load are amplified by the resonance.

There are three primary types of dynamic loading in FEA and each requires a separate solution type to solve for the proper response. The types are listed below.

- Transient response or time-dependent loading

- Frequency response or sinusoidal loading

- Random response

The first two types are commonly encountered in product design but a specialist should handle the third, random response. Chapter 18 discusses this in more detail, but a brief discussion of the concepts is included here.

Transient Response

A transient or time-dependent study is generally used for one or both of two reasons. If the duration or period of the event is so small that the system cannot respond quickly enough to fully deform before the load is removed, the actual stress levels may be significantly less than those reported by a static analysis. However, if the period of the impulse is short enough to excite a resonance in the system, a lightly damped and brittle structure may crack or shatter even though a gradually applied load of the same magnitude may have little effect.

Frequency Response

A frequency response analysis is also steady state, but differs from a static analysis in that both the magnitude and orientation of the load vary sinusoidally. As the frequency of the input approaches any natural frequency of the system, the difference between the static and dynamic responses diverges. The results from a frequency response analysis will behave nearly identically to a static solution with the same load magnitudes as the frequency approaches zero. However, if damping is high and resonance is not excited, the results may share some characteristics of a transient response in that the peak response, stress, or displacement, may be lower than if the load was applied gradually to full. The data used for damping, both material and structural, is critical to the correctness of a dynamic solution. If your system is subject to vibratory loading, it may be worth the time to check the static assumption by running a frequency response study.

Random Response

A random response study differs from a frequency response study in the nature of the loading. While loading in a frequency response solution is typically in pounds or pounds versus frequency, random response input is in the form of acceleration squared (G^2) versus frequency. These data are usually compiled in the form of a power spectral density (PSD) curve. PSD curves have been compiled for road noise in various cars on various surfaces and for different earthquakes or seismic events. The measured vibration is filtered to allow presentation in the PSD format. You can see from this description that random response input really is not random at all but a preapproved spectrum of excitations and frequencies. The proper application of random response requires that you have first mastered frequency response analysis. Note that whereas a frequency response analysis deals with one event at a time, a random response analysis deals with a spectrum of events simultaneously.

Using the Static Assumption

If the primary load input to your system is vibratory or of a short, impulse-like duration, the static assumption should always be checked against a dynamic solution. It is as simple as that. A static run may be used to debug a mesh and check boundary conditions but dynamic input should be solved with a dynamic solution.

Other Commonly Used Assumptions

In addition to the above main assumptions (geometry, properties, mesh, boundary conditions, linear, and static) which must be considered before every analysis, you will also need to identify and qualify many other assumptions that are applicable to your specific industry, parts, or current problem. The following lists should serve as a reference when reviewing your problem goals and setup. These lists are by no means comprehensive, but do cover a wide variety of situations which should prompt you to think of other assumptions applicable to your problem. Moreover, by no means do all of the following assumptions apply to all situations; in fact, some are actually contradictory. Be sure to use only those that apply to the analysis at hand.

Geometry

- The supplied CAD geometry adequately represents the physical part.

- Nonlinear geometric stiffening will not affect the behavior of the system.

- Displacements will be small such that a linear solution is valid.

- Stress behavior outside the area of interest is not important to this project such that geometric simplifications in those areas will not affect the outcome.

- Only internal fillets in the area of interest will be included in the solution.

- The thickness of the part is small enough relative to its width and length such that shell idealization is valid.

- The thicknesses of the walls are sufficiently constant to justify constant thickness shell elements.

- Local behavior at the corners and intersection of thin surfaces is not of primary interest such that no special modeling of these areas is required.

- The primary members of the structure are long and thin such that a beam idealization is valid.

- Local behavior at the joints of beams or other discontinuities are not of primary interest such that no special modeling of these areas is required.

- Decorative external features will be assumed insignificant to the stiffness and performance of the part and will be omitted from the model.

- The variation in mass due to suppressed features is negligible.

- A 2D axisymmetric solution will be used. It will be assumed that any part features that are not axisymmetric will have no impact on the behavior of interest.

- A 2D (plane stress or plane strain) solution will be used. It is assumed that any part features that violate the planar assumption will have no impact on the behavior of interest.

- Reflective symmetry will be used. The geometry and boundary conditions are, or can be assumed to be, equivalent across one or more planes.

Material Properties

- Material properties will remain in the linear regime. It is understood that either stress levels exceeding yield or excessive displacements will constitute a component failure, that is, nonlinear behavior cannot be accepted.

- Nominal material properties adequately represent the physical system.

- Material properties are not affected by the load rate.

- Material properties can be assumed isotropic (or orthotropic) and homogenous.

- The part is free of voids or surface imperfections that can produce stress risers and skew local results.

- Actual nonlinear behavior of the system can be extrapolated from the linear material results.

- Weld material and the heat affected zone will be assumed to have the same properties as the base metal.

- All simulations will assume room temperature although temperature variation may have a significant impact on the properties of the materials used. Unless otherwise specified, this change in properties will be neglected.

- The effects of relative humidity (RH) or water absorption on the materials used will be neglected.

- The material properties used assume dry, 50% RH, or 100% RH properties as specified in the manufacturer's data sheets.

- No compensation will be made to account for the effects of UV, chemicals, corrosives, wear, or other factors which may have an impact on the long-term structural integrity of the components.

- Material damping will be assumed negligible, constant across all frequencies of interest, and/or a published value or one determined from testing.

Boundary Conditions

- Displacements will be small such that the magnitude, orientation, and distribution of loading remain constant throughout the deformation.

- A static solution will be used. Loading rates are expected to be sufficiently low as to make this assumption valid.

- Frictional losses in the system will be considered negligible.

- All interfacing components will be assumed rigid.

- The portion of the structure being studied can be assumed decoupled from the rest of the system such that any reactions or inputs from the adjacent features can be neglected.

- Symmetry may be assumed to minimize model size and complexity.

- Load is to be assumed purely compressive, tensile, torsional, or thermal. No other load components are to be included in the study.

- Pressure loading will be assumed uniform across all loaded surfaces.

- The components modeled as pure forces will impart no additional rigidity in the actual system.

Fasteners

- Residual stress due to fabrication, preloading on bolts, welding, and/or other manufacturing or assembly processes will be neglected.

- Bolt loading is primarily axial in nature.

- Bolt head or washer surface torque loading is primarily axial in nature. Surface torque loading due to friction will produce only local effects.

- Stress relaxation of fasteners or other assembly components will not be considered. Failure is assumed to be early in the service life of the assembly.

- Load on the threaded portion of the part is evenly distributed on engaged threads.

- The bolts, spot welds, welds, rivets, and/or fasteners are numerous and stiff such that the bond between the two components can be considered perfect.

- All welds between components will be considered ideal and continuous.

- The failure of fasteners will not be considered.

General

- Only the results in the area of interest are important and mesh convergence will be limited to this area.

- No slippage between interfacing components will be assumed.

- Any sliding contact interfaces will be assumed frictionless.

- System damping will be assumed negligible, constant across all frequencies of interest, and/or a published value or one determined from testing.

- Stiffness of bearings, radially or axially, will be considered infinite or rigid.

- Elements with poor, or less than optimal geometry, are only allowed in areas not expected to be of concern and do not affect the overall performance of the model.

Summary

Is FEA a powerful tool for evaluating the structural integrity of a design and making comparisons between various alternatives? Absolutely! Is FEA the end-all that replaces testing and will allow users to design products with less understanding of sound engineering principles? Absolutely not! Dr. Bruce MacNeal of The MacNeal-Schwendler Corporation once commented, in the way you might expect a Dr. Bruce MacNeal to comment, "The intellectual participation of the user is a must." For all that FEA can provide to a design engineer and a design engineering organization, failure to understand the inherent assumptions in the modeling of systems will cause hours of time to be wasted and may possibly cost thousands of dollars to correct.

The qualification of assumptions is the key to successful use of FEA in product design. To be capable of qualifying assumptions, you must first understand the mechanics of the materials being modeled, the failure modes the products might encounter, and the manufacturing and operating environment expected for the product. Moreover, you must understand the capabilities and the limitations of each element type you choose to use and each solution method employed. There is an assumption behind every decision you make in finite element modeling. Once you accept this and work with this knowledge accordingly, you are soundly on the road to becoming a successful analyst. Once you have fully grasped the potential of the technology, you will become properly hooked.

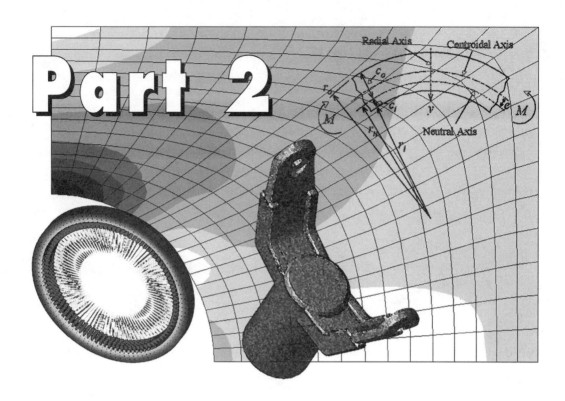

Finite Element
Modeling Basics

4

Common Model and Element Types

Once you have determined the goals of the analysis and the general category of the solution, such as linear versus nonlinear, static versus dynamic, and so on, thought must be given to the modeling type and element type best suited for the problem at hand. While a 3D automesh may seem like the easiest route, it can be overkill for some problems and, if inappropriate, could lead to incorrect results. The choice of modeling type will depend on factors such as general geometry style, time and resources available, and the goals of the study. If the immediate goal is to evaluate the effect of changing a fillet radius in a cantilevered feature, a simple plane stress analysis will be the most efficient approach. However, if exact stresses are required for failure verification, a more detailed 3D study might be warranted.

The choice of element type is somewhat dependent upon the modeling type selected. Obviously, a 3D solid tetrahedron or "tet" model would not be appropriate in an axisymmetric model. However, it is common practice to use two or more element types in a single model. This chapter will provide general guidelines for determining which modeling type is appropriate for various conditions and then review the commonly used elements and discuss their applicability for each modeling type.

Note that this chapter is meant to provide an introduction to the available modeling types and associated elements. Specific details about the properties of these elements will be covered more fully in Chapter 6 and the modeling needs of each element type will be covered in Chapter 7. In addition, the availability of some of these techniques and elements is code dependent, so keep your software documentation handy as you proceed through this chapter. Remember that there are many ways to model most situations, so your challenge will be to select the best option with the tools available in your software solution so that you may obtain the most efficient and accurate data.

Common Modeling Types

The most common modeling types are listed below.

- Planar simulations
  - Plane stress
  - Plane strain
  - Axisymmetric
- 3D simulation and modeling
  - Beam simulation
  - Symmetry or anti-symmetry
  - Plate or shell models
  - Solid models

As a rule of thumb, try to use the planar techniques, all of which are equally simple, and then move to the more computationally complicated 3D types as needed. Generally, simpler models are more accurate, both from a solution standpoint and from the reality that there is less chance of input error. Diverse modeling options have been developed to provide the analyst with choices to ensure more efficient and accurate solutions. The general descriptions that follow only briefly touch on the potential of each technique. When trend analysis or test models are the order of the day, creativity is the only limitation to the scope of information to be gained from simplifications or idealizations. If a nonlinear solution is required, as in the case of the clip shown in Fig. 4.2, the use of a planar solution can provide excellent convergence in a fraction of the solution time a 3D model would, thereby promoting improved accuracy.

Plane Stress Modeling

Plane stress modeling may be one of the most commonly used idealizations for which you never knew the name. A simple cantilevered beam analysis, as shown schematically in Fig. 4.1, is a plane stress idealization. The definition of plane stress requires that the behavior of interest occurs in such a manner that there is no stress component normal to the plane of action. This means that one of the three principal stresses is zero.

Fig. 4.1. Plane stress idealization of simply supported beam.

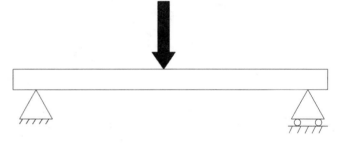

Identifying Plane Stress Models

A practical method for identifying plane stress opportunities is to look for geometry or systems that are essentially extrusions of a group of planar cross sections. The loads and constraints must be defined such that all resulting deflections allow the planar cross sections to remain in their initial plane. Externally induced deflection must be differentiated from Poisson's ratio related strain. If the ends of the so-called extrusions are not fixed in position, in-plane strain will cause normal strain due to Poisson's ratio. This will not necessarily generate measurable normal stress.

Stating the use of plane stress as an assumption implies other assumptions. It can be assumed that the behavior of interest remains planar and that various geometric inconsistencies which might violate the assumption of plane stress are negligible. Typically, the depth of a plane stress model is small compared to the size of the cross section, but this is not required. Figs. 4.2 and 4.3 show some problems that were solved using plane stress idealizations.

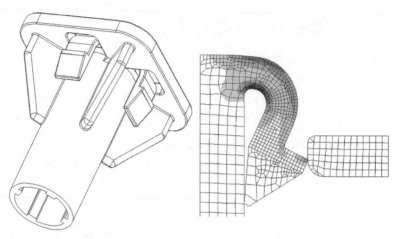

Fig. 4.2. Plane stress model of plastic clip feature.

Fig. 4.3. Plane stress model of cable retainer with multiple wall thicknesses.

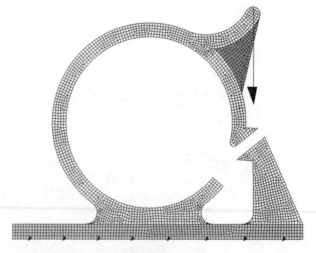

Geometry in a Plane Stress Model

If plane stress is valid, it can be assumed that any cross section or slice parallel to the generating cross sections would have the same stress distribution as any other. Consequently, the geometry for the model can be generated by cutting a solid model or assembly with an appropriately oriented plane. In many cases, it will be easier to work with 2D geometry imported from a drawing or created specifically for the analysis than with a 3D solid. The depth of the model, or the extrusion, must be specified in a plane stress idealization and the total loading should be applied, as if it were a 3D model. Multiple thicknesses can be assigned in a plane stress model although the user must understand how the solution interprets thickness variations. First, assume that the thickness is applied equally on both sides of the actual cross section. Therefore, all thickness will be centered on the plane of the model. Second, remember that the transition between the thicknesses is a sharp corner.

The snap clip in Fig. 4.3 has three thicknesses: the body of the clip, the web at the top, and the base. The solver will interpret these thicknesses as shown in Fig. 4.5. If your preprocessor does not have a plane stress option, the condition may be modeled by constructing a 2D mesh of plate elements, preferably with in-plane rotational stiffness, and constraining translation normal to the plane. If this is somewhat confusing, it will be clarified later in the book. In addition, many preprocessors will require that a plane stress model be constructed in a specific global coordinate system plane, usually the X-Y plane. Check your software's documentation to confirm.

Fig. 4.4. Plane stress model of a spur gear mesh.

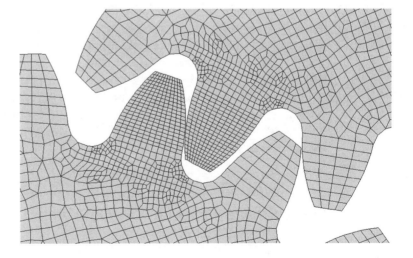

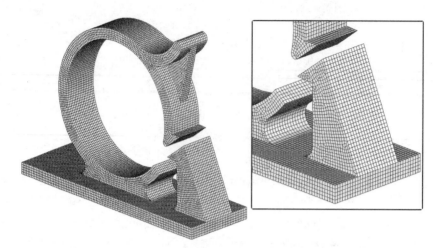

Fig. 4.5. Equivalent solid model of the retainer shown in Fig. 4.3. (Image courtesy of Tyton-Hellermann.)

Plane Strain Modeling

Many of the concepts discussed previously for plane stress hold true for *plane strain* with some notable exceptions. While the depth of the plane stress model is usually small compared to cross-sectional size, the depth of a plane strain model must be large in comparison to the section. In fact, it is common to assume an infinite depth so that any effects from end conditions are so far removed from the modeled cross section they can be ignored. Moreover, due to the large depth of the system, it can be assumed that strain normal to the modeled cross section is zero. A shorter model with perfectly constrained ends will also allow a plane strain idealization as normal strain is eliminated by the boundary conditions. Consequently, the behavior of the system local to the modeled cross section has only planar or biaxial stress and strain.

If an assumption of infinite depth is valid, then the specification of model depth becomes trivial. Therefore, the depth of the model is not a required input and loading is applied in units of load per depth. A plane strain problem is often referred to as a *unit depth* problem.

Fig. 4.6 shows a typical plane strain problem, a long pipe under uniform static pressure. If it can be assumed that the end effects caused by bends, fittings, or valves are far from the center of the pipe, and that the displacement has stabilized for a significant portion of the pipe, the assumption of plane strain is valid. Plane strain problems are common in civil engineering and are used to model retaining walls or dams.

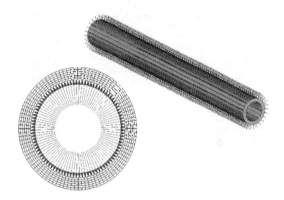

Fig. 4.6. Plane strain idealization of long pipe under constant pressure.

Axisymmetric Modeling

In an *axisymmetric model,* the geometry and boundary conditions are or can be assumed to be revolved 360° about an axis. Some common uses of axisymmetry are pressure vessels and tanks, as seen in Fig. 4.7. Axisymmetry problems are planar models in which the solver understands that the modeled half of the cross section is revolved 360°. By using a planar model instead of a full 3D solid or some combination of solids with symmetry, many more degrees of freedom can be introduced in critical areas without a significant gain in run time. If contact or another nonlinear solution is required, the benefits of using axisymmetry are huge, as in the battery model shown in Fig. 4.8. Some codes allow axisymmetric shells as well as the more common axisymmetric solids. The former appear as line elements in an axisymmetric model and represent thin-walled shells of revolution to the solver. Other uses of axisymmetry are illustrated in Figs. 4.9 and 4.10.

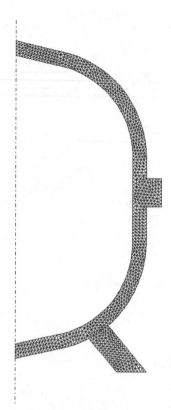

Fig. 4.7. Axisymmetric model of a pressure vessel.

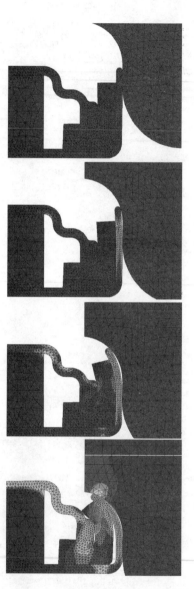

Fig. 4.8. Axisymmetric model of the forming of a calculator battery assembly.

Fig. 4.9. Axisymmetric model of an engine valve stem.

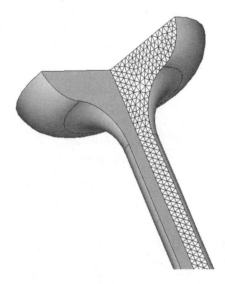

Fig. 4.10. Axisymmetric model of a hydraulic cylinder piston.

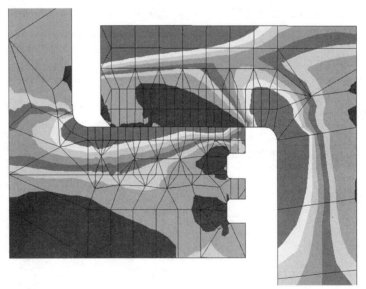

Orientation of an Axisymmetric Model

Axisymmetric models are typically required to use a global or system axis as the axis of symmetry. Again, the choice of this axis depends on the code used. Some require the axis to be the Y axis and some the Z.

Some codes even require that the axisymmetric model be built in a certain quadrant or on a particular side of the axis of revolution. Check your software's documentation.

Identifying an Axisymmetric Model

When attempting to identify axisymmetric problems, the most obvious means of qualification is that the base feature of the part should be a solid of revolution. In many cases, axisymmetry is still valid if there are small features that break up the revolution. In the example of the punch shown in Fig. 4.11, a double-D flat configuration has been machined into the shank and a hole for stabilizing the part while the chuck is tightened has been drilled into the body. Engineering intuition should suggest that with compression loading at the shoulder and a constraint at the tip, any deviation of the shank geometry should have no effect on the results of interest. The hole in the body cannot be written off so easily. A common practice, described in subsequent chapters, is to bracket the solution to determine if a feature contributes to the response of interest. In this case, it would still be faster to converge two axisymmetric models than a solid model with the required hole. The two bracketing conditions would be first, a punch with no hole, and second, a cross section with the full cutout of the hole.

Fig. 4.11. Solid CAD model of steel hole punch.

These models are shown in Fig. 4.12. It can be seen from the stress distribution that the cutout has little impact on the region of maximum stress and this is a significantly worse approximation than the actual condition. Why? Because the axisymmetric model understood the cross section of (b) to be a punch with a groove turned about its body 360°. It should be obvious that the geometry from (a) is much stronger than (b).

Fig. 4.12. Axisymmetric
analyses of hole punch (a)
without hole feature and (b) with
hole cutout.

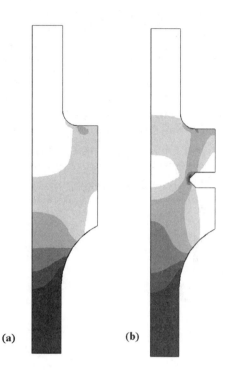

(a)　　　　　(b)

Loading

As mentioned previously, the loading must be consistent about the axis of the part. Loading is applied to an axisymmetric model differently in different solvers. Some codes allow you to apply the actual, total load to the model, while others force you to divide it by 2π, or 6.28, before applying the load to the model. Check your particular software's documentation. As the degree of approximation increases in the model, the need for checks on the model increases. A couple of quick checks on the punch model described earlier can be made to ensure that the loading was applied correctly. The first check is a simple stress prediction. All the applied load must flow through the narrow section at the tip of the punch. Therefore, the stress in that portion of the model should exactly equal the applied load divided by the cross-sectional area. If this does not match, something is incorrect. A second quick check is to sum the constraint forces at the tip. The reaction force should equal the applied load.

Constraints

As mentioned previously, the constraints must be constant around the part's axis. Axisymmetric models only require constraints parallel to the axis of revolution. Being a planar approximation, no out-of-plane translation or rotation is permitted by definition. In addition, the model is constrained automatically by its axis of revolution. Since all geometry and boundary conditions must be revolved, the results must be axisymmetric as well. Consequently, the pressure vessel lid shown in Fig. 4.13 only needs to be constrained in the Z direction. Do not fall into the trap of believing that this constraint requires less thought because your options are reduced.

Fig. 4.13. CAD solid model of pressure vessel lid.

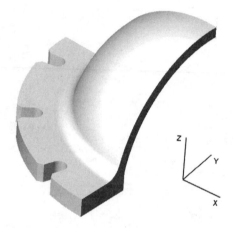

Fig. 4.14 shows four possible constraint schemes with four very different results.

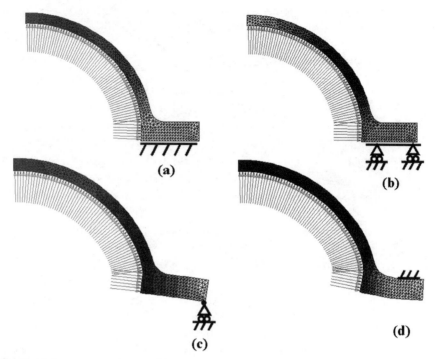

Fig. 4.14. Four constraint conditions and resulting displacement plots for the pressure vessel lid.

Although not universally implemented, some codes provide additional axisymmetric capabilities. ANSYS allows an axial torque load to be applied to axisymmetric geometry and COSMOS/M allows non-axisymmetric, radical loading on these models.

Symmetry

Symmetry modeling is an option that can be used with essentially any modeling type with great success. In fact, if the option of symmetry exists in a static problem, it should be used. The only exception to this might be a smaller problem where the time spent explaining the wonders of symmetry in FEA to an unenlightened reviewer of your results will take more effort than actually modeling the whole system. In most cases, utilizing as many planes of symmetry as are allowed by the problem will result in shorter run times, more accurate boundary conditions, and more accurate solutions deriving from the previous two benefits. There are two primary types of symmetry used in three-dimensional FEMs: planar and cyclic.

Reflective Symmetry

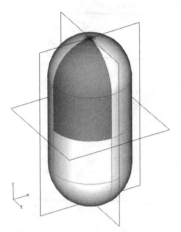

Reflective symmetry is more commonly used and available in all FEA packages because it requires no special programming, just correct application of standard boundary conditions. Any 3D model can have a maximum of three orthogonal planes of symmetry in which the geometry, properties, and boundary conditions are equivalent across these planes. By using all three, the domed pressure vessel of Fig. 4.15 can be captured with only one eighth of the geometry. Symmetry conditions require that the geometry and boundary conditions are, or can be approximated as being, equal across one, two, or three planes. Symmetry used in modal or subsequent dynamic analyses is not recommended. While a combination of symmetry and anti-symmetry can be used to find all the mode shapes, this combination is not feasible for most dynamic studies. Utilization of symmetry for a modal analysis will result in the computation of only symmetrical modes. You may miss many modes of interest that are not symmetrical about the planes you defined. This will be reviewed in more detail in Chapter 17, "Modal Analysis."

Fig. 4.15. Three symmetry planes for a domed pressure vessel with the section required for an internal pressure load.

Near-symmetry

In many cases, slightly asymmetric geometry can be initially modeled as symmetric by analyzing the less rigid half under the assumption that if the weaker half is acceptable, the stronger half will be too. The flange shown in Fig. 4.16 is a good example. The part actually has two planes of near symmetry, X-Y and Y-Z. The hole on the right side of the part violates the symmetry but an analysis of that quarter of the model shows acceptable displacement and stress. Therefore, a part with a hole in all four quadrants will be acceptable and, in all likelihood, the part with only one hole will be fine. Your engineering judgment is critical here because this assumption of near-symmetry is not valid in all cases. The asymmetry of this part may cause eccentric deformation of the main hole, which could lead to other system problems. Use care with this and other assumptions.

Fig. 4.16. CAD solid model of stamped flange.

Loading

The total load applied to a symmetric model should be divided by the number of symmetry planes used. One symmetry plane, or half symmetry, requires half the total load. Two symmetry planes, or quarter symmetry, requires one fourth of the total load, and so on. The exception is a pressure load or other area dependent load. These loads will automatically halve or quarter themselves due to the available surface area in the model. Use a check on total load and other simple checks described later in this book to ensure that the load was correctly applied.

Constraints

The constraints on a reflective symmetry model define the symmetry to the solver. The constraints on a solid model must prevent translation through the plane of symmetry on the entire cut face and the constraints on beam and shell models must also prevent rotation in the components parallel to the cut planes. These constraints ensure tangency and continuity at the cut plane, just as the other half of the model would if it existed. The boundary conditions for symmetry will be discussed in more depth in Chapter 8.

Cyclic Symmetry

Cyclic symmetry is a more specialized condition where features that are repeated about an axis can be modeled by a single instance of that feature. Common applications of cyclic symmetry are fan blades, turbine blades, flywheels, and motor rotors.

Fig. 4.17. Fan blade and hub model.

Fig. 4.18. Turbine blade model with cyclic symmetry instance exploded.

Fig. 4.19. Flywheels undergo windup in the transient portion of start-up and are good candidates for cyclic symmetry.

Fig. 4.20. Cyclic symmetry can be used on an instance of a motor rotor.

In addition to the geometric definition, which is subject to the near-symmetric assumption made earlier, constraints and loading must fit the cyclic symmetry requirements.

Boundary Conditions

In cyclic symmetry, each instance of the feature must see the same boundary conditions in its respective frame of reference. Acceptable loading might be centrifugal forces, radial displacement due to a press fit, or uniform wind or fluid resistance due to spinning. While the geometry of a motorcycle wheel or gear may fall into the realm of cyclic symmetry, the constantly oriented vertical weight of the bike or the tooth loading on the gear will invalidate cyclic symmetry.

Using Reflective Symmetry to Approximate Cyclic Symmetry

Cyclic symmetry is not available in all codes and some preprocessors do not provide tools for using it, even if your solver supports it. Cyclic symmetry can be approximated with planar symmetry in some cases. Because planar symmetry is available in all systems, this subset of cyclic symmetry is always available. The restrictions on using planar symmetry to model cyclic symmetry are that the geometry conforms to the definition of cyclic symmetry and that the only loading and resultant displacement are radial or coaxial. Figs. 4.21 and 4.22 show examples.

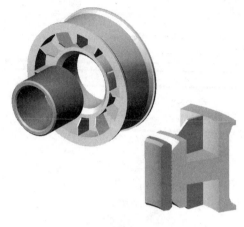

Fig. 4.21. Cyclic symmetry concepts can be simulated with planar symmetry in this press fit model.

Fig. 4.22. An odd number of features in an otherwise symmetric model can be simulated with radial cut planes as long as the loading is radial.

To use this method, the instance of the repeated feature is isolated using cutting planes, and planar symmetry constraints are applied to the cut surfaces. This usually involves the creation of a user-defined Cartesian coordinate system with an axis normal to the cut surface. The same constraint can be applied on a solid model by using a cylindrical coordinate system which is coaxial with the part, and applying a translational constraint in the theta, or angular degree of freedom. This procedure is not as effective for shell models because the rotational constraints cannot be simply defined in a cylindrical coordinate system.

If your code does not support cyclic symmetry, you must model the entire part. However, some time savings may be achieved by recalling that the results on each instance must still be identical. Therefore, only one instance requires a fine, converged mesh. The mesh on the rest of the part must be good enough to capture global deformation and stiffness, but not necessarily stress.

Anti-symmetry

Another type of symmetry modeling is available but not widely used by design analysts, primarily because it is not as intuitive as the previously described symmetry conditions. This technique, called anti-symmetry, can be used when geometry conforms to planar symmetry but the loading does not. Anti-symmetry exists due to the ability to utilize linear superposition on linear FEA results. The boundary conditions on the cut plane of an anti-symmetric model are exactly opposite those of symmetry.

Fig. 4.23 shows a simple example of the construction of results using anti-symmetry. On their own, the results from an anti-symmetry solution are meaningless. However, when combined correctly with symmetry results, large problems with symmetric geometry and asymmetric loading can be solved more quickly.

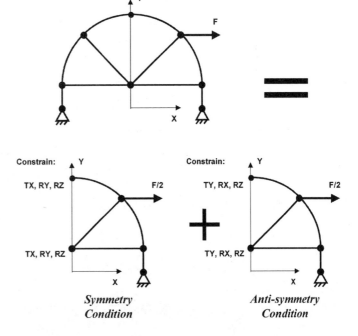

Fig. 4.23. Construction of
anti-symmetry results.

Additional Benefits of Symmetry

As mentioned previously in this section, symmetry can provide numerous benefits beyond simply reducing model size and run time. First, many symmetric problems cannot be effectively modeled without symmetry boundary conditions. Consider the simple ring under internal pressure shown in Fig. 4.24.

Fig. 4.24. Ring under internal pressure with a point constraint at the bottom and symmetry constraints. The smaller internal ring is the undeformed model for reference.

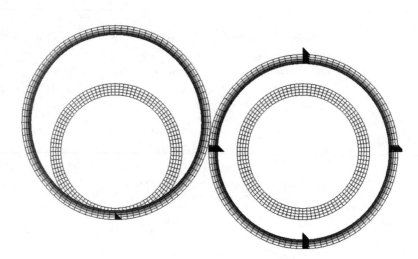

Without using quarter symmetry, fictitious constraints must be applied to prevent rigid body motion. This will result in counterintuitive displacement as well as high localized stress in the vicinity of the constraints. However, the quarter symmetry model is free to displace naturally without any questionable anomalies, which will require investigation and/or explanation.

When you believe that a full model must be shown in a report or presentation for clarity or simply to avoid explanation which might detract from the issue at hand, symmetry conditions can still be used to allow for more accurate modeling. Simply "bury" the symmetry constraints on appropriate cut planes, which divide the full model into halves, quarters, or eighths. The flange example shown previously has been modeled as a full model but the nodes at the edges of each quadrant have appropriate symmetry conditions in Fig. 4.25.

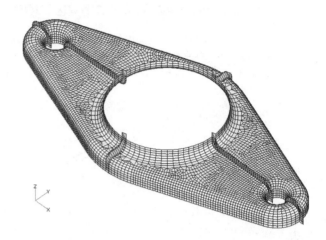

Fig. 4.25. Full shell model of stamped flange with symmetry constraints on the appropriate planes.

When modeling full models with buried symmetry, take advantage of the opportunity to gauge convergence by ensuring that the results in each portion of the model are identical. In a part with symmetric geometry and symmetric loading, the results must be symmetric. Nonsymmetric results in this case should be a red flag to even nonFEA users that something is wrong with the model.

Beam Models

Beam structures are the simplest, geometrically speaking, and the most efficient finite element models but they are the most complex from a user standpoint. Mastering beam modeling is one of those moments in an analyst's career that indicates growth from casual user to power user. Typically, beams should be used to model beam-like structures. The structural members of the system should be long compared to their cross-sectional area and should undergo relatively consistent bending, torsion, and/or tension compression. An immediate disqualifier for beam modeling is localized behavior of interest at a joint or feature on a portion of the beam. Beam models should be thought of as providing gross or general system performance information. From the beam model, reaction forces, moments, deflections, and interactions can be documented and used to generate inputs for more detailed submodeling. Beam models can utilize symmetry, although if the plane of symmetry cuts down the axis of a beam, the cross-sectional properties of the elements at the cut must be reduced accordingly.

Two Basic Types of Beam Elements

Beam elements typically fall into two broad categories. While the names used by different preprocessors and solvers may vary, most beams can be categorized as able to transmit moments or not able to transmit moments.

Rod Elements

The second case is actually a special instance of the more general beam element. Common names for beam elements which cannot carry moments are rod, bar, or truss elements. They can be thought of as long rods with ball joints on the ends. Because they cannot bend (a moment would be required to cause bending), the entire length of the modeled component can be captured with a single element. This member will transmit axial loads only and can be defined simply by a material and a cross-sectional area. In its most basic usage, it behaves as a spring, and in many cases can be used to model a spring, thus allowing for nonlinear or temperature dependent spring properties.

Fig. 4.26. Schematic representation of a rod element.

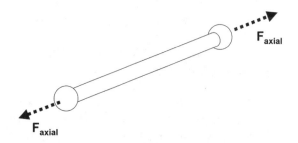

Beam Elements

The most general line element is a *beam.* Beam elements are defined by the geometric position of the end points, a material, a cross-sectional area, an orientation vector, the area moments of inertia, and torsional stiffness. (Refer to Chapter 2 for a review of these terms.) A restriction on beam elements is that the cross sections specified remain planar and perpendicular to the axis throughout the solution. In most FEA codes, a general beam can have a tapered cross section, the neutral axis can be specified to be offset from its geometric position, and any degree of freedom or load carrying component, can be turned off or released. In fact, the mechanism for executing this procedure is called a *beam release.*

COMSOL

Konrad Juethner
Technical Engineer

COMSOL Inc.
8 New England Executive Park
Suite 310
Burlington, Massachusetts 01803
Tel: (781)-273-3322
Fax: (781)-273-6603

Thus, by using beam releases, the more general beam element can be reduced to the rod element described previously.

Beam Coordinate System

Each beam element has its own coordinate system, which is independent of any spatial or model coordinate system. In most codes, the beam X axis is oriented along the length of the beam. The cross-sectional axes, Y and Z, are then oriented normal to the beam axis and orthogonal to each other. The two bending moments of inertia relate to the Y and Z axes and are sometimes called I_{yy} and I_{zz}. Some codes specify the moments of inertia as I_1 and I_2 and allow the user to arbitrarily assign them to the Y or Z axis. This can result in second guessing and double checking of the proper orientation. If the cross section is not symmetric about either axis, y or z, an additional term, I_{yz} or I_{12}, will be required. (See Chapter 6 for more detailed discussion.)

Fig. 4.27. Beam element coordinate system.

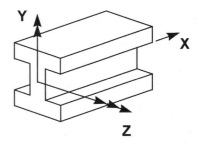

The neutral axis of a beam is, by default, on the geometric definition of the line element or its X axis. The term "line element" can be somewhat of a misnomer in that it really only applies to a linear or first order element, which is geometrically defined by its end points. A higher order line element may have a mid-side node, or a second order polynomial definition, such that its initial shape can be a conic curve. This higher order curve is usually required to be planar in its initial state although subsequent deformation during the solve may alter its planarity.

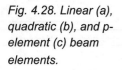

Fig. 4.28. Linear (a), quadratic (b), and p-element (c) beam elements.

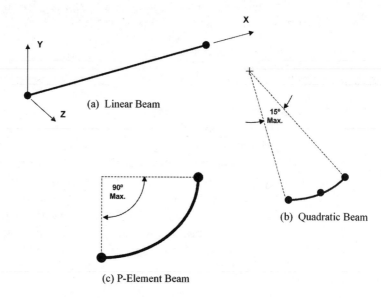

(a) Linear Beam

(b) Quadratic Beam

(c) P-Element Beam

Stress Recovery in Beams

Some preprocessors provide automated tools for calculating beam cross-sectional properties, ranging from predefined libraries of common cross sections to sketchers which calculate the properties on any drawn section. Regardless of the scope of the property automation tool, it is important to remember that in the end, the solver only knows "I"s, "K"s, and "A"s (inertias, torsional stiffnesses, and areas). Consequently, if the Is, Ks, and As are the same, an I-beam cross section will deflect identically to a circular cross section. However, in reference to Eq. 2.29, calculating the stress generated by a beam in bending requires one more piece of information: the distance from the neutral axis, or y. Because two shapes can have the same cross-sectional properties but significantly different dimensions, one more input to the beam definition is the *stress recovery point*. Most codes allow for up to four stress recovery points, which are typically defined at the outer corners of a bounding rectangle. Bending stress is calculated only at these points. Tensile and compressive stress is calculated for the entire beam, but reported bending stress and stress from torsion will depend on your choice of stress recovery points.

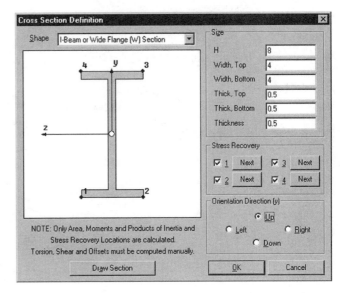

Fig. 4.29. Stress recovery points for an I-beam cross section in FEMAP.

Beams in Torsion

Beams in torsion also require the specification of a torsional constant. For circular cross sections, the torsional constant, K, equals the polar moment of inertia, J. Most codes still use the term J when referring to this property. Remember that there is a difference. The torsional constant is not automatically calculated by most cross-sectional property generators because the method for determining the torsional constant for a closed, circular cross section is vastly different from open cross sections or oddly shaped cross sections. Consider intuitively the torsional stiffness of the two cross sections shown in Fig. 4.30.

While the A_1, I_1, and I_2 values are nearly identical, the closed cross section is orders of magnitudes stiffer in torsion than the open section. The closed section K can be calculated by simply summing the I_1 and I_2 values (Eq. 2.9). The theoretical background for Eq. 2.9 requires that all sections remain planar. As seen in Fig. 4.30, the split cylinder is not capable of resisting much torque. The method to calculate K for the open cross section is much more involved. Extruding a solid or shell model into a beam and subjecting it to a moment is common practice. Using Eq. 2.36, with K in place of J, the K can be backed out of the angular rotation of the model. This method is valid for most beam models.

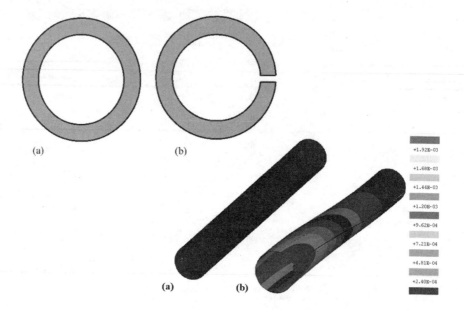

Fig. 4.30. A displacement plot showing the difference in torsional stiffness of a closed tube (a), and a split tube with the free end constrained to remain planar (b).

Convergence

While beam models will converge in most cases with fewer elements than most other modeling types, they should still be converged using the methods specified in Chapter 10. The lack of any truly local behavior in a beam is the primary reason for their efficiency and accuracy with fewer elements.

Section Orientation

In addition to geometric position and cross-sectional properties, the angular orientation of the cross section must be specified. While two I-beams rotated 90° about their axes from each other have the same cross-sectional properties, they obviously will not support the same load in the orientations depicted in Fig. 4.31. Consequently, some method of spatially orienting the cross section will be provided by your preprocessor and solver. Most codes use the straightforward technique of requesting an orientation vector for each element or mesh of elements. The solver then orients one of the principal axes of the cross section (Y or Z axis depending on the code) to the specified vector. Check your documentation to confirm which axis. Some codes use more complex methods to arrive at the same result. If this is the case with your solver, take some time to understand it if beam models are required. Do not be afraid to try some test models to ensure that you understand how to orient beams.

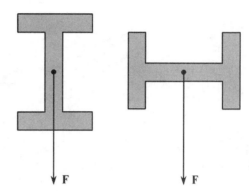

Fig. 4.31. The orientation of a section greatly affects its ability to carry a bending load.

Plate and Shell Modeling

The terms *plate* and *shell* are often used interchangeably and refer to surface-like elements used to represent thin-walled structures. While classical theory will differentiate between plate and shell calculations, for the purposes of this text 3D surface modeling will be referred to as "shell modeling." One differentiation that will be made refers to thick-shell elements, which are handled somewhat differently.

Shell Element Basics

Shell elements can be quadrilaterals or triangles. A quadrilateral mesh is usually more accurate than a mesh of similar density based on triangles. Most preprocessors can mesh a surface with quads only or apply a quad dominant mesh where triangles are used only when the mesher cannot resolve an area within specified element tolerances. Triangles are acceptable in regions of gradual transition. Linear or first order shell elements are normally planar and degrade in accuracy as their initial definition deviates from planar. This is an issue only for quad elements because a three-noded triangle must be planar. Higher order shell elements can provide accurate results with curved initial geometries. In fact, one of the benefits of using higher order elements is that positioning the mid-side nodes on the actual curved geometry increases the model's accuracy. P-elements can be defined to actually represent bidirectional curvature and can smoothly represent initial geometry.

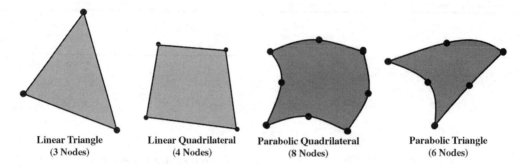

Linear Triangle	**Linear Quadrilateral**	**Parabolic Quadrilateral**	**Parabolic Triangle**
(3 Nodes)	(4 Nodes)	(8 Nodes)	(6 Nodes)

Fig. 4.32. Linear and quadratic shell element types.

Most first order triangle elements are only capable of calculating a single strain value across the entire element. Consequently, they are known as *constant strain elements*. This limitation can lead to overly stiff results under a given load as localized strain gradients will be difficult, if not impossible, to capture. They do provide adequate results when used on flat or gently curving surfaces with minimal strain variance across the span. Linear quad elements have a linear strain distribution from one node to the next so they are better at capturing localized stresses. Adding mid-side nodes to both these elements improves their strain distribution by a polynomial order.

Orientation

The default interpretation of a shell mesh is that it is centered at the mid-surface of the modeled geometry. Many codes offer the ability to numerically offset the mid-surface of selected elements to better represent the geometry. Shell meshes are usually a constant thickness but some codes also provide the option for tapering a shell mesh. Good preprocessors will allow for tapering an entire mesh over a surface in contrast to trying to specify the taper manually for each element.

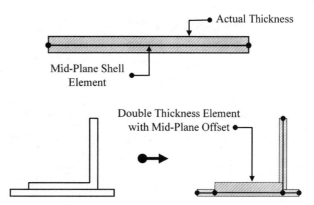

Fig. 4.33. Shell mid-planes can be offset to better capture thickness variations or a stack of plates.

The need to taper should be confirmed with test models before assuming that the extra work will bring sufficient increases in accuracy. In most cases, adjusting the wall to a constant, median thickness or using a stepped reduction may be sufficient.

Fig. 4.34. Tapered or drafted walls can be modeled with (i) tapered shell elements, (ii) a median thickness, or (iii) a stepped thickness reduction.

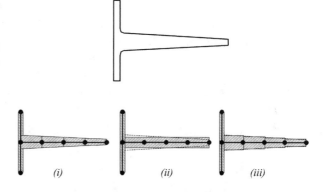

Shell elements have an orientation similar to beam elements. The x and y coordinate axes are oriented in the plane of the shell, and the z axis is normal to it. The element is defined with a top and bottom side. The two primary requirements for understanding the orientation of a shell mesh involve using pressure loads and evaluating bending stress. Pressure loads on shells are typically oriented from the bottom to the top.

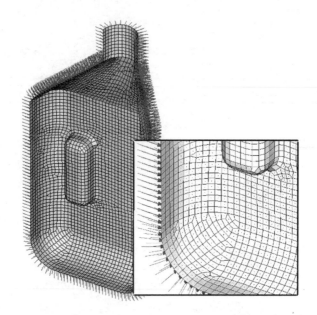

Fig. 4.35. Normals of shell elements in this bottle model point inward; therefore, a positive internal pressure would act as a vacuum.

Consequently, the top side of the shell mesh of a plastic bottle with consistently inward oriented normals is on the inside. A positive pressure on this mesh would result in a vacuum as interpreted by the solver. The resolution for this problem is to either reverse the normals of the entire mesh or to change the sign of the pressure load. While most preprocessors have handy tools for controlling shell normals, scaling the pressure load by -1 may be easier in many scenarios.

Stress Recovery

Similar to the discussion of stress extraction in beam elements, your solver interpolates the stress in the outer fibers of a shell because of the absence of geometry there. Some codes provide the ability to respecify the distance from the neutral surface at which stress is calculated.

Shell elements assume a linear stress distribution across the defined thickness. This should be a sufficient approximation for true shell problems, but it highlights the fact that there is variation in stress levels from the top to the bottom of a shell element.

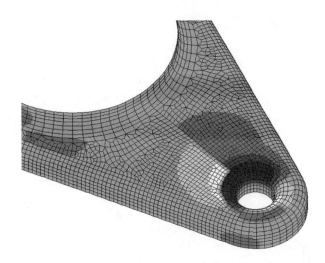

Fig. 4.36. If shell normals are not consistent, discontinuities in stress results may appear.

Recalling the tension-compression distribution of a beam or plate in bending should clear up any questions of the importance of this feature of shell elements. It is important to know which side of the element has the stresses of interest and to adjust the stress display accordingly. It is equally important that the shell normals be consistently aligned so that there are no annoying discontinuities in the contour results due to flipped shell normals. Some geometries lend themselves to continuous normal orientation. Other geometries have features that defy any sort of planning with respect to shell normals. Consequently, some postprocessors have provided an additional output feature that allows you to more quickly understand the results in a shell model. Most postprocessors provide the option of viewing results as calculated for the top or bottom faces of the shell. Some take it a step further and provide a results display of the maximum or minimum values of the element, regardless of orientation. This capability greatly simplifies the results interpretation of complex shell models.

Identifying Shell Model Candidates

The first indication that shell modeling is appropriate is that the wall thickness of the part or assembly is small compared to the overall size or surface area of the system. It is difficult to quantify "small," although various references consistently use the ratio of 10:1 surface area to wall thickness as the guideline for applicability of shells. Where the consistency stops is in reference to *"area of what."* Some references state element area to thickness and others state part surface area to thickness. However, it can be easy to find exceptions to both these guidelines in a perfectly valid shell model. One rule of thumb that is a little more qualitative and less quantitative can provide some quick guidance when evaluating various modeling methods. If the part would be understandable when modeled with zero thickness surfaces to someone unfamiliar with its actual form, a shell model is a likely candidate. If there are some bulky transitions or the thickness is so great that the relative position of the walls cannot be ascertained, then a shell model is probably not appropriate.

As with any idealization, the goals of the project should be clearly understood so that the limitations of the approximation can be properly evaluated. A shell model can provide excellent overall displacement results on most structures fitting the general requirements. Two shell elements that are continuous but not co-planar will be interpreted with some "virtual" overlap of their thicknesses. This can be ignored for gross displacement models but the stress at a corner will contain some error resulting from such overlap. It is good practice to attempt to transition a sharp corner using multiple elements following a fillet instead.

Fig. 4.37. "Virtual" overlap of the shell geometry as interpreted by the solver.

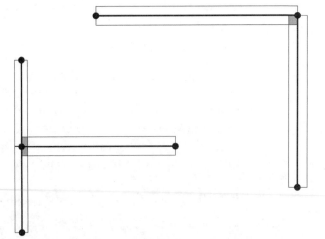

When bending is an issue, stresses calculated on continuous shapes or on large, unsupported areas will also be valid as long as the wall thickness is sufficiently thin. The inaccuracy inherent in the method is again at sharp transitions or corners in the model. The local model geometry will most likely deviate from the actual part in these areas, and this behavior is not generally well represented by the elements. However, as with many instances of local behavior, the stress levels away from the transition or joint are more likely acceptable as long as proper convergence methods are observed. It is not uncommon to build a large model using mainly shells for the initial pass of the study, and then single out more marginal components, from a shell applicability standpoint, and model them with solids.

Another good indicator of the validity of a shell mesh is in the modeling process itself. By default, a shell element is spatially located at the mid-surface of the geometry it is simulating. Fig. 4.38 shows the steps required to turn a simple I-beam into a shell model.

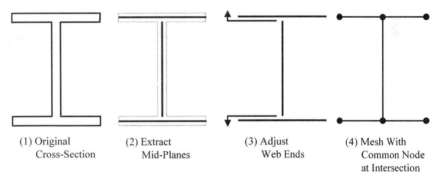

(1) Original
Cross-Section

(2) Extract
Mid-Planes

(3) Adjust
Web Ends

(4) Mesh With
Common Node
at Intersection

Fig. 4.38. Steps required to break an I-beam cross section into a shell model.

When the three key features of the I-beam are reduced to mid-surfaces, a gap of half the flange thickness exists between the surfaces. In a true shell situation where the thickness is much less than the other dimensions of the part, extending the web in both directions to meet with the flange surfaces should not measurably affect the overall behavior of the part.

However, in the weldment shown in the following illustration, making similar adjustments to the mid-surface geometry should cause you to sit back and question the move. The degree of pain or the volume of air sucked between your teeth as you tweak the model is inversely proportional to the applicability of the shell method.

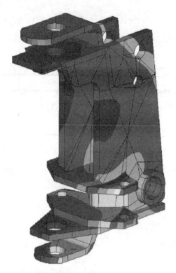

Concerns about adjustments should not automatically stop you from using shells for a particular problem. Again, the goals of the problem should guide all decisions regarding modeling. The cast cross brace shown in Fig. 4.40 has relatively thick transitions at the ends which would not intuitively suggest shells.

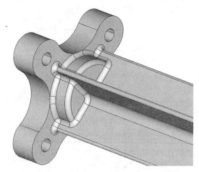

Fig. 4.39. Thick-walled weldments may not be good shell model candidates.

Fig. 4.40. Solid CAD model of a cast cross brace.

However, under the loading which resulted in the deformed shape in Fig. 4.41, the behavior of interest is far from questionable geometry and a solid model would not have significantly improved the accuracy.

Fig. 4.41. Stress results of a shell model of the cast cross brace.

Accuracy

Shell models, where applicable, may be significantly more accurate than solid models in bending with reasonable solution times. As the popularity of embedded meshers in solid modeling CAD systems increases, the tendency for new users to tetrahedral mesh a thin-walled part is growing. Figs. 4.42 through 4.44 show a thin-walled panel modeled three different ways.

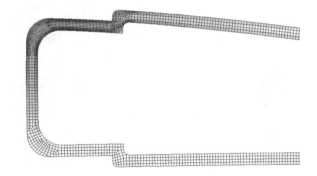

Fig. 4.42. Plane stress model of thin-walled panel.

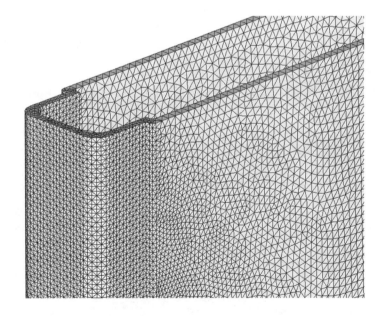

Fig. 4.43. Solid model of thin-walled panel.

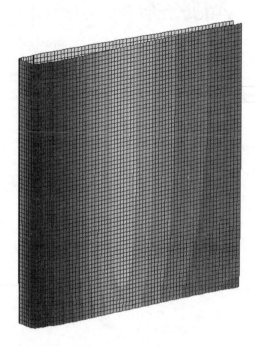

Fig. 4.44. Shell model of thin-walled panel.

A plane stress analysis indicated the need for a finer mesh at the end detail but due to out-of-plane bending, a 3D model was eventually required. An automeshed solid model became prohibitive for rapid design iterations at 200,000 degrees of freedom with only a fraction of the stress expected in the end feature. The above shell model modeled the collapsing behavior of the panel as well as the bending behavior of the wall. It solved in minutes and could be modified very quickly. In this case, a shell model, the appropriate modeling method, produced faster and more accurate results.

Special Shell Elements

Some specialized formations of shell elements can be found in various codes. These include a membrane-stiffness-only derivative, a shear panel, and a bending-only element, as well as the aforementioned thick shell formation. Thick shell elements have a different supporting set of equations that capture bending behavior more accurately for shells that, as you might guess, are thicker than normal shell elements.

Additional Benefits

Shell models can be used with planar symmetry, cyclic symmetry, and anti-symmetry. As described later in this chapter, they can be mixed with beam elements and/or solid elements. In addition to the accuracy benefits and the solution speed gained by using shell elements, the speed at which changes can be made and design iterations can be explored makes them superior to solids when geometry allows. First, it is a simple matter to delete the mesh on only one or two surfaces that might warrant topology changes. After adjusting the geometry, these surfaces can be meshed and merged with the previous mesh in a matter of minutes. Second, making a change to wall thickness is as simple as typing in a different number. Many combinations of wall thickness can be explored quickly with shell models. The use of a numerical input for wall thickness has made automatic optimization on this property popular.

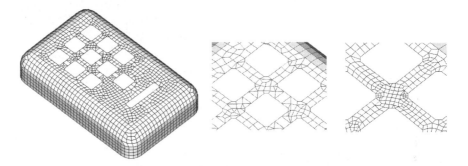

Fig. 4.45. Local mesh refinement of shell model near cutout.

Finally, achieving convergence by locally refining an existing shell mesh is much more straightforward than in a solid model. With many preprocessors, the mesh can be refined in selected areas with no more than a window pick. Most solid models require re-meshing the entire model to effect a convergence pass. It is commonly known that easier tasks are more likely to be done. Consequently, it is fair to say that an engineer in a schedule crunch will be more likely to take the time to converge a shell mesh than a solid mesh.

Solid Element Modeling

Solid element modeling is the most CPU intensive technique and should be considered when the other options cannot be applied. Converging a solid mesh or making a geometry modification usually requires a completely new model. So why is a tetrahedral automeshed solid rapidly becoming the model of choice for new users? The actual reason is somewhat of a catch 22. Design engineers are always looking for ways to make their jobs easier and automeshing tets requires the least amount of thought. Software developers are somewhat sensitive to the desires of the market so they gave people what they wanted. Coming full circle, design engineers see the increasing availability of these products and the marketing blitz behind them and assume that they must be the way to go because they seem so popular.

While improvements in automeshing coupled with fast and affordable solid CAD modeling have made solid FEA more accessible to the general engineering community, it is critical that you, the user, remember that it is just one of the available modeling techniques. It should only be used when it is appropriate, and the analyst must understand the capabilities and limitations of all solid element types and techniques in order to obtain the most accurate solution.

Identifying Solid Model Candidates

The ideal solid model is a bulky, low aspect ratio part. The term "potato shaped" is frequently used to describe these types of parts.

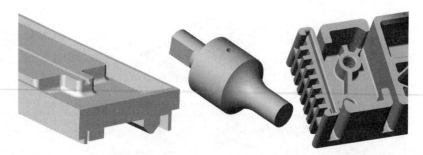

Fig. 4.46. Chunky solid CAD models are likely candidates for solid FEA models. (Figures courtesy of Tyton-Hellermann.)

As walls become thinner and bending behavior becomes significant, care must be taken to ensure that the correct element type is used and convergence is checked. Identifying a solid model is relatively straightforward. If the part or system cannot be modeled with one of the planar approximations, as a beam model, or as a shell model, it must be modeled as a solid. All the symmetry techniques discussed previously are applicable to solid models *and more so* due to the likely size of solid meshes.

Solid Element Basics

There are three types of solid elements commonly used: brick, wedge, and tetrahedron, otherwise known as the tet. Each of the elements can be modeled as first order or second order in most h-codes.

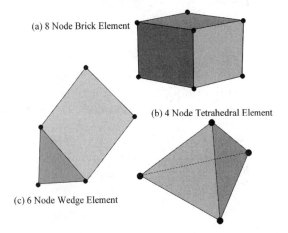

Fig. 4.47. Common solid element shapes.

(a) 8 Node Brick Element

(b) 4 Node Tetrahedral Element

(c) 6 Node Wedge Element

P-element codes use the same element types but the edge definition is, naturally, more complex. First order tetrahedrons in h-element codes are also constant strain elements and tend to behave in an overly stiff fashion compared to their higher order counterparts. Consequently, many more linear tets are required to approach the accuracy of a second order tet mesh and even then may not be enough. In many cases, first order tets will converge to a different answer than higher order elements or brick elements. While tetrahedral meshes met with some resistance from specialists as they became more available, most users will acknowledge that a well-converged parabolic tet mesh can provide acceptable results in most cases. When the time savings in tet meshing compared to building a brick/wedge mesh is taken into consideration, it is clear why they have moved into the mainstream for many analysts.

Tet versus Brick Meshes

Fig. 4.48. Comparison of an automeshed valve stem to a revolved brick mesh.

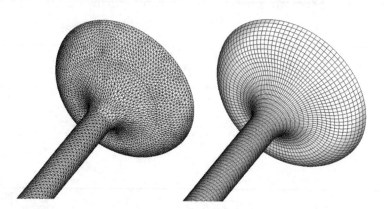

Despite marketing hype, no automatic brick mesher for solid CAD geometry has gained broad acceptance. Some smaller software providers have developed semi-automatic technology that has been billed as automatic, but the general consensus of the analysis industry is that the current offerings are not reliable from a general use or an element quality standpoint. Chapter 7 presents the pros and cons of brick meshing in more detail. Suffice it to say here that solid geometry which can easily be extruded or revolved into brick/wedge meshes without any loss of geometric integrity will probably be more computationally efficient and more accurate than a comparable tet mesh density. However, if the geometry cannot be easily broken into components for brick meshing, a tetrahedral mesh is the only generally accepted means to fill a volume that can be used by nearly all popular structural FEA codes.

Solid Modeling Tips

When considering solid models, remember that they may often be combined with other element types to allow for more complex boundary conditions. Building assemblies using solids can be more complex than with shells or beams. However, a solid element has fewer required assumptions and is inherently less ambiguous. The actual geometry can be modeled, including welds, fillets, chamfers, bosses, and transitions. There is no issue of top or bottom sides or stress recovery points, so the resulting data are simpler to interpret.

Special Elements

Previous discussion has focused on single element type models using common elements found in nearly every FEA package. There are other special elements available with most codes that facilitate complex boundary conditions, assembly idealizations and modeling, and *transitional*, or mixed element type models. The latter two conditions, assemblies and transitional meshing, will be discussed in more detail later in the chapter.

Spring Elements

One of the most commonly used special element is a *spring*. Most codes incorporate springs as simple 1D (line elements) connecting two nodes. Springs are typically used to represent...springs. However, in assembly modeling or when contact conditions exist, springs can provide a great deal of additional modeling flexibility. These uses will be discussed in the appropriate modeling sections. Springs can be axial or torsional. In the discussion of beam elements, it was suggested that the rod type, uniaxial line element could easily be used as a spring element with the proper modulus and area. Most codes, however, include a spring element which allows the user to simply enter a spring rate, K. In addition to the basic point-to-point springs, some codes make available a *spring-to-ground* element that automatically and mathematically fixes one end of the spring in all degrees of freedom.

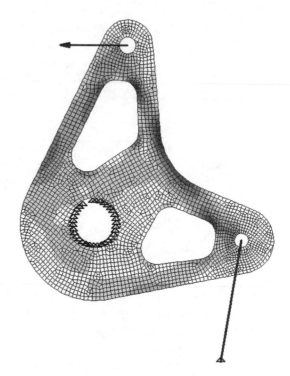

Fig. 4.49. Spring elements can be used in a simulation in the same way as they are used in the actual part.

In reality, there is not another "end" to these springs, but that is the most straightforward way to look at them. Another type is a *degree of freedom (DOF) spring*. The DOF spring differs from a regular spring element by providing stiffness in only specific directions or degrees of freedom. Whereas a general spring element derives its force from changes in spatial point-to-point separation, a DOF spring only looks at translation or rotation in the specified DOF. Some codes allow for preloaded springs and nonlinear springs. All the types mentioned can be valuable if machine or assembly modeling is likely to be a key part of your FEA challenge. Consult your software documentation or your support person to confirm which options are available.

Damper Elements

Damper elements provide dashpot type damping for dynamic models only. With units of force per velocity, they are rarely used in a static analysis.

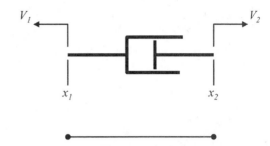

Fig. 4.50. Schematic representation of damper element.

Dampers can be axial or torsional and should be used primarily in place of an external damping device. Most codes that support dynamic analyses provide other means to represent material and structural damping.

Mass Elements

The primary use of a mass element is to idealize the mass of a component that provides a contribution to the loading of the part being studied, but is much more rigid and/or too complex to include as a mesh. Mass elements are used to represent engines in cars or motorcycles, display tubes in televisions or monitors, and pumps and motors on models of machinery. Mass elements are typically single node elements with no geometric properties. Consequently, a static analysis with no gravity or acceleration load will see no benefit from the inclusion of mass elements. Being attached to only one node, they cannot by themselves affect rigidity.

Fig. 4.51. Mass element tied to vertical plate with rigid links.

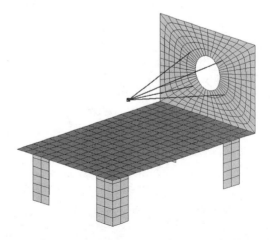

Because modal and dynamic analyses derive a significant portion of their solution from mass distribution, mass elements are commonly used in these studies. If gravitational or acceleration loads are applied to a static model, using concentrated masses to improve mass distribution may also have a major impact on results. Mass elements are either attached directly to nodes on the mesh or positioned in space at the correct center of gravity of the part and attached to the model with rigid elements.

Mass Moment of Inertia

In a dynamic or modal analysis, the mass moments of inertia should always be considered. A mass element itself is interpreted by the solver as a point mass. However, the shape of the part will affect the rotational inertia of the system.

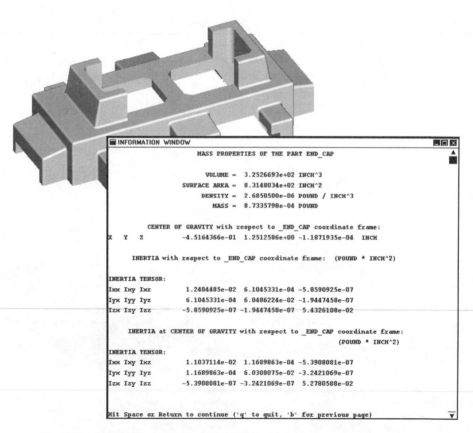

Fig. 4.52. Pro/ENGINEER mass property listing for solid part.

These inertias may be difficult to determine without testing. A CAD solid model of the part or of a simplified version could provide the required data as well.

Rigid Elements

Rigid elements may go by the terms *rigid links, links,* or *multi-point constraints (MPCs)*. In simple terms, they connect the degrees of freedom of one node or entity to the degrees of freedom of one or more other nodes or entities. The relationship between the first (independent) node and its dependent nodes is somewhat like "follow the leader." A node tied to the X translational degrees of freedom of another node is mathematically constrained to translate an equivalent X distance for any X displacement of the independent node. Another way to look at it is that their relative position in the X direction is fixed. A similar link between rotations or any combination of translations and rotations can be made. No mass is introduced to the system with rigid elements, but they can affect the overall mode stiffness. They actually reduce the overall model size somewhat as they remove unknown degrees of freedom from the system.

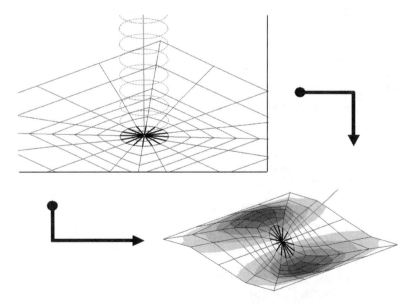

Fig. 4.53. Use rigid elements to distribute beam moments at the correct footprint on a shell model.

Uses of Rigid Elements

Rigid elements are extremely valuable in assembly modeling because you can easily tie together meshes that do not touch or lack aligned nodes. As mentioned previously, rigid elements can tie a mass element to the rest of the model in one or more degrees of freedom.

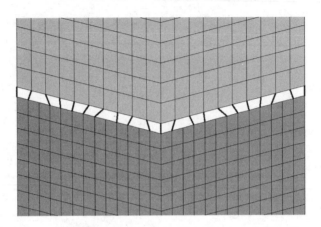

Fig. 4.54. Rigid elements enable meshes of dissimilar densities to be quickly connected.

Rigid elements are also critical in transitional meshing due to the inherent incompatibility of different element types. These elements can tie the rotational degrees of freedom of a shell element to the translational degrees of freedom of an adjacent solid.

While rigid elements are versatile and handy, care must be taken to avoid overstiffening a model due to overuse of rigid connectivity. Chapter 7 will discuss the "dos" and "don'ts" of rigid element usage in more detail.

Contact Modeling

Contact modeling is a more complex, nonlinear technique that allows for two parts or different portions of the same part to touch, slide, bounce off, and/or react against each other. Because it is a nonlinear technique, the analyst should verify with a linear, noncontact model that the parts in question actually contact. Mesh debugging and converging should also be taken as far as possible in a linear solution before introducing the contact condition as nonlinear runs can take two to 20 times longer than the initial linear run. Chapter 15 discusses this in more detail and also shares some techniques for faking contact using linear elements and methods.

There are three primary types of contact elements: *gap elements, slide line elements,* and *general contact elements* that allow for arbitrary surface-to-surface or curve-to-curve contact.

Gap Elements

Gap elements are point-to-point connections that behave as conditional spring dampers.

Fig. 4.55. Gap element orientation.

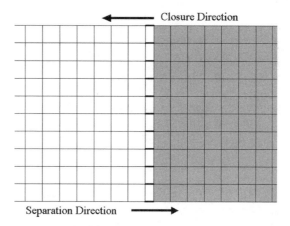

Their initial direction or assembly indicates the vector corresponding to *separation.* If the initial node-to-node separation becomes more positive, the gap has no stiffness and is, for all intents and purposes, dormant. As the separation becomes less positive and approaches zero or some predefined gap distance, the element "wakes up" and becomes a stiff spring which resists further closure. Gap element stiffness controls the numerical stability of the solution and the resultant penetration of the node pair. If the stiffness is too high, the element may "bounce" as the load is incremented near zero separation.

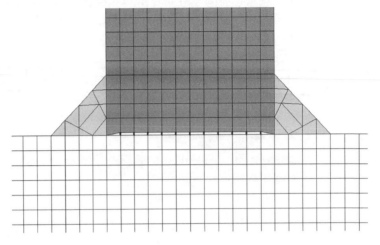

*Fig. 4.56. Gap elements
ensure that the load
transfer beyond the root
of the weld is correct.*

If the stiffness is too low, external forces may actually drive one node through the other. Many FEA codes employ an adaptive gap element that can adjust its stiffness internally to allow a predetermined penetration tolerance without wreaking havoc on solution stability.

Gap elements should be used when very little relative displacement or sliding normal to the direction of closure is expected between the nodes on one part and the corresponding nodes on the other.

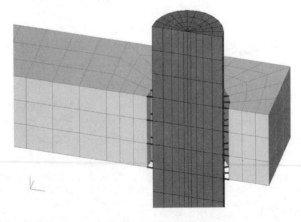

*Fig. 4.57. Gap elements
enable shaft to transmit load
to retaining bore.*

The contacting meshes should be built so that the nodes are directly aligned with their counterparts in the direction of expected contact. This can be difficult in automeshed solids and may require significant manual intervention. All codes have some tolerance in which the contacting nodes can be misaligned without noticeable degradation in element performance. However, you should try your best to align the nodes and save what tolerance there is for any displacement normal to the gap element resulting from model solution.

Gap Elements as Cables

Gap elements can also be set up to work in the opposite direction such that they behave as cables instead of compression springs. By specifying a separation distance corresponding to the "tight" condition of a stretched cable, the element can turn on or off in tension depending on relative displacement in the separation direction.

Slide Line Elements

Slide line elements are essentially contact curves that will allow significant relative sliding between the contacting parts. A slide line is created via the connectivity of two selected sets of curves or nodes that define the curves.

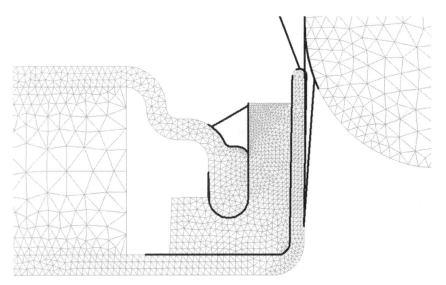

Fig. 4.58. Slide line contact regions required to model battery assembly.

A means of specifying the contact direction versus the separation direction must be provided. If this cannot be determined automatically by your code, check the documentation for the proper procedure. When set up properly, a slide line will resist penetration, regardless of the relative position of the nodes or elements on the opposite sides of the contacting regions.

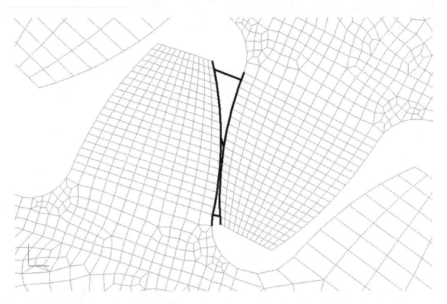

Fig. 4.59. Slide line elements allow spur gear teeth to load each other naturally.

The stiffness considerations for gap elements apply to slide lines and they only work in compression.

Slide Lines in 3D Contact

While slide lines are defined as planar contact entities, they can be used for 3D contact if two or more corresponding slide lines are positioned on the contacting parts such that they remain relatively coplanar throughout model deformation. Setting these up requires some advance planning at the geometry stage to ensure that curves or lines of nodes are properly oriented.

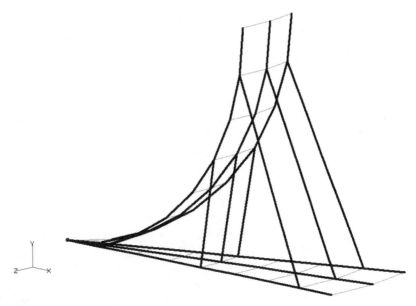

Fig. 4.60. This sliced hemisphere can contact the plate in three dimensions using radially oriented slide lines.

Fig. 4.61. This automotive bumper can contact the pole correctly using parallel slide lines.

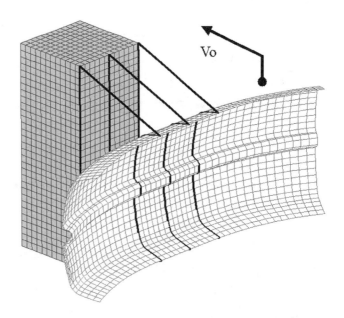

General Contact Elements

General contact elements are conceptually the simplest, but the most computationally intensive. The user will typically define a contact pair consisting of two surfaces, curves, surface meshes, or edge meshes.

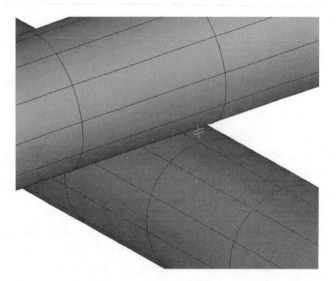

Fig. 4.62. General 3D contact regions are required to model impact between crossing cylinders.

The contact stiffness is usually derived from the material properties of the elements in the contact pair—typically Young's modulus and, in some codes, material damping. The contact and separation directions are also derived from the underlying elements, using their normals or boundaries. These contact entities provide the smoothest and most realistic contact pressure distribution but the price is paid in solve time.

Friction in Contact Elements

Static and dynamic friction can be specified for all contact elements described. However, the availability of this option is code dependent. The friction coefficients used by FEA contact elements should not be assumed to have a direct correlation to the friction coefficients pulled from data sheets or empirical testing not specifically designed to develop FEA properties. If friction is required, a simple test model should be correlated to an actual friction test designed to provide these coefficients. A handy test setup involves a mass on an inclined board supported by a spring.

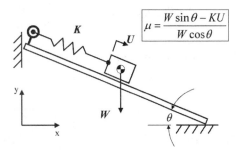

Fig. 4.63. Schematic representation of friction correlation fixture.

$$\mu = \frac{W\sin\theta - KU}{W\cos\theta}$$

This minor modification to the standard coefficient of friction test will eliminate model failure due to rigid body motion as the block begins to slide. Once a friction coefficient has been correlated between test and analysis for a specific material pair, it should be reliable for general use.

Learning to use all contact options available to you is a worthwhile exercise, even if their use is not immediately recognized. Some thought should always be given to linear approximations that provide contact-like behavior. However, in many scenarios contact is the only way to accurately model the interaction between two parts. As hardware gets faster, some situations will solve quickly enough with nonlinear contact, thereby making the effort spent on a creative workaround needless. This is where experience with the tools at hand becomes important.

Crack Tip Elements

Elements have been developed in some codes to capture the singularity at a crack tip. They go by the names of quarter point, crack tip, or singularity elements. Crack propagation is a highly specialized analysis technique and should only be undertaken after you are extremely confident in linear and nonlinear modeling and results interpretation.

Part versus Assembly Modeling

The modeling techniques described thus far contain all tools for modeling parts or assemblies. In some cases, such as some weldments, the part and the assembly do not require different modeling methods because they behave as a single continuous entity. However, most analysts, at one time or another, will have to model an assembly of rigidly or semi-rigidly attached components. The techniques for attaching parts are as varied as the techniques for modeling parts themselves. Chapter 13 provides greater detail on assembly interactions and techniques. In this introduction to assembly modeling, a few key points are mentioned that should be made clear before this type of assembly modeling is even considered.

Fig. 4.64. Pulley and mount assembly.

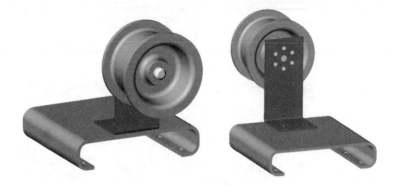

Component Contribution Analysis

Try not to start your project with an assembly, even if you are pretty sure it is going to go there eventually. There are enough areas of uncertainty in the modeling of a single part to have to worry about first without introducing uncertainty due to assembly interaction. A process called *component contribution analysis* is recommended when evaluating the structural behavior of an assembly. This method involves isolating each part or rigidly connected subassembly for a separate initial analysis. The interaction of the mating components should be accounted for using boundary conditions, or in extreme cases, simplified representations of the other parts and, possibly, contact. The goal is to develop a simple model of the individual part and to focus on its particular analysis requirements.

Fig. 4.65. Solid element analysis of shaft component.

Turning the One Disadvantage into an Advantage

There are many advantages to the component contribution technique and one glaring disadvantage. First, the bad news: it takes longer to examine each part individually before building the assembly model. It delays the excitement of showing off your results on a complicated, multi-part, full system model to your boss or peers and finally garnering the respect you know you deserve. The disadvantage is directly related to the first major advantage. While delaying the presentation, it minimizes the chance that the spectacular model you hang your hat on is wrong! As mentioned previously, FEA is not a tool for those who crave instant gratification. While the results are usually obtained more quickly than in a physical test, a disciplined, methodical approach is the only thing protecting you from a screen full of garbage.

Considering Interactions in Terms of Boundary Conditions

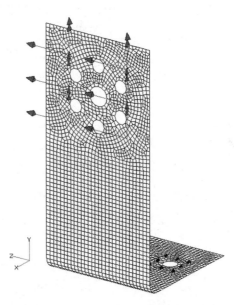

Fig. 4.66. Shaft loading will produce a 'Y' moment and a 'Z' lateral shear load on the flange.

The second major advantage to component contribution analysis is that it forces you to dedicate some serious thought to the interactions of the external loads with each of the parts in the assembly and prepares you to be able to evaluate the end results with confidence. A lateral force on an assembly will impart both a force and a moment on all parts away from the point of application that are in the load path to ground. Your component model should have a correctly calculated moment and reaction if it is expected to behave as it will in the assembly. If the results of the component models do not compare favorably with the assembly model, something is wrong with one or both of them. If they do correlate, it is likely that the interaction was modeled correctly.

Isolating the Performance of Each Part

The third major advantage is that the component contribution technique forces you to consider each component on its own merits and allows you early identification of the weak link or links in the assembly. Conversely, it may suggest that a part is significantly more rigid or less prone to failure and it can be substituted for in the final assembly model with boundary conditions or a simplified representation. Assume that all energy from the applied load goes into the part in question. If the stress and displacement are acceptable in this scenario, the part will most likely perform even better when load is shared by multiple elastic components. However, if the component study suggests that either stress or displacement or both may be a problem, you will then know to take great care in the modeling and results interpretation of this part in the assembly model. Use your understanding of the part, the system, and the physics of the problem to determine if the part can be further improved or optimized in the component model. Perhaps a small adjustment to the simplified model is all that is needed to better represent the mating bodies and allow a design decision to be made on the part without going as far as a full assembly model.

Fig. 4.67. Load path to base causes moments in both 'Y' and 'Z' with the 'X' oriented lateral shear.

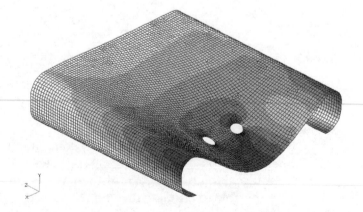

Keep It Simple...

A final point regarding assemblies has a parallel in single part modeling. Look for the simplest means of tying parts together to minimize the error introduced. Always keep the goals of the study in mind and do not get carried away with building a complex assembly just because you can. It is much easier to emotionally justify adding complexity to a joint or interface than to simplify it. Most new users will tend to make a complex interaction even more complex if the results are not quite as expected. It is likely in these cases that the modeling techniques or the physical system being represented were not well understood in the first place.

Transitional Meshing

Fig. 4.68. Special handling is required to transition solid shaft to shell flange mesh.

One of the more widely used techniques for joining parts of dissimilar shapes and sizes is to mesh each with the best element for the geometry and then attach them afterwards. If the interface between the two parts is rigid and continuous, it is common practice to merge aligned nodes of the two meshes. This creates a continuous transition between element types that are not necessarily compatible at a common set of nodes. Transitional meshing is also used when one part has the characteristics of two or more element types. Meshing a part with multiple element types should not be entered into casually. At and near the transition between types, the results are always suspect. Consequently, an estimate

must be made up front as to the proper location for the transition so
that the behavior of interest is not affected by the local error.

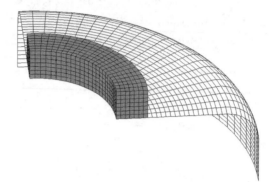

*Fig. 4.69. Depending on its
geometry, a single part may need
to transition from one element type
to another.*

In addition, elements of different types have different degrees of freedom (load carrying/transferring capabilities) at respective nodes. A
solid node by itself cannot transfer rotational forces and requires the
couple created by the other nodes on the same face to create the
moment. A shell element connected to only two of a solid's nodes will
hang as if hinged on the solid unless special techniques are employed
to correct the load transfer at this transition. The techniques used to
make this correction are simple in theory but can be complex in practice. These techniques are discussed in Chapter 7. Always debug the
interaction between dissimilar elements with test models so that you
can be confident about their contribution to the overall structure.

*Fig. 4.70. Pressure loading
on the gusset will cause
significantly different
displacements if the shell
solid transition is ignored (a)
and correctly adjusted for
load transfer (b).*

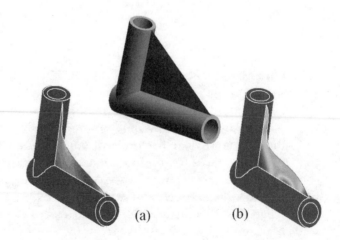

(a)   (b)

Use Test Models to Debug Idealizations

Assess the quality or validity of a transitional mesh by modeling a test problem with the transition desired and then an equivalent model of a single element type. If a beam/shell interface is considered, try modeling the whole joint area as a shell model. Comparing this to the more idealized version will provide valuable data on the extent of the localized transition error and allow you to mentally adjust the results of the final model as necessary.

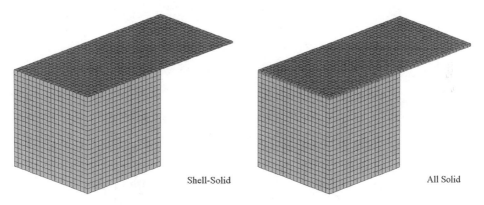

Shell-Solid   All Solid

Fig. 4.71. Test model of transitioning mesh as a continuous solid can aid in results interpretation.

Utilizing multiple element types is an extremely valuable capability found in full-feature analysis tools. As with any new techniques, learn them with simple test models, rather than a complex problem under a tight schedule.

Summary

Choosing the appropriate modeling type and using it effectively is one of the skills that will come with experience in the technology. Few analysts are fortunate (or unfortunate) enough to truly require only one modeling method. Consequently, it is worth your time to learn each technique and element with simple test models early in your education so that you will be prepared when the inevitable need hits. This chapter has provided some guidelines for identifying key features of the various planar or 3D models. "If it looks like a duck..." is one way to put it. However, as soon as a guideline is adopted, an exception will be found. Seasoned analysts have forced solids into extremely thin-walled parts and have used beams for short, stumpy members. Let the goals of the project, your experience, and the experience of your peers and coaches guide you in the selection of modeling types. Part of the fun of FEA modeling is the problem solving process and using all the options at your disposal. Consequently, do not try to avoid these more idealized simulations. Your creativity and willingness to learn should be your only hindrances to solving even the most complex problem.

5

CAD Modeling for FEA

In an ideal scenario, CAD and FEA activities are coordinated to minimize the duplication of effort as analysis is made an integral part of the design process. The geometry built by the design team will ideally be usable for FEA and all downstream applications. However, in the real world, some CAD models cannot be used for analysis and some models should not be used for analysis. This chapter will sort through criteria that will help you identify occasions when CAD and FEA should work together and when they should not.

It is the responsibility of both the analyst and the designer, or *geometry provider*, to plan projects such that the optimal level of coordination between CAD and FEA is achieved. The recommended vehicle for this coordination is the initial design review. Designs and geometry initiated by the design group and passed to the analysis group should be differentiated from geometry developed specifically for analysis and returned to the designers for documentation. There are distinct differences in the ways geometry is handled for the following four typical scenarios seen in product design.

- CAD models prepared by the design group for eventual FEA.

- CAD models prepared without consideration of the analyst's needs.

- CAD models unsuitable for use in analysis due to the amount of rework required.

- Analytical geometry developed by or for the analyst for the sole purpose of FEA.

One important distinction made in this chapter is the difference between CAD (design) geometry and analytical geometry. CAD geometry will refer, hereafter, to the models or files intended for the final documentation of the design. These are models which will be utilized by all downstream applications including detailing, manufacturing, prototyping, and tolerance analysis, as well as FEA. Analytical geometry, on the other hand, refers to geometry developed for use with FEA in final analyses and test models or initial part development.

This chapter will outline techniques required for each of the above scenarios to make the most out of existing data and minimize model building efforts. Because there will be some overlap in the options discussed, reference might be made to an earlier description. Due to this overlap, it is important to understand and be comfortable with all techniques.

Design versus Analytical Model

From the introduction of CAD in the 1970s to the present, analysts have relied on existing CAD geometry to reduce meshing time. Because of the investment in CAD geometry and CAD software, as well as the relative quality of these data, the ability to employ existing CAD data is typically a prerequisite for selecting analysis systems.

As design analysts seek ways to complete all assigned tasks including simulation, direct meshing and analysis of existing CAD data have become a customer-driven, FEA industry priority. In fact, numerous products from major players in the analysis software market have been developed solely to address this need. The marketing programs of many such products have even suggested that by bridging the gap between CAD and FEA, and providing highly reliable meshing of geometry within the CAD system, the difficult part of FEA is over. They position their products as complete solutions providing designers and engineers an effortless entry into the world of FEA. While this image is often believed, the importance of mixed element models, thoughtful boundary conditions, and informed results interpretation are just as frequently ignored or downplayed. Ignoring the latter tends to obscure the fact that the mesh is only one part, and often a small one, of overall solution quality, and that in many cases, the design model and the analysis model may be, should be, or must be quite different.

Before attempting to consider the merits of using the design model as the analysis model, the conditions listed below must be met.

1. Design models are built in 3D solids or surfaces that fully enclose volumes.

2. The part can and should be meshed with tetrahedrons, or is simple enough to provide the foundation for solid mapped brick meshing or mid-plane surface extraction for building shell models.

3. The CAD model exists at the time the analysis is to be performed.

The third point deserves further discussion. When the design has progressed to the point that a detailed CAD model exists, the analyst must evaluate the possibility of saving time by using it. If the first two conditions are met and the geometry was constructed with the proper practices to be discussed later in this chapter, the analyst should take advantage of it. However, if the need for analysis precedes the geometry, the analyst should direct the development of the model for analysis requirements alone. There are a couple of reasons for taking this approach.

First, the request for analysis indicates that the integrity of the proposed part has been questioned and the analysis will most likely suggest modifications and alternatives, which may not be easily incorporated into an existing CAD model.

Second, at this stage of the project, there is little time invested or emotional commitment to a specific design. Consequently, the cost factors and inertia against modification are small. Fig. 1.5, which represents the traditional product development process, illustrates the cost of change at various stages of the design cycle. The benefits of FEA and "predictive engineering" will be reduced if an investment is made in a detailed CAD model at this stage. Let the analysis results drive the design.

The concept of "emotional commitment" may be one of the major, intangible contributors limiting the value analysis can bring to the table. Prior to high end solid modeling, the drafting or CAD portion of the design process was more of a documentation task. An E-size layout with five sheets of Mylar overlays did not evoke the same level of excitement that a photorealistic solid model does. Today, solid modeling has caused the CAD task to feel more like creation than documentation. Consequently, after seeing their ideas transform into a part over the course of a few days or a week, designers are much more likely to embrace evolutionary improvements versus revolutionary or radical changes as may be suggested by the analysis. Beginning the simulation process with simplified, conceptual geometry reduces the risk of becoming emotionally tied to a concept.

In reality, only a small portion of parts or problems can be addressed directly from CAD models in a general sense. Some industries or part types, such as low aspect ratio or "potato-shaped" castings or machined parts, are natural fits for the CAD to FEA link. Sheet metal parts or assemblies, and thin-walled castings and plastic parts or large assemblies are usually best addressed with an analysis-specific geometry approach.

Fig. 5.1. Potato-shaped parts usually mesh cleanly and solve well with automeshed tetrahedrons.

How should you treat existing CAD geometry that meets the three criteria established above but was not built to the practices to be defined in the next section? The subsequent section presents ways to work with this data, as well as indicators which will suggest if proceeding with these models is worthwhile. The bottom line is that if working with existing

CAD data will take longer than building an analysis model from scratch, or if it introduces error due to inconsistencies in the resulting mesh, it is best to start from a clean sheet of paper, figuratively speaking. Considerations for developing geometry specifically for the simulation appear later in this chapter. Many efficiencies can be gained in the geometry creation when it is not expected to go any further than performance studies. Finally, techniques for using automated optimization routines based on CAD model dimensions are presented. As these tools become more popular, proper geometry preparation will maximize their effectiveness in developing optimal designs.

Building CAD Models for Eventual FEA Use

It should be clear by now that the analyst's job will be made much simpler if the CAD geometry exists in a usable state. Many design analysts have the luxury of being responsible for the entire design. Consequently, they have control over the development and quality of the geometry to be used for FEA. While design analysts may feel overburdened by all the responsibility, they have the power to minimize surprises and make their lives easier. However, most analysts are dependent, in part or in whole, on geometry developed by others. In these circumstances, it becomes important to educate the geometry providers, in addition to the analysts, in the proper practices of preparing geometry for FEA or any other downstream application. True integration of simulation in the design process will require a team approach. The use of design reviews to plan out parts requiring analysis is discussed in detail in Chapter 20. Suffice it to say at this point that whenever possible, parts expected to pass through the analysis process should be built in CAD using the practices described here. This means that geometry providers as well as downstream users should be aware of the inconsistencies and issues which cause problems later. They should be shown the results of bad geometry and made familiar with the consequences of poor planning.

Solid parts to be automeshed with solid tetrahedrons are the best candidates for time savings with a clean transfer between CAD and FEA. As mentioned previously, these are parts that satisfy criteria number one for using the design model in the analysis process. Consequently, they will serve as the basis for much of the discussion. In addition, this discussion assumes that the analysis under consideration is intended for design verification or qualification only. When the design is being developed using simulation driven data, the geometry should be built specifically for that purpose.

Written guidelines based on the practices described herein should be distributed and "policed." Tools are being developed to run within CAD programs which will flag problem areas such as sliver faces and fragile dependencies. These tools can serve as a teaching method, as well as a reminder to the person making the model to observe proper practices.

Solid Chunky Parts

Why is it that most FEA software demonstrations tend to utilize thick-walled, low aspect ratio parts such as castings or machined blocks? It is primarily because these parts mesh cleanly directly off the CAD models. Automeshed parts containing a single tetrahedron through a wall thickness will most likely result in spotty stress contours, require extremely large meshes to avoid high aspect ratio elements, and raise issues of nonlinearities. However, solid, potato-shaped parts typically undergo small displacements and easily mesh with good-looking elements relatively quickly. Parts meeting this description, like those shown in Figs. 5.1 and 5.2, provide the best opportunity to save time and cost by using the design model as the analysis model. Observe proper modeling techniques and the designer and the analyst will coexist in harmony. There are several areas of modeling which can improve the transfer of data between applications that must be known by the geometry providers. These modeling issues fall primarily into two categories: (1) clean geometry and (2) fragile "parent-child" or dependency relations. Each will be equally frustrating to the analyst in its own way and is defined in the following pages.

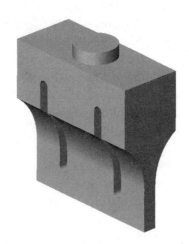

Fig. 5.2. Parts similar to this one must clearly be modeled with solid elements. Using a reliable automesher is the most efficient approach for such projects.

Building Clean Geometry

In short, clean geometry can be defined as a solid CAD model that maximizes the possibility for a mesh which in turn captures the features required for correct results. Two key points are made in this statement. First, the geometric features must not prevent the mesh from being created and must also contain surfaces of consistent size and shape ratios to prevent forcing high aspect ratio elements and/or transitions between element edges that may compromise accuracy. Second, simplification or manipulation of features in an attempt to clean up the geometry should not reduce the structural integrity of the part. The best mesh is the smallest model that yields correct data. Consequently, manipulation of the model, either by adjusting dimensions or suppressing features far from any area of interest, is acceptable as effects local to the simplification will most likely not affect the global behavior of the system. However, care should be taken when adjusting a model near an area of concern.

The best safeguard against the need to clean up geometry near an area of interest is to not create a problem in the first place. Many designers make dimensions and feature size choices based on convenience. Consequently, it is not surprising that short edges or sliver surfaces appear randomly in a model. If the feature choices are made with the knowledge of downstream needs, many of the modeling issues that plague the automeshers and analysts can be avoided. Fig. 5.3 shows an excellent example. The rounded rib on the inside of the piston has a thickness of

0.30 and a radius of 0.145 in. This radius has produced a flat surface of the dimensions 0.010 by 2.563. A mesh size no greater than 0.050 will be required to avoid highly distorted elements. This results in a model size of 284,353 nodes with a second order tetrahedral mesh. Adjusting the fillet radius by 0.005 to 0.150 eliminates the sliver surface and allows for a nominal mesh in this area of 0.120. The resulting mesh of 33,476 nodes is 88.2% smaller and the resulting structural behavior of one model versus the other is indistinguishable. Local mesh refinement near the sliver surface would have reduced the model size somewhat but that still requires extra time which would not have been required with some forethought in geometry creation.

Fig. 5.3. Designer's choice of feature size controls the creation of dirty geometry.

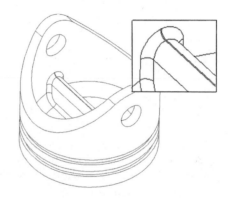

Short Edges

Fig. 5.4. Short edges on large faces can cause highly distorted elements or a failed mesh.

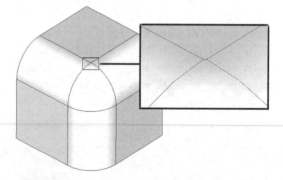

Edges that are short compared to either model size or, more importantly, nominal mesh size, will introduce error into the model by forcing one edge of a triangle or quadrilateral to be the length of that edge.

While this will be covered in more detail in Chapter 7, some insight into the way most automeshers work may be helpful here.

Fig. 5.5. This short edge was created by the proximity of two nearly aligned surfaces.

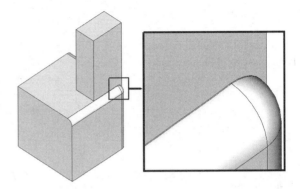

Essentially, automeshers pave or seed the outer surfaces of a part with triangles, and then fill towards the center of the volume. Unless local mesh refinement is employed, the automesher will try to space nodes on edges first and then within surfaces at the defined nominal element size. However, most meshers are constrained to use every point or edge on the part. Consequently, when a short edge is encountered, it will space the two legs of the triangle that are not on the offending edge to full element size and limit the edge length of the third to the physical geometric edge. This short edge will affect at least two elements in a tet model and may affect more depending on your tool's algorithms for transitioning.

Limiting the size of small edges to no less than one-third of the expected nominal element size is good practice. This is great on paper, but difficult in practice. Educating geometry providers on the needs of FEA will help. Planning and evaluating features as they are created will also help in this case.

The primary causes of short edges are the misalignment of features and the proximity of fillet edges to other edges. Some commonly created short edges that could have been eliminated with proper planning of geometry are shown in the three illustrations appearing in this section.

*Fig. 5.6. Sloppy
geometry creation often
causes small cracks or
gaps in faces that are
difficult to visually detect.*

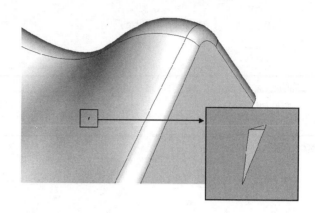

Sliver Surfaces

Sliver surfaces are faces on a part with a high aspect ratio. Short edges
and sliver surfaces often accompany one another. Figs. 5.7 and 5.8 show
a short edge with no associated sliver surface, and a sliver surface with
no short edges.

*Fig. 5.7. These two
cylindrical features are
offset only slightly. While
this may not create detailing
problems, an automesher
will struggle with the short
edge caused by the*

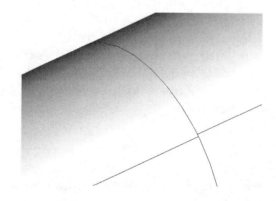

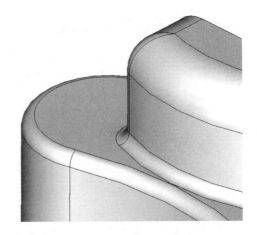

Fig. 5.8. As the two curved edges approach each other, a surface dimension much smaller than any part feature is created. Resolving the results will be difficult for the mesher.

Checking for both problems is required. The same modeling practices that create short edges create sliver surfaces. The end result on the mesh is the same. The mesher may create a short edge between the smallest distance between two surface edges or it may create an extremely flat element. Each of these conditions can lead to inaccuracies in the model. Figs. 5.9 to 5.11 show common sliver surfaces.

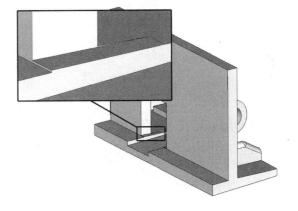

Fig. 5.9. More thought on aligning these two features would have eliminated the sliver surface.

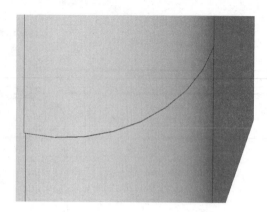

Fig. 5.10. Sliver surface caused by a slightly undersized fillet.

Fig. 5.11. A fillet across this shallow angle has created a difficult meshing situation.

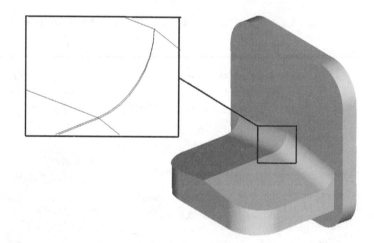

Voids in Solids

These geometry problems are difficult to spot and it is impossible to predict their effects on the resulting model. Most voids are the result of misalignments and tend to contain short edges or sliver surfaces. However, many voids are larger with low aspect ratio faces. If the voids are small and the local behavior is far from an area of interest, results may not be affected.

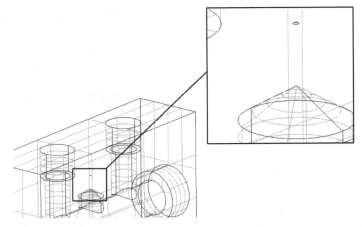

Fig. 5.12. It is extremely easy to accidentally create voids in a solid model while designing a new part.

However, voids will force incorrect mass properties which could affect dynamic analyses and, if large enough, may seriously impact the stiffness of the part. Voids are often the result of poor modeling practices where holes or cuts are filled in with "plugs" instead of corrected with feature redefinition. Once again, poor planning is ultimately responsible for these problems. Fig. 5.12 shows a typical void in a solid part that may affect results.

High Order Surfaces and Edges

Styling is becoming easier to incorporate into CAD models with the power of surfacing tools and the ability to stitch surfaces into solid bodies. The industrial designer is primarily concerned with outward appearance and may not think twice about complex lofts or blends. The more complex the surface or edge definition, the greater the chance it will contain inconsistencies in the form of gaps and waves. Most h-element automeshers will ignore small geometric disturbances and simply pave over them. However, p-element meshers that map elements directly to geometry definition will get hung up on these surfaces. Fig. 5.13 shows the response by Parametric Technology Corporation's Pro/MECHANICA to a high order surface abnormality.

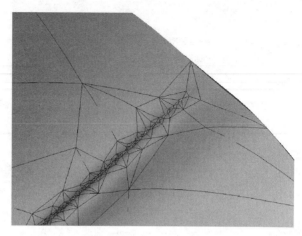

Fig. 5.13. A p-element mesher which maps the element definition to the geometry definition is highly sensitive to any type of inconsistency.

Identifying these problems prior to meshing can be difficult. A visual examination of a shaded image may assist by pinpointing inconsistencies in the shading. A wireframe view of surfaces with grid or surface definition lines visible could also highlight problems. Creation of these problems can be minimized by using the simplest geometric definition possible to create a feature. Whenever possible, use extrusions and revolutions instead of sweeps or blends. Avoid building a volume out of surface patches when standard solid modeling operations would suffice. Use lines and arcs in profiles when possible. Many splines can be approximated with lines and arcs to a high degree of accuracy. Explore your CAD tool capabilities for simplifying feature definition. While geometry creation may take more thought and time, the savings in downstream applications will be worth it.

Corrupt Geometric Definition

Geometry tools and interfaces are becoming more robust every year. Standards such as ACIS, Parasolids, VDA, and STEP have greatly reduced the problems typically associated with model conversion. IGES and DXF will continue to cause headaches, primarily because of the disparity between standard interpretations and level implementation. While the latest IGES standard provides for advanced geometry definition, there is no guarantee that both software packages involved in the transfer use the same revision or entity designation. Even though standards such as ACIS and Parasolids minimize incompatibility, multiple revisions of these standards exist. One of the surest ways to corrupt geometry is through multiple transfers and translations. Every translation introduces the likelihood that a face or edge will be redefined incorrectly. Any error introduced will propagate through future translations. Many companies rely on outsourcing of design work, which often entails conversion of data between CAD systems. The impact on downstream geometry quality can be minimized by following the steps listed below.

1. Try to use the same CAD system for all components in a design.

2. When the above is not possible, translate geometry through kernel based tools such as ACIS or Parasolids. Using standards based (i.e., IGES, DXF, or VDA) translations may lead to problems.

3. Visually and systematically inspect the quality of imported geometry before it is incorporated into the product database. Do not assume it is clean.

4. When possible, avoid modifications of imported geometry in a second CAD system. Recreating the part in a native system may be preferable if modifications cannot be made in the original system.

5. Use the original geometry for analysis when available. If the native CAD geometry cannot be used directly, use a translation directly from the original model. Minimize the iterations of translation to reduce the error which can be introduced by these manipulations.

Parent-child Relationships

Many companies have switched to parametric or history based CAD systems which define features off other CAD features. An example is the wall dimensioned off a fillet feature shown in Fig. 5.14. This dependency is often called a "parent-child" relationship. In this case, the wall is a child of the parent fillet.

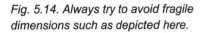

Fig. 5.14. Always try to avoid fragile dimensions such as depicted here.

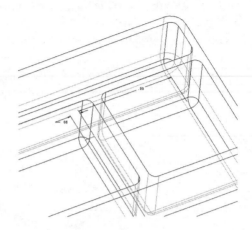

Parent-child relationships are of interest to the analyst because many features such as outside corner rounds and small holes are insignificant to the structural behavior of the part and only complicate the model. One of the most useful features of parametric and/or history based CAD systems is the user's ability to turn off or suppress features to simplify the geometry for downstream applications. Unfortunately, suppressing features which have children can range from annoying to impossible. As the degree of dependency increases, the user's choices for redefining the parent feature diminish.

It may be difficult to vow never to make a fillet or draft feature a parent of downstream features. At times, it is clearly unavoidable. However, good modeling practices can minimize the impact of parent-child relations on the analyst's job.

Ideally, the geometry is being driven by analysis results. If this is the case, insignificant features would not be created prior to the simulation phase. However, most companies still work in the mode where the geometry is developed in the product design process. It is then provided to analysts for optimization or verification.

As discussed previously, one of the best tools to minimize the impact of parent-child relationships is the initial design review and subsequent interaction between the analyst and geometry provider. If parts to be analyzed are identified in the design review, the important features of these parts can be noted and the geometry can be created to ensure that they are not children of insignificant features.

Education of the geometry providers is the most important consideration when attempting to minimize poorly structured geometry. The needs of downstream applications should always be communicated to the geometry providers. In addition, the impact of improperly constructed relationships should be illustrated. One means of illustrating this problem is to make the geometry providers execute the simplifications themselves. Under the direction of the analyst, they will learn the difficulties of preparing a part for simulation.

Guidelines for Part Simplification

It is important to note before proceeding that the following guidelines are not intended as ironclad rules. Every problem is different and various loading scenarios on the same part may require different levels of simplification. It is not safe to assume that all small fillets or even outside corner rounds should be suppressed by default before analysis. While this is often the case, they must be evaluated on a problem-by-problem basis. When deciding to suppress a feature, consciously acknowledge the impact of the feature and its absence on the end result. If there is hesitation due to potential effects, be conservative and retain the feature. In general, the features listed below should be considered for suppression.

- Outside corner breaks or rounds.

- Small inside fillets far from areas of interest.

- Screw threads or spline features unless they are specifically being studied.

- Small holes outside the load path.

- Small protrusions outside the load path.

- Decorative or identification features such as company logos or part numbers. A caveat to this occurs when verification analysis is required to examine a failure near a decorative feature suspected of causing a stress riser.

- Large sections of geometry that are essentially decoupled from the behavior of interest and not required for their mass contribution. A couple of simulation iterations may be required to determine which areas are actually safe to suppress or delete.

Guidelines for Geometry Planning

When constructing geometry, the needs of the analyst or any other downstream user of the geometry should be considered first and foremost. In addition, good modeling practice would suggest the geometry tips listed below.

1. Delay inclusion of fillets and chamfers as long as possible.

2. Delay addition of draft as long as possible.

3. Try to use permanent datums as references where possible to minimize dependencies altogether.

4. Avoid using fillet or draft developed edges as references for other features.

5. Never bury a feature in your model. Delete or redefine unwanted or incorrect features. An example of a buried feature is a hole which has been plugged with a cylinder rather than deleted. Buried features can accidentally become parent features.

Additional Considerations

While the previous section focused primarily on solid parts expected to be automeshed, many thin-walled parts can be constructed to facilitate compression to mid-plane surfaces for shell element meshing. Because few CAD tools provide this functionality, it is important that you understand the capabilities and limitations of your system. If the capability exists, both analyst and geometry provider should learn how best to create parts for mid-plane extraction. Remember that creating such parts usually requires a good deal of planning, because geometry creation techniques may be vastly different than those for faster, more efficient model building. Next, remember that most thin-walled parts or assemblies such as plastic parts may be difficult to plan for automatic extraction and might best be constructed independently only for the analysis.

Planning the geometry for analysis requires a little extra time but can save a tremendous amount of time and effort down the road. As more companies strive to develop a single geometry database, proper construction of features in the geometry will become essential. Many CAD packages have other features which facilitate use of downstream applications. Family tables or table driven configurations can allow for the creation of a design part, analysis part, assembly or visualization part, and so on. The ability to define a simplified representation or named, predefined configurations will allow the various users of the part to quickly access the appropriate version. Explore the capabilities of your particular CAD tool to maximize efficiency when building parts for eventual use in analysis.

Working with Existing Geometry

Even the most careful planning will not eliminate the possibility that you will need to analyze geometry that was not developed with FEA requirements in mind. Attempting to work with this geometry can be as time consuming (or even more so) as the analysis setup itself. This section reviews techniques for identifying and correcting problems and provides tips for preparing geometry, planned for FEA or not, to create a mesh. It is important that the analyst be familiar with the guidelines for geometry creation discussed previously before proceeding with this section. This section covers techniques for working with geometry in the native CAD environment as well as the analysis preprocessor.

Geometry Guidelines for Working with Existing Geometry

These guidelines apply to geometry setup in general. They should be considered regardless of whether or not the part is constructed for eventual analysis.

Symmetry

When symmetry is allowed, it should be used. The rules for symmetry usage will be discussed in greater detail in Chapter 8. Symmetric FEA modeling will always result in faster runs and may result in more accurate or accurate looking results. Within the native CAD environment, a symmetry cut should be the last feature on the model. The cut should utilize independent datums or global references. The cut should be deletable and/or suppressible and should never have dependent features. Some preprocessors provide geometry tools as well. By the time geometry is brought into a preprocessor not integrated within the CAD system, links to parametric features and the history tree are broken. Consequently, the part should be split with the most convenient means. If solid operations are permitted on a solid database, simply cut away half the part. If splitting the part in two with a plane is the fastest method of achieving a symmetry model, it is best to discard the unwanted portion so that it will not accidentally be used.

When solid operations are not available or the part is not a solid, wireframe techniques must be employed. Two quick ways to develop geometry at the cut plane are to use surface-surface intersections or midpoints. While all preprocessors are different, all addressed in this text provide some combination of these two methods. If curves can be created from surface-surface intersections, create a rectangular surface that is larger than the projection of the part onto the surface. Use this surface to create the curves that outline the desired cross section. Then, depending on the surfacing capabilities of your tools, either trim the affected surfaces or delete them and rebuild to the new cross section. Another method for developing this cross section is to create points at the mid-point of all curves that cross the symmetry boundary. Note that if a curve that crosses the cut plane does not have a mid-point on the cut plane, the part may not be symmetrical. How these points are used is again dependent upon the surfacing capabilities of your system. Some preprocessors allow for the creation of a curve projected onto the surface between two points. If this technique is not available, you may be forced to delete the crossing surfaces and recreate the missing geometry by hand. It should be clear from this brief discussion that using the native CAD environment to split the part before transferring to FEA is desirable.

Boundary Condition Adjustments

A *patch* is the term used to describe a logical break in a surface or curve for more precise placement of boundary conditions. Patches are also used to provide more control over the mesh, possibly near a high stress gradient where transition from very small elements is required. Figs. 5.15 and 5.16 present selected common uses of boundary condition patches.

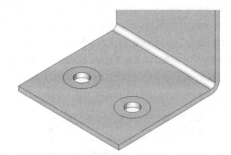

Fig. 5.15. Distributing load across bolt head or washer diameter can be achieved with patches.

Means for creation of patches are dependent on available geometry tools. In general, however, applying patches in the native CAD system is usually the easiest. When possible, physically break the geometry at a patch instead of simply associating the patch to the surface or curve. This patch will be less likely to be deleted accidentally and will allow boundary conditions to be placed on that surface instead of trying to use the nodes or elements contained within the patch. When possible, consider the need for patches in the creation of geometry. (See "High Order Surfaces and Edges" above.) You should verify that the application of a patch does not create any of the "dirty" geometry conditions defined in the "Building Clean Geometry" section.

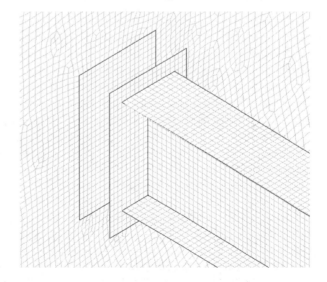

Fig. 5.16. While the actual location of the supporting beam represented by the rectangular patch might be off the part edge by a wall thickness or two, this small adjustment should not affect the reliability of the shell model.

Being comfortable with the level of assumptions in your model will greatly facilitate the creation of patches. If there is great uncertainty or understood inaccuracy in a model, adjusting a patch size or location slightly to ease the meshing process may be warranted. If, on the other hand, pains have been taken to ensure that the properties, loads, and geometry are correct for an analysis, the patch should be created in a similar manner.

Another means to locate boundary conditions is to utilize the orientation of *face cracking* or splitting on a cylindrical surface. Fig. 5.17 shows an example of a cylindrical hole with a split surface on the piston discussed earlier. Most CAD systems define a cylinder as two 180-degree, four-sided cylindrical surface halves. If a load is to be applied to this piston to simulate a pin-to-hole or bearing load, proper orientation of the two halves is required.

Fig. 5.17. Proper orientation of split surfaces facilitates the application of loads.

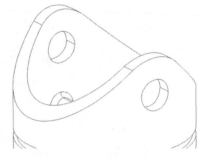

You may have control of this orientation in your CAD system or you may have to adjust it in your preprocessor. If orientation of a cylinder is important, it is best to take care of it during the geometry creation process because adjusting orientation may entail recreating geometry. If the surface is not split, it may be split using the techniques described above for modeling symmetry in a surface environment. Most meshers will handle a split cylindrical surface better than a continuous one and some older meshers may choke on continuous cylindrical surfaces. This is equally true for spherical and conical surfaces.

Feature Suppression versus Submodeling

Feature suppression was discussed briefly in the "Building Clean Geometry" section from the standpoint of geometry planning. Choosing the features to remove or suppress requires engineering judgment. It is not good practice to assume that all small features or fillets may be suppressed. Feature suppression will be differentiated from submodeling in that submodeling involves the isolation of certain portions of the geometry which can be assumed decoupled from the rest of the part. For purposes of this discussion, feature suppression is defined as removal of detail in or near the area of interest.

The assumption of decoupling warrants further discussion. Depending on the model, it may not be easy to determine if decoupling is possible, and if so, where to break the model.

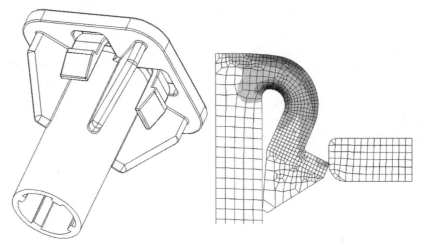

Fig. 5.18. The mounting tab on this connector can easily be decoupled for submodeling.

The connector shown in Fig. 5.18 has a relatively rigid body and a relatively flexible mounting tab. It is fairly safe to assume that geometry beyond the local mounting area will have no effect on the insertion behavior of the tab. However, it may not be so readily apparent that the connecting rod shown in Fig. 5.19 can be decoupled or split as shown. Because connecting rods act as two-force members, each can be modeled separately in a static analysis. Symmetry conditions would be used at the cut plane. This part also has a true symmetry plane parallel to axis plane X-Y as shown in Fig. 5.19.

Fig. 5.19. Connecting rod can be divided into two separate studies due to uniaxial nature of loading. (Image courtesy of Briggs & Stratton Corporation.)

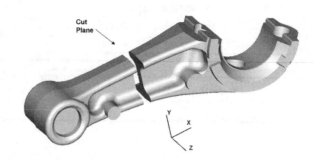

Feature Suppression

The ability to suppress, delete, or neglect to add certain details is a key enabling assumption to efficient finite element modeling. Depending on the goal of the analysis, great liberties can be taken with details in a part. Typically, displacement, modal, and trend analyses require less detail than a stress analysis. This is because local phenomena will have little effect on global behavior. Even detailed stress analyses may allow defeaturing away from the area of interest. Hopefully, the value of planning described in Chapter 1 has begun to sink in. Essentially, the goals of the project will drive the level of detail required in the model. Initial behavioral studies might allow prismatic, coarse geometry, whereas focused stress or fatigue studies will require more detail. It is not uncommon to include more detail in the first few iterations of a stress analysis. As you become more familiar with the stress distribution in the part, convergence runs on minor modifications may see an equal amount of mesh coarsening as mesh refinement. This is equally true for feature details. When results suggest that an area may be of concern, more detail and precision might be added to capture the behavior in the area. As other areas prove to be of little interest, the detail may be further reduced. The series of images in Figs. 5.20 to 5.22 illustrate this process.

In Fig. 5.20, the bracket has been modeled with moderate detail, because the behavior of the part is unknown.

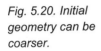

Fig. 5.20. Initial geometry can be coarser.

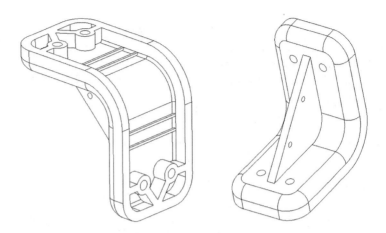

Fig. 5.21 reflects the addition of fillets where the high stress was found, and the removal of fillets and holes where no activity was noted. A corresponding adjustment in mesh refinement would accompany this geometry modification.

Fig. 5.21. As behavior of part is clearer, fillets are added to ensure that the stress distribution captured is correct.

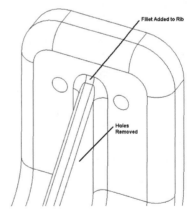

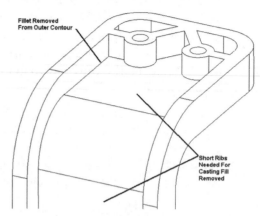

Fig. 5.22. The final analysis can have a much tighter mesh because a considerable amount of detail was removed from areas with no structural concerns.

Listed below are three primary rules guiding feature suppression and removal.

1. Consciously acknowledge the effect that the removal of the feature will have on the quantity of interest or the goal of the study. While most clearly removable features will have a negligible effect on the results, removing a feature that may have a noticeable effect is still valid. Your understanding of this effect determines whether the suppression can be made.

2. If you are unsure of the effect a feature removal may have, leave it in. More detail will only cost you run time. If the initial runs indicate that there is little behavior of interest near the feature in question, the feature may be removed later.

3. Balance the cost (i.e., time) of feature removal with the cost of retaining it. It is very common to get caught up in cleaning up a model and spend more of your time on feature removal than you would actually save on run time. While it will be difficult to estimate the tradeoffs when you are learning FEA, a useful guideline is that the run time difference over the life of the model should be four or more times greater than the time required to implement the simplification. Because many models will be run two, four, and even a dozen times, the sum of the time savings should be considered. An expected one-time verification run is not typically worth much simplification time.

A caveat to rule three involves balancing model detail against software resource limitations. Many codes are purchased with a fixed node limit. Similarly, some models will push the capacity of your system's RAM or hard disk. In these cases, feature suppression away from an area of interest may be the only way to get the model to run.

Adherence to rule number 3 must also take into account the modifiability of the supplied geometry. If the part is available as a parametric, history based solid built to the guidelines defined in the "Building Clean Geometry" section, features can be evaluated mostly on their merit alone. However, if dependencies are confused and complex, or the geometry is available as a featureless "lump" due to cross-platform transfer, it may be very costly to make any change to the model. In a nonparametric model, fillets and rounds are easier to add than to remove. Holes and slots are usually added and removed with equal cost. Ribs may be difficult to add or remove, depending on end conditions. As stated previously, geometry simplifications are nearly always more efficient if (1) the model was built to the guidelines for downstream use, and (2) simplifications can be made in the native CAD system.

Submodeling

This section will refer to submodeling as defined previously. You may find the term submodeling used to describe a technique where large models can be divided into logical substructures for simultaneous modeling. In that case the boundaries between the substructures are defined early in the design process so that the behavior at one substructure's boundary can be used as input to an adjacent structure. This technique is heavily used in the automotive and aerospace industry. It will not be addressed any further in this text. Contact your software technical support organization for more details if you believe that such technique may be helpful in your simulation efforts.

Submodeling here simply refers to performing a detailed analysis on only a portion of the overall geometry. Typically, submodeling refers to breaking up a single part. Yet some assemblies, specifically weldments, behave as a single continuous structure and might be candidates for submodeling. Pulling a part out of a nonrigid assembly for detailed study, however, falls into the category of creative boundary condition application.

The validity of submodeling is entirely dependent on the analyst's ability to determine where behavior local to the area of interest ceases to be affected by remote features or results. In many cases, a full model run may be required to determine the load path before submodeling at an abrupt change in the rigidity of the part (as in the connector from the previous illustration) or large distances between independent loads. When evaluating the results from the full model run, look for stress levels to approach zero between areas of loading or constraints indicating decoupling, or a constant stress distribution which can be attributed to predictable conditions, such as the force over area (uniaxial stress) behavior of the connecting rod from Fig. 5.19.

Of equal importance to the existence and location of a submodeling cut are the boundary conditions at the cut plane. The following four choices for these boundary conditions appear below.

- *Fixed constraints.* Used when the portion of the model being studied is attached to a much more rigid structure.

- *Symmetry constraints.* Used when the cut plane experiences reaction forces similar to a uniaxial condition.

- *Enforced displacements.* Used when the full model suggests a known displacement, translation, and/or rotation at the cut plane. This condition assumes that changes to the geometry in the submodel will not significantly affect the displacement input.

- *Reaction loads.* Used when the reactions at the cut plane can be calculated or extracted from the full model study. This condition also assumes that changes to the submodel will not affect input loads.

Because most submodeling operations require cutting the original geometry at a plane or a simple profile, a postprocessor with good solid or surface modeling tools will be as efficient as the native CAD environment. If your preprocessor lacks these capabilities, it is best to modify the CAD model before submitting it to the analysis software. If this is not possible, the techniques described previously for symmetry modification should be applied.

Mid-plane Extraction

As mentioned in the "Building Clean Geometry" section, some CAD integrated preprocessors allow for mid-plane extraction to facilitate shell model construction. More often than not, this will not be automatic. Some manual adjustment of the geometry or resultant mid-plane surface will be required. Techniques will vary according to the software. Another commonly used technique is to simply use the inner or outer surfaces of a thin-walled solid as the shell model surface. While these surfaces are offset from the mid-plane by half of the wall thickness, large ratios of surface area or part size to wall thickness can reduce the error introduced to nearly nothing. The automotive body panels shown in Figs. 5.23 and 5.24 will not show a perceptible difference if modeled at the true mid-plane or the inner/outer surfaces.

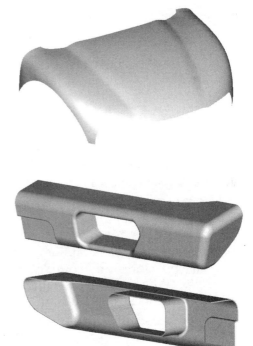

Fig. 5.23. The overall size of this part is so large in comparison to its wall thickness that a mid-plane model will behave nearly the same as a mesh on the outer surface.

Fig. 5.24. The time spent developing a mid-plane surface of this geometry would probably not pay off.

However, the thicker, walled weldment shown in Fig. 5.25 will most likely exhibit different stress results when modeled at the mid-plane ver-

sus the outer surface. If you are uncertain as to whether an exterior/
interior surface is a valid modeling assumption, try simple test models
which represent the problem being approached.

*Fig. 5.25. This thick-walled part is
questionable from a shell modeling
standpoint, but must certainly be
modeled at mid-plane in order to
achieve any level of accuracy.*

Some codes allow you to offset the neutral surface of the shell elements
used to model mid-plane surfaces. Chapter 6 discusses this property
adjustment. Offsetting the neutral surface may allow you to correct for
the physical offset made due to surface choice.

Adjusting Features

As stated previously, the level of uncertainty in a model due to properties and boundary conditions will typically outweigh inaccuracies due to
small variations in geometry. This is why nominal dimensions are typically used in place of detailed examination of the worst case tolerance
stack-up. If a detailed failure analysis or verification is required, this
approach may not be valid. However, most of the time it is a good
assumption. In these cases, making small adjustments to features to
facilitate meshing is valid. Corrections to eliminate "dirty" features as
described in the "Building Clean Geometry" section are usually excellent ways to ensure a more efficient model. However, rule number 3 for
feature suppression is applicable here as well. If you have access to the

native CAD system, and model adjustments do not conflict with dependent features, these modifications are warranted. However, if the geometry tools are not easily accessed, it may be better to spend your time applying mesh control techniques to ensure that degenerate features do not cause inaccuracies. Typically, your time is more valuable than run time. Keep in mind, however, that the cost of "dirty" geometry is additive when multiple runs or iterations are required.

Knowing When to Bail...

Knowing when to stop can be as valuable as knowing when to start when it comes to geometry preparation and simplification. While more applicable to design analysts, even specialists have time constraints on their projects. It is not advisable to spend valuable hours or days cleaning up geometry when starting from scratch or running with a slightly larger mesh will take less time overall and will allow you to reach the end faster.

Building beam models usually derives little value from a CAD database. It is almost always faster to build the wireframe neutral axis geometry from prints and sketches. Shell models will require decisions on a case-by-case basis. Typically, as the complexity of the part or assembly increases, the likelihood that starting from scratch will be faster increases. Weldments quickly increase the degree of complexity. When you decide that a new model is the best bet, develop the surface model to the degree of complexity dictated by the goals of the analysis. Guidelines in this area are included in the next section.

Deciding when to bail on a solid model is the opposite of the thought process for shell models. As the degree of complexity in the solid increases, the time required to remodel, even in the native CAD system becomes prohibitive. The "cleanliness" of the geometry, as defined in the "Building Clean Geometry" section, becomes the major decision making factor. If it is difficult or impossible to adjust or suppress features which significantly impact the accuracy or efficiency of the model, rebuilding the part specifically for the analysis may be warranted. The "Building Geometry Specifically for Analysis" section addresses guidelines for building these models.

Working with Limited Geometry Tools

Unfortunately, many readers of the previous discussions may be struck by the reality that native CAD or even solid modeling operations for geometry preparation and cleanup are not options. When this is the case, a different mind set must prevail. It is extremely important that you know the capabilities and limitations of the geometry tools at your disposal. The previous sections described techniques for cutting a part at a symmetry plane using wireframe and surfacing operations. Good preprocessors, such as those discussed later in the book, provide limited surfacing and some solid modeling tools. Surfacing and wireframe techniques can be used effectively to rebuild "dirty" portions of an imported model. These techniques do require more patience and planning compared to solid model based CAD systems, but will help provide cleaner, more efficient geometry.

Surfacing tools may be required to adjust or stitch surface edges on a CAD import. A quick surface mesh with triangular elements and a subsequent free-edge check will usually identify problem faces or surfaces. This technique is covered in more detail in Chapter 7, "Automeshing of Solids." The decision to fix the geometry versus simply fixing the mesh depends on how many times the same part will be remeshed. A one-time task of geometry cleanup can provide major savings if multiple convergence or optimization runs will be performed on the same model. Use your preprocessor's surfacing tools to stitch adjacent surfaces, rebuild corrupt or "dirty" features or to defeature a part. Fig. 5.26 shows the surface make-up of a complex part, passing through IGES translation.

Fig. 5.26. Solid model corrupted by IGES transfer.

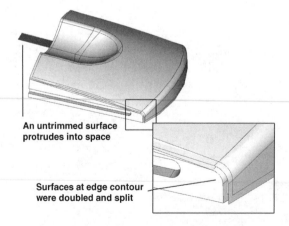

An untrimmed surface
protrudes into space

Surfaces at edge contour
were doubled and split

Deleting and rebuilding the surfaces in the area shown can yield a much more efficient geometry, as seen in Fig. 5.27.

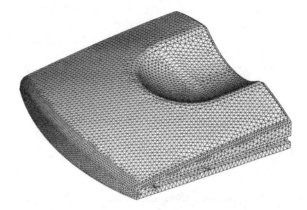

Fig. 5.27. With corrupted surfaces removed and rebuilt, the part is ready to mesh.

Many preprocessors provide you with the capability of ignoring features when automeshing. If this technique is available to you, practice on some test models to verify that you do not get unexpected results on a time-crunched project. These tools work by identifying loops at the base or face of features on a larger surface. By specifying these features in this way, the automesher can simply pave past them as if they did not exist. Significant time savings can be gained over modeling the local geometry if the resultant mesh is acceptable.

If the time to clean up geometry in your preprocessor becomes excessive because the native CAD environment is not available, it may be worthwhile to reconstruct the needed geometry in a CAD tool to which you have access. Again, balancing hands-on time versus run time must be considered. If multiple optimization runs will be required, having access to parametric geometry for solid models can be extremely valuable. Simpler solids and most shell/surface models can be reconstructed relatively rapidly in most preprocessors. Before choosing this route, consider the need for modifications and communicating developed geometry with clients or vendors. If your preprocessor will not support export of the geometry you develop, it may be best to take the extra step of building the model in a CAD system that supports at least IGES or DXF and then importing this database into your preprocessor.

Developing Geometry Specifically for Analysis

As mentioned previously, the ideal use of FEA in product design is at the concept stage of the process. When used in this manner, the geometry can be developed specifically for the analysis and the analysis can drive the geometry. The key to making the simulation results work for you is to start with simple conceptual geometry and work toward the more complex details as your understanding of the system grows. Liken this method to the understanding gap shown Fig. 1.1. Spend as little time as required to gain the maximum amount of knowledge in the concept stage.

Start Simple

When the analyst has control of the geometry from the start, many efficiencies can be built into the study. Depending on the goals of the analysis, it is usually advantageous to start with very coarse models of various structural designs. This is the ideal stage to utilize topology optimization software to suggest configurations with a balanced load path. Chapter 12 suggests tools to accomplish this task.

The value of starting simple is illustrated by the images in Figs. 5.28 and 5.29. The backboard shown can be constructed with several vastly different rib schemes for stiffening.

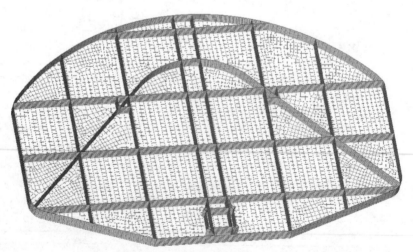

Fig. 5.28. One rib pattern may appear visually to provide the best stiffness.

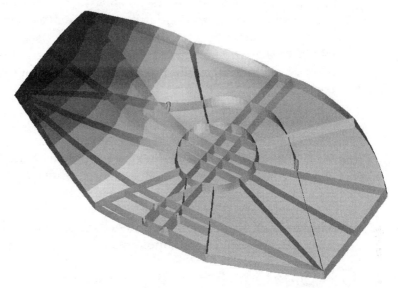

Fig. 5.29. However, a drastically different pattern might actually be the better choice.

The choice of the optimal design is not intuitive. Quick shell models can point out winners and losers without much work. In addition, the initial studies might yield two or more comparable designs which could be presented to manufacturing or marketing to include nonstructural factors in the decision. At this level of conceptualization, all parties involved will be more willing to suggest and accept change.

The Burden of Commitment

Two things happen–*emotional commitment* and *concurrent commitment*– when the suggestions mentioned previously are ignored and a detailed design progresses before the simulation is initiated. Either type of commitment can tie the hands of the analyst and restrict improvements suggested by the simulation to those of the minor evolutionary kind.

Emotional commitment occurs due to the feeling of ownership designers take upon completion of a part. The degree of emotional commitment is proportional to the time spent on CAD modeling. While always present, this phenomenon is more prevalent in the solid modeling world. Few designers are immune to this. However, this bond does not usually develop at the brainstorming concept sketch phase. Initial geometric models for concept prove-out should be thought of as sketches.

Concurrent commitment refers to the fact that few systems are designed sequentially, component by component. While the details of one component are being worked out, the details of the mating or surrounding parts are being worked out as well. If the design is essentially complete by the time it is submitted to analysis, it is safe to say the mating components will be nearing completion too. At this point, decisions cannot be made solely for the benefit of the component under study. A compromise between the optimal part configuration and the one which will have the minimum impact on the mating parts must be struck. If the guidelines described in this text are followed, critical components which are candidates for structural and cost optimization will have been identified in the initial design reviews. These will undergo a simulation and optimization cycle before the interaction with the rest of the system is determined. A verification run should still be made after all features are finalized, but the general topology should be correct.

Guidelines for Analytical Geometry

These guidelines are aimed at geometry to be used solely for analysis. In some isolated cases, this geometry might be converted into the final CAD model but that is not the intent. If it is important that the analysis geometry be used for design, refer to the "Building Clean Geometry" section, which approaches the problem from the design end. Typically, analytical geometry will be built to facilitate meshing and topology changes to be studied. Feature construction may vary a great deal from that suggested by good CAD practices. In addition, most of these models are meant to represent or suggest the final design. This approach liberates the analyst from the burden of painstaking geometry detail.

Elements and Model Types

The first analytical models should be constructed using the approach which will yield the most data in the shortest period of time. To that end, the analyst must consider plane stress, plane strain, and axisymmetric models in addition to beams or shells. Initial geometry must reflect this consideration.

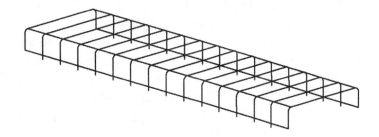

Fig. 5.30. A simple beam model can quickly allow you to evaluate options, whereas a solid model of the wire rack would be nearly impossible to mesh and solve.

If a planar representation can provide initial behavioral details, do not spend time on a 3D model until data from the 2D version has been exhausted. Similarly, it may be understood that a beam model will yield only a snapshot of the gross behavior and that a shell or solid model will be required. However, the gross behavior may be fine tuned over several iterations in minutes so that the time spent on a detailed model can focus on the details. Figs. 5.30 to 5.32 indicate the relative benefits of coarse initial models.

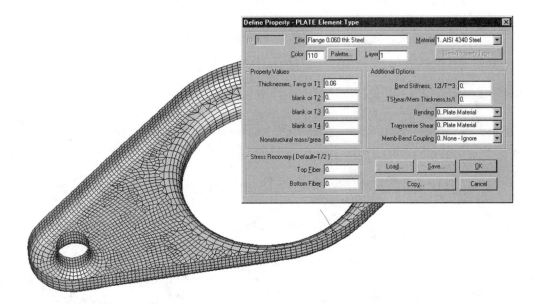

Fig. 5.31. This shell model can be modified quickly and easily in pusuit of an optimal design.

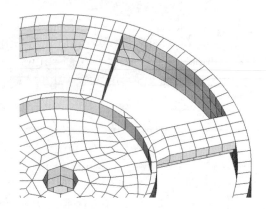

*Fig. 5.32. A simple
representation of a solid can
be meshed with brick elements
for a quick and reliable study.*

When no simplifications seem appropriate and solids appear to be the only way to proceed, the same thought process should prevail. Build the solid model with the minimum amount of geometry to serve the function of the part. Build the geometry to facilitate the meshing. If run time is an issue due to tight deadlines, consider building the part with extrusions and revolutions yielding a more efficient brick and wedge mesh. See Chapter 7 for discussion of the benefits and techniques required to effectively develop these models.

From the minimum geometric model, begin to add reinforcement as required to bring simulation results within specifications. Spend your time on the areas that suggest stress risers or marginal results. Add detailed fillets and features only to verify actual behavior where required. Remember that these geometric guidelines must be accompanied by proper meshing and convergence practices. The areas on which you have focused your study should most likely be flagged for tighter tolerances or controlled manufacturing procedures after the design is complete.

Plan for Boundary Conditions

As mentioned previously, the loading and constraint schemes should be mapped out before beginning more complete models. Test models should be used to confirm these schemes, and boundary conditions to be bracketed should be identified. Because the method of boundary condition application should be known, build geometric features to accommodate the method. Refer to the discussion of load patches earlier in this chapter.

Plan for Assembly Features

Many analyses will be performed on assemblies. Consequently, as parts are being constructed, build in features to allow accurate placement of assembly features. Fig. 5.33 shows a structure with two stiffening cross braces. Note that the side frame has a mesh patch matching that of the end plate of the cross brace.

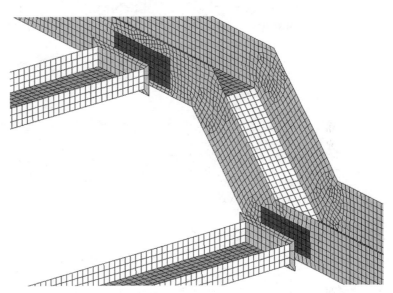

Fig. 5.33. Use surface patches to aid in aligning assembly components.

This planning provides meaningful spots to attach beam or rigid elements to model the bolts, as well as springs or contact elements to provide more accurate load transfer. It is much easier to plan for assembly features when the initial geometry is being developed than to modify the solid or surface model as an afterthought.

Plan for Automatic Optimization

The speed of engineering workstations, solution algorithms, and automatic meshing tools have made geometric optimization not only feasible but common in integrated CAD/FEA systems. While many tools for geometry optimization exist, the type which works directly off CAD dimensions holds the most promise for widespread use by design engineers. These systems allow you to specify a range of values on a model dimension that can be varied in conjunction with other variables to determine an optimal part configuration that satisfies design criteria.

While this technology can be powerful and relatively easy, its benefits may be limited by the modeling and dimensioning schemes used to develop the initial geometry. In fact, the best modeling and dimensioning for use with optimization may not be the best for detailing or future design modifications. Consequently, it may actually be beneficial to optimize on analytical geometry as opposed to the final design model to make the most out of the project. The simple example shown below highlights many of the modeling issues affecting optimization. In this example, the original plate dimensions shown in Fig. 5.34 represent the maximum geometric envelope for a wall-mounted bracket that must support a cantilevered load. In addition to outer perimeter optimization, lightening holes will be allowed to further reduce weight.

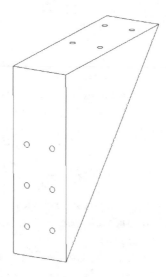

Fig. 5.34. New bracket design must have mounting features as shown and fit within the specified envelope.

Topology Constraints

Because the tools being discussed only allow design variables on existing model dimensions, it is unlikely that the optimizer will be able to add holes or ribs in pursuit of the optimal design. Consequently, several optimization models might be planned which represent possible topology configurations. Fig. 5.35 shows three different concepts for the bracket which would be best approached from different optimization studies.

Fig. 5.35. Evaluate
drastically different
approaches independently
instead of trying to develop
a single model that can
capture all possibilities.

Upon completion of the studies of each concept, the best of the best can be chosen for the final design. One trick for getting around topology constraints is to use buried or overlapping features. Fig. 5.36 shows two configurations of three holes or solid "cuts" which can combine in an overlapping condition to make a single larger hole.

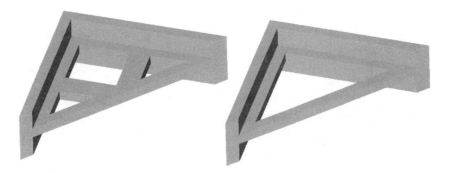

Fig. 5.36. Holes shown in configuration at left can "merge" to create the larger opening shown at right if dimensioned properly.

Using the above techniques, holes and ribs can be made to "appear" or "disappear" in the optimization process. Fig. 5.37 shows a rib dimensioned in such a way that it can completely bury itself inside the plate thickness.

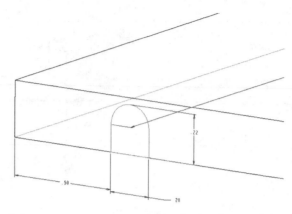

Fig. 5.37. By making the rib height less than plate thickness, the rib can be made to "disappear" in an optimization study.

It is also recommended that you use automatic optimization routines to simply refine a concept, rather than to develop one. This means that simple concept models should be used to prove out and qualify designs based on more manual methods. When one or two designs show promise for more detailed study, automatic routines can be applied to optimize the features of these concepts.

Relationship Constraints

One feature that has made sketching cross sections and profiles in parametric CAD systems faster and easier is dynamic, intelligent relations. This feature will recognize situations such as parallel and perpendicular lines, tangent arcs, concentric circles, and equal size entities, and automatically constrain them as such. This procedure is intended to anticipate design intent. However, it can also limit the shapes available to the optimization routine. If two lines are constrained to be parallel, they must always remain parallel. It is best to disable these constraints prior to geometry creation or at least prior to optimization. If they cannot be disabled, the geometry should be created to avoid their automatic creation. The method used to create dimensions can also force these types of constraints. Line-to-line or plane-to-line dimensions will force the entities involved to remain parallel. Point-to-point dimensions provide the greatest degree of flexibility when considering optimization. Fig. 5.38 shows an initial geometry configuration and dimensioning scheme that will allow an optimization routine to evaluate many possible configurations.

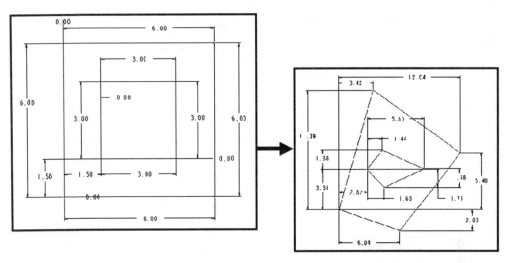

Fig. 5.38. A simple plate with a rectangular hole may require more complex dimensioning for an optimization study than for detailing a single design.

Test the Limits of Geometry Variables

It is possible to apply design variables which can combine to produce an illegal shape or a configuration that will not regenerate. It is also highly likely that more than three or four variables will conflict in at least one combination. If the optimizer encounters one of these combinations, it may crash out of the routine. Hopefully, this crash will indicate the problem area. The biggest disadvantage of an error in these studies is that most optimizations are extremely lengthy processes and many hours of computer time could be lost due to poor preliminary planning. To minimize the chance of conflicting dimensional variables, test your model at all combinations of extreme values. Some software packages provide tools to review these extremes by providing a table or a form in which to enter values. If these aids are not available, use the regular geometry manipulation tools to explore the combinations which might cause model failure. If a conflict is identified, adjust the geometry or the variable definition to eliminate the problem. It may be sufficient to simply adjust feature size, suppress unnecessary features, or reorder them so that the conflict does not exist. In Fig. 5.39, the fillet feature might fail when the hole penetrates it if the fillet was created after the hole, but would be unaffected by the design variable if it was created before the hole.

*Fig. 5.39. Order of creation
becomes important if an
optimizer is allowed to adjust the
position of overlapping features.*

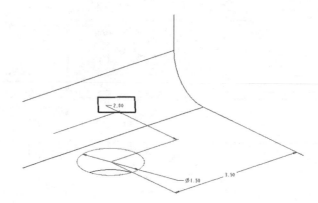

Summary

Today's CAD tools are a mixed blessing for the design analyst. While the combination of solid modeling and more robust automeshing has made FEA more accessible, the availability of ready-made models has limited many users' scope of this broad technology. Companies wishing to make FEA an integral part of the design process must learn how to bend the geometry tools to the needs of the analysis. This approach is somewhat opposed to the current attitude which seeks to set up analysis functionality to fit within the limitations of CAD systems.

Geometry and FEA are inextricably attached. As new analysts must learn the capabilities of their simulation tools, so must they understand the most efficient ways to utilize the geometry tools at hand. Project planning and interaction with geometry providers will significantly improve the situation. When working with CAD models, try to work in the native CAD environment. CAD models built with downstream applications in mind will certainly make meshing easier and models more accurate. Finally, remember that the FEA solver does not ultimately understand geometry. In the end, it is the quality of the mesh that represents geometric accuracy. If the available geometry will allow fast construction of an accurate and efficient mesh, use it. If it will not, do not be afraid to remodel. When wearing the analyst "hat," focus on the needs of the task at hand.

6

Assigning Properties

Once the finite element model (FEM) geometry has been developed, you should begin considering the assignment of all model properties, including material specifications and other properties governing the local behavior of each finite element. Never take this portion of the model building process lightly. Your FEA solution can only be as good as your representation of the real system. Remember that materials in FEA are perfect, and so are shell thicknesses and beam cross-sectional dimensions. Before proceeding with the analysis, you should understand exactly what it is that you are assuming when you enter a property value. These assumptions must be properly documented so that possible discrepancies may be easily accounted for later in the correlation phase. These are essential concepts for which you need to develop a solid understanding. Hence, this chapter will discuss in great detail the nature and implication of each property required and/or available in your FEM. You may also wish to refer to Chapters 2 and 3 for more in-depth presentations of property fundamentals and the assumptions associated with their application.

Introductory Concepts

In the current state of FEA, most codes have made it possible to assign properties to the underlying geometry of the model. If it is possible in your code, it is recommended that you do this before proceeding with the mesh. For one thing, it is easier to manipulate a model before it is overwhelmed with elements. It is also far simpler to select a large geometric entity than to pick each small element within it. In addition, if a volume, surface, or curve has consistent properties, by assigning them to the geometry you can guarantee their application to every element within, even if the elements must be modified or recreated later. This last point becomes very relevant in possible iterations of your FEM, where simple changes to the geometry may not be supported by the existing mesh.

Fig. 6.1. Examples of good (a) and not-so-good (b) habits for shell property entries in Pro/MECHANICA.

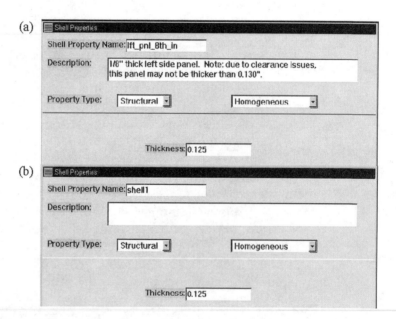

Property Names

It is also likely that your code will allow user-specified names for each property. Be sure to take advantage of this powerful capability. If the component or system you are modeling has distinct sections, such as different parts or different wall thicknesses in the same part, it is useful to name these differently, even if their property value is the same. This way, if a design revision or a sensitivity study is required for a particular section, its properties can be easily accessed and controlled. Detailed property names also provide great means for bookkeeping. Let each name remind you not only of the section to which a property is assigned, but also of its value. For example, *shell1* does not nearly convey as much as "eighth inch left side panel." Some codes will restrict the number and type of characters that may be used in property names. If so, learn to get creative with the characters you can use. As shown in Fig. 6.1, for a code that limits the property name to 16 alphanumeric characters, starting with a letter and no spaces in between, a possibility may be *lft_pnl_8th_in*.

Comments and Descriptions

Some codes even supply a comment or description box to accompany each property, as in Fig. 6.1. Writing an entire report is not necessary, but noting pertinent information that may be easily accessed later is recommended. If you used an equation to determine a shell's thickness or a beam's torsional rigidity term, it is useful to write it in the comment box for future reference. The same can be said for material properties that derive from a book, fax, phone call, or data sheet; their origin may soon be forgotten if an easy-to-find note is not made that will stay with the model. If your code allows you to add properties to a library for use in other models, property names and comments gain even more importance as efficient means of accurately tracking information. This is especially true of codes that are used in a company network; being able to share the properties that others have spent time deriving or locating is extremely useful, as long as the source(s) can be quickly confirmed.

Colors

Another feature that may be found in your code is the possibility of assigning different display colors to each property. Being able to visually inspect a model's properties serves as a quick and powerful check on the model. These display colors are often related to groups or layers that can be turned on or off, providing you with additional flexibility and clarity when working with larger models.

Remember that properties and their related names, comments, and groups are "free." By splitting up your model's specifications into clear and complete assignments, you will not pay with disk space or run time. You will, however, gain much in terms of information that will always be attached to the FEM and serve as an important means of communicating assumptions and intent.

Material Properties

Probably the most difficult of all properties to fully appreciate, material properties and their underlying assumptions are the foundations of the FEM. For example, selecting the property set for "steel" in the code materials library is easy enough. However, are you aware of the assumptions made with this selection? As mentioned previously, materials in FEA are perfect. Their properties hold constant throughout the assigned entity. Localized changes due to heat or other processing effects are not accounted for. Any impurities present in the parent material are commonly neglected. Hence, simply picking "steel" works, as long as you have considered all associated assumptions and adjusted your interpretation of results accordingly. Otherwise, you must allocate resources for obtaining material property values specific to the application under analysis. If it is impractical or impossible to do this, you must at least bracket your solution using the extremes. Refer to Chapter 3 for more information on this subject.

Types of Materials

As mentioned in Chapter 2, there are basically two types of materials: isotropic and anisotropic. *Isotropic* materials have properties that are independent of geometric orientation. On the other hand, *anisotropic* material properties always require definition of material orientation. A special subtype of anisotropic materials is known as orthotropic, which exhibit material properties that are load direction dependent in three orthogonal planes of extreme values. Most analyses assume properties that are isotropic and homogeneous. The homogeneous assumption means that property values remain constant throughout the volume. Because very few materials behave perfectly, both of these approximations must be kept in mind when presenting or correlating FEA results with reality.

*Fig. 6.2. FEMAP
material property
entry window.*

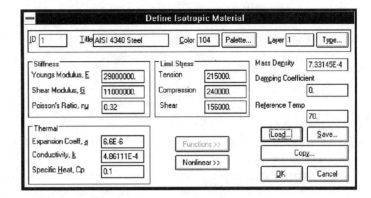

Stiffness Properties

The only material properties that are generally required by an isotropic, linear static FEA are its Young's modulus, Poisson's ratio, and shear modulus. Because these three properties are related by a single equation (Eq. 2.28), you should enter only two. Entering the three of them opens the door to the possibility that their values do not satisfy the equation. If a discrepancy is found, a warning message may be issued and you may lose control over which properties are actually used by the solver. Most codes and solutions require that you enter Young's modulus and Poisson's ratio, while the system calculates the shear modulus.

Other Properties

A thermal expansion analysis also requires a coefficient of thermal expansion, while more advanced thermal simulations ask for conductivity and specific heat values as well. Orthotropic studies request values for all structural and thermal quantities in each of three orthogonal directions. Modal analyses make use of a mass density, and dynamic studies allow for the input of a material damping coefficient as well. This property is not necessary if an overall structural damping coefficient is determined to be applicable for the entire system. It becomes important, however, if certain components in an assembly have much higher damping than others, such as a cast iron or rubber components in a steel weldment assembly. Codes that include optimization routines might add other optional properties such as a cost per unit mass. This property allows you to optimize on the more costly materials of a component or assembly instead of simply optimizing on the overall mass.

Units

Regardless of the properties you are asked to enter, it is extremely important that your units are consistent with those of the rest of the model. Be especially aware of the mass density entry. In the inch-pounds-second system the units for this quantity are sometimes referred to as slinches/in^3, which are essentially obtained by dividing the weight density (lb/in^3) by 386.4. In the mm-Newton-sec system these are in metric tonnes/mm^3 or 1000 kg/mm^3. It is always a good idea to check the units used by your code by comparing common material properties from the code's library against equivalent textbook properties and looking out for gross numerical differences. Refer to the last two tables in Chapter 8 for common units usage and conversion factors.

Nonlinear Material Properties

Additional property definitions are necessary for problems that require a nonlinear material model. The linear, elastic properties mentioned earlier must be specified in addition to the nonlinear stress-strain relationship that will govern behavior in the plastic region, where strains exceed the yield stress as indicated by you, the analyst.

Nonlinear material models can be input in one of several manners. Included are bilinear, trilinear, or multilinear models. A bilinear model simply requires a plastic modulus and a transition stress to identify when to switch the element stiffness definition from the elastic Young's modulus to its plastic counterpart. A multilinear model requires the input of stress-strain data pairs to essentially communicate the stress-strain curve from material suppliers or testing to the FE model.

In addition to nonlinear models of standard engineering materials, other types of nonlinear solutions require special material models and solver algorithms. A hyperelastic solution might be used to accurately model highly deformable, low stiffness, incompressible materials, such as rubber and other synthetic elastomers. This type of material definition requires distortional and volumetric Mooney-Rivlin constants or a more complete set of tensile, compressive, and shear force versus stretch curves. A creep (viscoelastic) analysis requires time and temperature dependent creep properties for long-term effects under near constant stress. Plastic parts are extremely sensitive to this phenomenon.

Another property that is sometimes required, and is often useful, is the definition of material orientation. This allows you to define anisotropic material properties and/or review general FEA results in terms of specific system- or user-defined directions. The default material orientation is typically that of the world coordinate system, but may be redefined in terms of a user coordinate system as necessary. Referring to Chapter 2 for a more in-depth review of material properties is recommended.

General Element Properties

Beams

In addition to a material property specification, beam elements have other required and optional properties. The most basic is its cross-sectional entry, which is input relative to its principal coordinate system. Generally, the x axis of this system is parallel to the axis of its definition curve and requires the user to specify the orientation of the other two axes using an orientation node or vector. For the purpose of this discussion, this x axis definition is assumed. Next, your code may allow you to define two offsets and a rotation angle that place the beam section away from its definition curve, although this curve continues to define the axis of load application and transmissibility.

Once you have successfully oriented and located a beam element in space, its cross-sectional property definition requires that the quantities below be calculated and entered.

- Cross-sectional area, A.

- Principal area moments of inertia, I_{yy} and I_{zz}. For beams with no transverse axis of symmetry, I_{yz} will also be required. Refer to Chapter 2 for a full description of the principal area moments.

- Stress recovery points, C_y and C_z. These points define the distances along the principal y and z directions and away from the neutral axis (principal x-axis) at which you want bending stresses to be calculated.

- Torsional stiffness factor, K. Often the most confusing of all cross-sectional quantities, the confusion arises from the nomenclature used by FEA codes. In engineering, J has historically been used to denote the polar moment of inertia of a section, which, in this case, is simply equal to $I_{yy} + I_{zz}$. However, in the world of FEA, the same J is often used to define the section's ability to resist torsion. For a circular section, both of these quantities are equal. For most other sections, however, they can be very different. A great source for equations that define a torsional stiffness factor for a variety of sections is *Roark's Formulas for Stress and Strain* by Warren C. Young (McGraw Hill, 1989), Chapter 9, Table 20. Alternatively, this quantity may be obtained through FEA by building a straight solid model of length l of the section under consideration ($l >>$ section dimensions), applying a torsional load T to it, and backing it out from the equation $K = Tl/G\theta$, where θ is the resulting angular deflection and G is the shear modulus of the material.

Some codes also allow for the definition of both y and z shear factors in terms of ratios of effective shear area to true cross-sectional area. The input of these quantities permits the examination of shear effects in the principal y and z directions of a beam. Regarding shear, note that most codes assume that the shear center of the beam element coincides with its neutral axis. Because this is true only for sections that are symmetric about one or both of their y and z principal axes, reported displacements and stress results may not be correct. Adjust your model if the cross section does not fit this description to ensure that all displacements are calculated correctly.

It should be pointed out that most FEA codes have internal libraries of common section shapes. Moreover, a few allow graphical input and use shape calculators to obtain the section properties. CAD systems will also provide most of this sectional information. Many engineering textbooks tabulate these properties and/or provide equations for them as well. However, when you use a method external to your code for calculating section properties, you must be careful with the torsional stiffness factor definition. When in doubt, build yourself a solid test model as outlined above.

⁃⁌ **NOTE:** *SDRC provides a general section sketcher that determines properties from an internal FE analysis using the section, similar to the means described previously for determining K.*

Advanced Beam Properties

In addition to the properties described above, beams offer a great degree of flexibility in most solvers, thereby allowing even more detail in the simulation of structures. Neutral axis offset and shear center off-set were mentioned previously. These properties allow you to physically attach the beam element in a convenient location, but force the model to interpret the beam as if it were in a different spot. Fig. 6.3 shows a plate stiffened with I-beams. In an FEM of this structure, placing the beam elements directly on the plate is most convenient. However, the bending stiffness of the plate is much different if you use the default interpretation of the beam placement–buried halfway in the plate so that the neutral axis is aligned with the actual element locations–rather than offsetting the beams to the correct location. Utilizing a neutral axis offset, the model understands that the I-beam neutral axis is located at half of the height of the beam off the plate, thereby correct-ing the overall stiffness of the system.

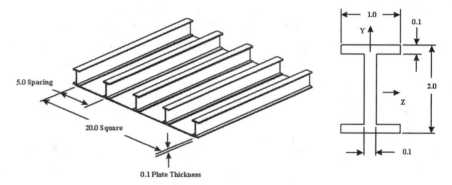

Fig. 6.3. Plate with stiffeners.

Another advanced property that can be assigned to beam elements is a linear taper from one end of the beam to the other. This property is not supported in all codes but can have a significant impact on the stiffness of a structure when actual taper is involved. The hard way to specify taper across a set of beams is to manually calculate and input each end size to each beam. While a simple model with one or two beam ele-ments might not take much time, this would be prohibitive in a larger model. That is why some preprocessors allow you to select a group of beams and specify the two end conditions. The program calculates the actual element taper based on a linear slope between the ends.

A final beam property that is frequently required in design analysis is a *beam end release.* This property disables one or more degrees of freedom (DOFs) on the end of a beam so that the beam cannot carry any load in the disabled DOFs. Rotational DOFs can be released to allow a beam-to-beam connection to behave like a pinned joint, or the axial DOF can be released to allow a beam to slide as if it were in a journal. This additional flexibility allows for simple, yet realistic, beam models of more complex systems.

Shells

Shell elements require less mathematical definition than beams because they have more geometric definition than beams. In other words, the mathematical section property of a shell need only specify thickness, the visually absent dimension. Once the thickness is input, geometric representation of the other two dimensions will complete the shell definition. Of course, the material orientation, as outlined previously, will affect orthotropic properties (if used) and result representation. In addition, for the examination of shell results, you can select the orientation of their normal vector to tell the system which side is "up" and which is "down." Ensure that the shell orientation is consistent across every region of the model. This is extremely important when making use of pressure loads, because these are understood by the system in terms of the shell normals. Refer to Chapter 8 for more information on pressure loading.

Advanced Shell Properties

Like beams, shell elements can be used at a higher level than simply adding thickness to a surface element. The physical location of a shell element, by default, defines the neutral surface of the element. However, like beams, the convenient placement of the shell in the FEM might not be the true location of the neutral surface. Good examples of this phenomenon are a relatively flat surface with thickness that varies on only one of its sides and a doubler plate in a weldment. To accurately model these types of systems, some codes allow that a shell element be defined so that its neutral surface is offset from its physical location or tapers in the same manner that beam elements can. Ideally, if the latter property variation is important to your design challenges, your preprocessor should allow the assignment of taper across a group of elements and not force you to enter the taper element by element.

Similar to end releases in beam elements, the bending, membrane, or shear properties of a shell may be released or isolated by some codes. A membrane shell might represent a thin skin on an airplane wing, whereas a bending-only element might be used to simulate a complex composite or linked panel that, due to its assembly interaction, can resist bending but pull freely in tension. These specialty uses of shells should be learned with test models before attempting to incorporate them into a larger, more complex model.

Solids

Taking the argument for shell elements one step further, solids do not require any mathematical definition because they are completely defined by the geometry. The only solid properties that can be prescribed are those pertaining to material and orientation, as reviewed in the material properties section above.

Special Element Properties

Mass Elements

Mass elements that act as inertia simulation entities in your FEM must be fully defined. In a static analysis, this definition may simply consist of a mass value. However, in a modal or dynamic analysis, you should not forget the moment of inertia contribution of a lumped mass representation. Recall from Chapter 2 that just as a body's translation is governed by its mass, its rotation is regulated by its mass moments of inertia. Hence, if a mass element is used to simulate a nonmodeled component of a size comparable to or larger than that of the modeled structure, its moments of inertia are likely to have a significant effect on the overall dynamic response of the system. Fig. 6.4 shows a very good example of this phenomenon in a modal analysis intended to ensure that the first mode of a switch assembly system be higher than 100 Hz. Fig. 6.4(a) uses a point-mass representation of a switch body attached to its mounting bracket. Its mass, but not its moments of inertia, is input into the element definition form. Fig. 6.4(b) shows the results of a full model of the system. Fig. 6.4(b) shows the results of a full model of the system. Note that had one used (a) to provide a design solution, the answer would have been 73% higher than that indicated by (b) and nonconservative in nature; the analysis would have incorrectly reported having met the design criterion.

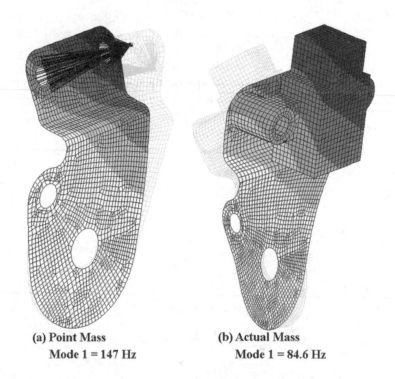

Fig. 6.4. Inertia modeling comparison: point-mass model without MMOI input (a) versus the full model (b). (Image courtesy of Buell Motorcycle Company.)

(a) Point Mass   **(b) Actual Mass**
Mode 1 = 147 Hz   **Mode 1 = 84.6 Hz**

Mass moments of inertia for simple structures are tabulated in most engineering textbooks or can be calculated using the equations presented in Chapter 2. For more complex shapes, these inertia values may be obtained directly from a CAD solid model. Alternatively, if the part was not modeled in solid geometry or the CAD program itself does not provide this capability, most FEA codes will usually provide this information as a simple model query or as additional output at the end of a run. Hence, to obtain these properties with FEA, you must create a simple model representative of the geometry under consideration. You can then either query your code for its inertia properties or solve a simple analysis, such as a thermal expansion simulation with $\Delta T = 0$. Most codes will allow you to define a coordinate system about which you want the mass moments of inertia calculated. However, certain codes provide this output only about the model's world coordinate system or its center of mass.

Springs

Spring elements prove very useful as constraint entities in many types of simulations or to simply represent a physical spring in an assembly. If you make use of spring elements in a model, you must make sure that their properties are properly defined. Whether a spring is used between two element nodes or to connect a single node to ground, there are basically two types of stiffness properties that must be considered: extensional and torsional. Although both types are used to limit motion, extensional stiffness opposes linear translation while torsional stiffness fights back angular rotation. Point-to-point springs simply require a spring constant, either torsional or axial. A more complex spring type allows you to enter a different stiffness in each of the six degrees of freedom. These stiffness quantities are always input in terms of coordinate directions. Hence, to define them, a spring orientation must be known or declared. Of course, the default orientation of a degree of freedom spring in most codes is that which aligns with the world coordinate system.

Dampers

Damper elements are line elements that, in their most basic use, represent dashpot-like devices in a system. The units of a damper element are *force/velocity* (or force-time/length). Such units are of little use in a static analysis and only generate a force when velocities are calculated. Consequently, if there are discrete damping devices in the physical system that you are modeling in a modal or dynamic analysis, damper elements should be used. Damper elements can also be used to model vibration absorbers, or if used to connect all nodes on a surface to another surface or ground, can represent a damping continuum in which the damping is known in units of force/velocity. As with any new element type, you should experiment with them in test models.

Contact Elements

The property definition of contact elements depends on the technique(s) your code allows for creating contact pairs. If surface-to-surface or curve-to-curve contact is allowed, you may not have any properties to enter because the contact behavior is derived from the material properties of the elements on each side of the contact pair.

However, most general purpose codes continue to require the use of slide line or gap elements to define a contact condition. Consequently, some level of input is necessary, depending on the capabilities of the element and the degree of automation programmed into the system.

Gap Elements

All gap elements will require two properties and some have others that you can adjust as well. The two basic properties of a gap element are its *compressive stiffness* and *initial gap*. The compressive stiffness must simulate the local stiffness of the elements in the contact pair. Unless two elements share a common node, they are not aware of each other. The gap element provides a means to specify interaction between two otherwise independent nodes. Therefore, the stiffness property is a spring constant that allows the gap to resist one node from crossing the spatial location of the other. The point at which the gap begins to resist this displacement is defined by the initial gap. Consider two plates modeled at the mid-plane. Due to the mid-plane modeling, these two plates that might rest on each other in the real system would, by definition, be separated by half the sum of their thicknesses. By bridging this separation with gap elements assigned an initial gap of zero, the solver knows that this initially assembled position represents two parts already in contact.

A cautionary note when specifying gap stiffness: if the assigned stiffness is too low, some unacceptable penetration will be observed between the contact pairs. If the property is too stiff, the solver may experience what is best described as *bounce*. Numerically, the solver will not be able to reconcile the forces causing the contact to occur with the reaction forces from the gap/spring. The process of choosing a gap stiffness may be an iterative one as an initially soft gap is stiffened to the point where penetration is minimal but the nonlinear solution still converges. Some codes have implemented adaptive gap stiffness technology that will adjust the gap property internally to achieve convergence without penetration within some defined tolerance. In an adaptive solution, the method for specifying an initial gap stiffness is the opposite of the manual, iterative process. Choose a stiffer gap property and let the software soften it for you as is required numerically. This will be more efficient than the other way around.

Gap elements can alternatively be assigned a tensile stiffness in place of a compressive stiffness. In this mode, they behave like stiff cables. The initial gap property represents the fully extended length of the cable. The tension stiffness is simply the spring constant of the cable member.

Slide Line Elements

Slide line elements require a compressive stiffness property just like gap elements. However, slide lines cannot support tension. The choice of a slide line stiffness is more complex than for gap elements. In a gap element, you can estimate the spring constant of local elements using the equation, $K=AE/L$. This is not the case with slide lines. Your software documentation should provide guidelines for choosing stiffness of slide line elements. Some codes have default properties and adaptive methods that can greatly simplify the use of these elements.

Friction

Most codes allow for some type of frictional interaction across a contact element. This property should be interpreted in the same light as a coefficient of friction between any two physical bodies. However, the magnitude of the FEA coefficients may not necessarily correspond to the physical coefficients. Use testing and correlation analyses to determine static and dynamic friction coefficients between two materials in your FEM as necessary.

Summary

Assigning properties to an FEM has become an easier task in recent years. As more and more codes have moved in the direction of geometry based, finite element meshing, the application of properties has been greatly simplified through the utilization of the underlying model geometry. Making use of this capability is recommended to eliminate sources of error due to original selection or later modification of the elements themselves. However, you are cautioned not to embrace this user-friendly technology simply to rush through the property assignment step. You are also advised to take advantage of other features provided for your convenience, such as user-defined property names, comments, and groups.

Property assignment in FEA is an extremely important part of the overall analysis process. Both material and elemental properties must be fully understood in terms of their direct effects and implications. The results of any simulation must always be viewed in light of the parameters that were utilized for its execution. You must always ensure that you understand exactly what is being asked in each property entry form and keep in mind all assumptions made in the application of each property. For instance, it makes no sense to imply a 10% error in your results when the property variability, material and/or elemental, is higher than 10%. Similar to all other assumptions, you should always try to quantify property approximations to get a better feel for the safety factor necessary to build into your analysis.

7

Finite Element Model Building

Typically considered the most tedious part of the FEA process, model meshing is the subject of much automation development by all major players in the industry. While it is true that meshing is still a necessary evil for most modeling efforts, some people find completing a complex model satisfying in the same way as overcoming a tough, character building challenge. It is unrealistic to expect that every design analyst will learn to enjoy meshing as much as they seem to enjoy building parts in a CAD solid modeler. It is hoped, however, that you will at least learn to appreciate the importance of every meshing decision and resist the temptation to simply blast an automesh at a part and let the pieces fall where they may.

In this chapter, the basic concepts and techniques for both manual meshing and automeshing are presented. The concept of element accuracy is discussed with examples of the effects of distorted elements. Basic guidelines are discussed for choosing to mesh a part manually versus automatically and techniques for combining dissimilar meshes. In brief, this chapter is focused on the basic tools in a finite element modeler's toolbox. It is important that you understand each tool's capabilities and limitations so that you can make the best modeling decisions toward achieving accuracy and efficiency.

Setting Up the Model

The first step in any mesh is the planning stage. At the time that your goals are reviewed and your boundary conditions are ironed out, you should make the determination of which idealization to use. Should a thin-walled part be modeled with shells? Should a planar idealization be used? If this determination can be made prior to beginning geometry, you will be one leg up. If not, discuss with your peers or coach how to make the best use of what you are given.

Plan geometry as discussed in Chapter 5 to make your modeling tasks easier. Plan for all required coordinate systems. Adding coordinate systems when the model database is small and before the mesh is built will be more efficient than waiting until the last minute.

Grouping and Layering

It is always easier to create elements or other model entities in the groups and layers they belong to than attempting to move them later. Grouping will help organize a model into logical sections. The various parts of an assembly model should be organized in separate groups to assist in model building. Such organization will also facilitate results viewing because each component can be displayed individually. Most codes provide means for automatically grouping model entities by material, property, or entity type. You may wish to place all shell elements in a single group and all beams in another for quick access to the entities for checking and editing tasks. Planning groups and layers up front and using the automation tools in your preprocessor can be a great time saver.

Resource Requirements

The time to think about memory and disk requirements is before the modeling starts, not when it is finished. If you know from experience that you only have enough RAM for 30,000 shell nodes or enough disk space for 75,000 solid nodes, plan your model accordingly. Due to the speed of today's systems, it is often more convenient to overmesh than to take the time to be judicious. However, when you know you might be running into a resource crunch, you can utilize mesh control to focus the mesh density where you need it. This will save time compared to deleting a complex solid automesh and starting over; yes, even automeshes can be complex.

Element Selection

The big decisions in element selection are made in terms of the idealization deemed appropriate for the problem. Use solids for solid parts, and beams for beam-like parts, and so on. However, beyond these principles you will be faced with choosing triangular or rectangular elements. Again, this decision is made for you if you accept the defaults. Modern preprocessors will default to quadrilateral elements or a quad-dominant mesh for a shell model and triangular tetrahedrons for a solid model. The former choice is made for accuracy, and the latter for convenience.

Rectangular elements provide a linear strain distribution across the edges or volume. First order triangular elements only capture a single strain value; they are often called *constant strain elements.* Therefore, you will need many more triangular elements relative to quads to capture a high gradient. However, few solid automeshes have been offered in the market that can provide a high quality solid brick automesh. Consequently, less accurate tetrahedrons are used for solid automeshes, primarily because there is no alternative. The bright side is that in most cases, second order elements or parabolic tetrahedrons can capture more complex local strain gradients and provide reasonable results with proper convergence methods.

There will still be a few instances where element selection falls into your lap. The decision to build a manual or semi-automatic mesh will usually be determined by the need to use more accurate bricks over the less accurate tets. Other more code dependent reasons to choose against the defaults are when an axisymmetric model cannot support anything but triangles or a nonlinear solution cannot support second order tetrahedrons. In the case of a nonlinear solution, it might be prudent to take the time to model in more accurate bricks for the run time savings alone. A fair estimate is that you will need five tetrahedrons for every brick element in a model to get the same results. Because nonlinear runs generally tend to be more time consuming, a smaller model may allow you to make more design iterations within the time allotted.

Manual versus Automatic Meshing

If accuracy and speed were equal, few design analysts or even analysis specialists would dispute that automeshing is the way to go. The goal of FEA is not to build a mesh but to get performance data. Unfortunately, manual meshes and automatic meshes are not equivalent. Some clarification is probably required before proceeding. When manual meshing is mentioned, images of long days and nights typing in the coordinates of nodes and connecting the dots by hand come to mind. While the occasion will arise to connect nodes for making a quadrilateral or a triangle when cleaning up a model, most cases of manual meshing involve some combination of automatic and semi-automatic methods. Given the power of today's preprocessors, the need to manually mesh a shell model should never arise. If a surface model can be developed in either a CAD system or the FEM tool, a little preparation can allow you to automesh nearly the entire model. The issue of manual versus automatic arises most often in the context of solid models. Even in solid models, a typical solid "manual" mesh consists of revolving or extruding automatic or semi-automatic surface meshes. The real task is in planning the extrusions or revolutions so that the mesh matches up at the seams between the different steps. In the end, your challenge is to determine when manual meshing a solid model as defined is beneficial. More detail on this issue appears later in this chapter.

Some of the key differences between equivalent automeshed and manual meshed solid models are discussed below.

Modeling Speed

There is no way to sugarcoat it–manual meshing is very time consuming on even moderately complex solid parts. On extremely complex parts, it can be prohibitive in a design environment. Automeshing, on the other hand is the hands-down speed champion. However, while an automesh might get you to the run button faster, the excess of elements required to achieve the same degree of accuracy might cause the solution to take far longer. This is not the norm, however, in modern software.

Solution Speed

A single model of comparable convergence levels will solve more quickly if built entirely of wedges and bricks rather than of second order tetrahedrons. However, the solution speed of even large tetrahedral models is relatively fast for linear static analyses. Total modeling and solution time must be considered, which also means considering how many times a model must be remeshed and how many times a particular mesh must be solved. You must keep in mind that you will pay a high price for an excessive mesh in a nonlinear solution.

Accuracy

For a given mesh density, a brick mesh will provide more accurate answers closer to the converged solution than a second order tetrahedral mesh. A linear tet mesh should always be considered inaccurate unless the time is taken in test models to confirm that the stress change is gradual enough to allow linear tets to converge correctly. However, many studies have shown (and you should perform your own) that properly converged second order tets can provide the same accuracy as a linear brick mesh. Tets should still be avoided in thin-walled parts where shells are more appropriate. In most bending situations, two or more tetrahedrons are required through a wall thickness to properly capture the behavior. One accuracy issue that gets lost in the shuffle is the fact that many geometric simplifications are required to obtain a brick mesh. One could argue that the simplifications required to build a brick mesh cancel out any element accuracy issues when compared to a second order tet mesh with little or no simplification.

Convergence

It should be clear from the previous points that you will achieve better convergence with fewer brick elements than tetrahedrons. However, one intangible convergence issue pertinent to relatively manual meshing techniques is that if it takes a long time to complete a manual mesh, it will take a long time to refine it. In a hectic product design process, a design analyst might have time to run through four to five convergence passes in a day with an automesh, but it might take more than a day to make one convergence pass on a manual model. While run times for the manual mesh may be faster, the time required to modify it might be prohibitively long. The truth is that if convergence is difficult or time consuming, most design analysts will not invest the necessary time.

Perception

This intangible cannot be fully quantified. Many analysis specialists will immediately choose the more traditional brick mesh over a tet mesh. If you take the time to prove to yourself that a tetrahedral mesh is as accurate as a brick mesh for your particular problem, you should share and discuss these results with potential detractors. However, if you have not taken the time to perform this test, you should be prepared to take whatever is handed to you by traditional analysts. In any industry or specialty, you must put in a minimum amount of work to gain credibility. If you are willing to simply accept automesh results, you are not living up to your responsibility as an FEA user.

Manual and Automatic Mixed Meshes

Very few codes allow the combination of manual and automatic meshing on a single geometric entity. If you wish to mix the two to take advantage of the best of both worlds, you will need to break your geometry into parts. You will probably be responsible for ensuring that the two meshes line up. Remember that tetrahedrons cannot directly mate with brick elements. In certain limited cases, wedges can serve as interface elements. In most cases, you will have to transition these two dissimilar meshes with rigid links or multi-point constraints. The use of these elements is discussed later in this chapter.

P-elements and H-elements

Many codes provide the option to choose between p-elements and h-elements. P-elements are excellent for capturing high stress gradients. The additional resource requirements of these elements are warranted in these cases. For areas of gradual stress transition or away from any area of interest, h-elements are more efficient. H-elements can capture most stress conditions if enough degrees of freedom are placed in the area. However, when the option is available, choose to use the best element in the best location. Refer to your software's documentation to confirm that the option of mixing h and p elements is available and for information on specific usage techniques.

Meshing Beam Models

The simplest element type to model is a beam element. In contrast, beam elements are the most complex type to set up because configuring properties can be quite involved. Placing beam elements in a model is extremely easy. The best and easiest way to construct a beam model is to prepare CAD wireframe at the neutral axis of all beams. Take the time to split the wireframe at every joint or connection of two beams. In all shell and beam models, breaking the geometry into the smallest segments possible is beneficial because making changes to the model would involve disruption of the smallest number of entities. Some pre-processors provide tools for directly meshing beams between two points or nodes with a specified number of elements without underlying geometry. While this technique is extremely efficient for adding cross braces or struts in a large structure, it provides the least flexibility for modification. The line of elements must be treated as a cohesive group to permit adjusting the placement of the member represented by the beams. Use groups or layers to simplify changing elements associated with geometry.

Guidelines for Beam Element Size

Beams must be converged just as any other element type. You will learn, however, that beams converge with a minimal number of elements. The greater the dominance of bending, the larger the number of beams required by the model to capture the behavior. One guideline typically appearing in FEA references is that the length of the beam should be about ten times the maximum cross-sectional dimension. Typically, it is unclear whether such references are asserting that the overall beam—represented by the beam mesh—must be ten times the beam cross section, or that the length of each beam element must be ten times the maximum cross-sectional size. The best guideline for determining the applicability of a beam model is that if it looks like beams, or if a 2D or 3D wireframe representation conveys most of the structure with little ambiguity, then a beam model is probably appropriate. Use test models to confirm the validity of beams if a particular case is borderline. Once the decision is made to use beams, converge the model as you would any other finite element mesh. As the beam element length approaches the resulting displacement of the beam, it is possible to experience errors in a linear analysis which state that displacements are excessive. If your beam mesh is this dense, you have probably overconverged the model. However, a nonlinear large displacement solution will solve the small displacement problem for the same results as the linear solution with slightly larger elements, and can be used to complete the solution if the mesh is required.

Other Beam Modeling Issues

Multiple properties should be considered when developing a beam model. Cross-sectional orientation, neutral axis offsets, beam end releases, and taper should be included where appropriate. If the wireframe template was broken into a sufficient number of components, beam ends should line up and you should have a continuous model after merging the coincident nodes in the model. However, a visual check of the continuity of a beam model is difficult. Solid and shell elements can be checked with free face and free edge plots, respectively. There are no equivalent plots for beam models. Your best tool is diligence in the initial model building. If the model is complex and you receive an error message indicating insufficient constraints after kicking off the solution, use a modal analysis to identify the loose elements. This technique is described in more detail in Chapter 9.

Meshing Shell Models

Building a shell model requires mid-plane surfaces in one form or another. For single parts and simple sheet metal assemblies, some CAD systems that have a preprocessor integrated into the interface can automatically compress a solid model into a mid-plane surface model. However, the model must be constructed with just the right features to allow this to happen. Most thin-walled models will be too complex for automated CAD to mesh routines. A good technique for starting shell models is to sketch the part first to identify the key features required in the model. Most sheet metal assemblies have many inconsequential features and details that are not required in the mesh. In addition, when working directly off a print, identifying the dimensions to convert to mid-plane dimensions and those to be left to the actual dimensions is difficult. Fig. 7.1 shows a typical sketch of a sheet metal weldment. This model should be sufficient for gross deformations and it can be refined if necessary to improve the stress results.

Fig. 7.1. Sketch of a sheet metal weldment.

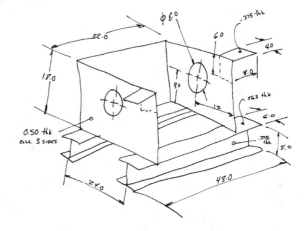

Figs. 7.2 through 7.6 show a sequence of the surface model development. As with beam models, a little up-front planning and geometry manipulation can greatly facilitate final model building. A well-prepared, underlying surface model will also improve the efficiency of making changes. This model was built in FEMAP v5 from a couple of initial curves. You can see that the curves are broken into smaller segments early in model development. It is good practice to try to break the geometry into four-sided patches with corner angles as close to 90° as possible. Surfaces are entered into the model when the curves in an area are completely defined. By Fig. 7.6, the entire surface model has been developed. It is common to backtrack, delete a couple of surfaces, and rebuild an area to obtain a better topology. When the surface model is complete and all edge curves of adjacent surfaces are merged or otherwise accounted for, mesh the entire model with a consistent element size. This technique will ensure that the nodes at the edges of the surfaces will align and can be merged cleanly. Choose a mesh size that you expect will provide a good representation of the geometry and response to the loads.

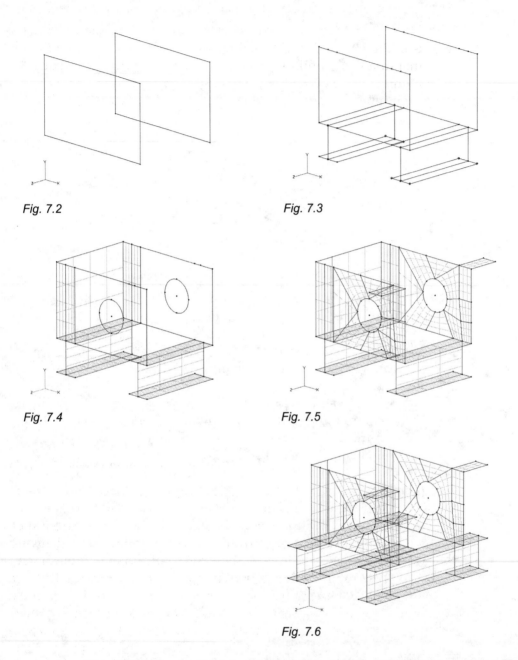

Fig. 7.2

Fig. 7.3

Fig. 7.4

Fig. 7.5

Fig. 7.6

Fig. 7.7 shows the initial mesh on the model. After merging the nodes on coincident surface edges, you should check a boundary or free edge plot of the model. This plot, as shown in Fig. 7.8, should show the edges of shell elements that are not connected to any other element. Look for cracks in the model that will show as free edges in the middle of a surface that should be continuous. Clean up the model by locally fixing the discontinuity or deleting the mesh on surfaces around the area of concern, adjusting the surfaces or mesh sizes, and remeshing.

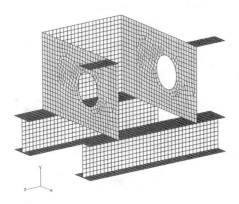

Fig. 7.7. Initial mesh on model.

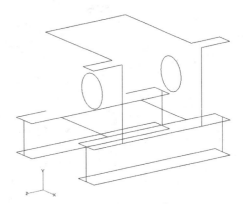

Fig. 7.8. Plot showing edges of shell elements not connected to other elements.

Once you have a continuous mesh, it is relatively easy to refine a portion of the model to improve the stress calculations. This can be done by deleting elements on surfaces near high stress regions and remeshing with a smaller element size. If your preprocessor supports this, you can also refine the mesh locally around specific areas of concern.

Other Methods of Creating Shell Models

There are three other common ways of building whole or partial shell models. First, curves can be extruded or revolved into shell elements. The I-beam extensions added to the previous example in Fig. 7.6 were made with extrusions of the I-beam cross sections built previously. Next, a shell mesh can be built between four points or nodes with a specified number of elements between the corners. This does not require geometry up front, but is harder to modify afterward. The third means of creating shell elements is to manually build them, node by node. While this may sound archaic with the preprocessing technology available today, rest assured that if you spend enough time building finite element models, especially shell models, you will need to stitch an occasional element in by hand. It is not nearly as daunting or painful as it sounds.

Element Shape Quality

The ideal shape for a triangular element or face is an equilateral triangle and the ideal shape for a quadrilateral element or face is a square. Elements with these shapes will be the most accurate; deviation from these shapes introduces error into the calculations. An element's robustness, or ability to provide nearly perfect results with a less than perfect shape, is an indication of the quality of the FEA code. Fig. 7.9 shows four element distortion types that must be controlled in a model, either by you or your meshing technology. Limits on these distortions vary with the solver you are using. For example, h-elements should have an aspect ratio less than 5:1, whereas p-elements can produce good results with an aspect ratio as high as 20:1. Typically, distorted elements only affect the results local to the bad element, although this should not be construed as a license to mesh sloppily in every area outside the immediate areas of concern. The overall stiffness of the model has a significant effect on local results. Mesh your parts with the goal of no bad elements and check the areas of high stress for distortion. If the elements in the area you are concerned about appear distorted or seem to affect the results in any way, take the time to clean up the mesh and re-run the solution.

Fig. 7.9. Four element distortion types to be controlled in the model.

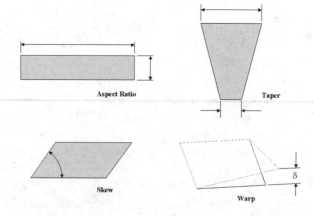

Mapped Meshing

Mapped meshing refers to specifying or forcing a particular mesh pattern by indicating the number of nodes on all the edges of a surface or volume. The edge description defines the area of volume fill. Consider mapping the element pattern on the surface patches developed for the model to provide a more uniform, better looking mesh. If all surfaces were perfectly rectangular, mapped meshing would not be as much of an issue, because most h-element meshers will fill a rectangle with uniformly shaped elements. However, meshing an irregular surface with a fixed nominal element size can yield unpredictable results. The simple trapezoid-like surface in Fig. 7.10 illustrates this concept.

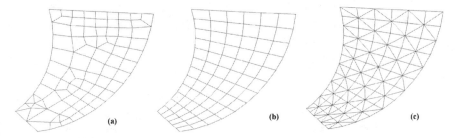

Fig. 7.10. Trapezoid-like surface illustrating meshing of irregular surface.

The mesh in (a) was applied to the geometry without consideration of the resulting distribution. You can see that the mesh is irregular with both triangles and quadrilaterals. This is primarily due to the differences in length of each side. Because the automesher will fit the nominal mesh size to each side, they may end up with different numbers of elements. The mesher then needs to fill in the area while maintaining the edge element count. The meshes in (b) and (c) show mapped meshes of the same surface using a quad pattern and an alternating tri pattern. A tri pattern may be handy if a surface mesh is being prepared for eventual fill with solids.

Biasing a Mesh

One technique for improving the mesh near an area of interest is to bias the mesh. This can be accomplished with either solids or shells, although it is easier with shells. Fig. 7.11 shows a rectangular surface mesh biased toward one corner (a), and a revolved mesh biased toward the axis of revolution (b). Biasing a revolved mesh also allows you to control the aspect ratio of the elements as they progress out from the center. Use mesh biasing to control the distribution of elements in a mesh and force smaller elements closer to a high stress area.

Fig. 7.11. Rectangular surface mesh biased toward one corner (a), and revolved mesh biased toward axis of revolution (b).

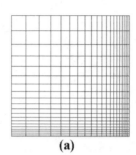 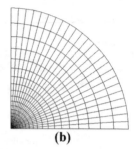

(a)   **(b)**

Transitioning Mesh Densities

Mesh biasing is a means of forcing smaller elements near an area of interest, while allowing larger elements in regions with a more gradual stress gradient. Another technique is to use automeshing of surfaces with the geometry broken into patches. A good rule of thumb for minimizing occurrence of high aspect-ratio elements is to limit transitions to ratios of 2:1 or less. The meshes in Fig. 7.12 show the use of patches to transition from a tight mesh to a loose mesh. In (a) the mesh transitions from a 0.05 nominal element size to a 0.50 nominal element size without control of the transition. If the area around the hole is important enough to be concerned about its local mesh size, the irregular mesh created by the transition will yield disappointing results. The mesh in (b) uses a series of regions to effect the transition. It is clear how much more gradual and uniform the elements in (b) are because of the extra work put into the mesh. The results around the hole in (b) have an excellent chance of providing accurate results.

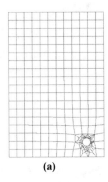 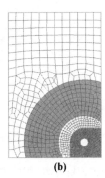

Fig. 7.12. Mesh showing use of patches to transition from a tight to a loose mesh.

                       (a)   (b)

Controlling a P-element Mesh

Underlying mathmatics explains why p-elements can be much larger than h-elements. P-element automeshers will attempt to fill the model with the largest elements possible, within a default or user-specified aspect ratio and edge or face angle tolerances, for solution efficiency. However, as mentioned previously, p-element solvers are not immune to element distortion. Tightening the element creation tolerances is the simplest way to improve the general mesh quality in a p-element mesh. Because such tightening can lead to excessive model sizes, means for locally controlling the mesh are often used.

One way to force a better mesh in local areas is to manually mesh areas of concern or *seed* the model with elements. If your system supports it, the rest of the geometry can then be automeshed without additional controls. A mesh can also be seeded with curves or points. Seeding a mesh forces a mesh pattern by specifying entities that must be used by the automesher. It is common to create points on edges or associated to surfaces in such a manner that smaller, low aspect ratio elements are created. This method should be employed near contact regoins or in anticipated areas of high strain gradients such as sharp internal corners or notches.

Manual Meshing of Solid Models

Occasionally, when the pros of manual meshing outweigh the cons, these techniques should indeed be used. Nonlinear analysis and the study of thin-walled solids are two of the biggest reasons to investigate manual meshing techniques. Remember, the longer the run times or the more times a single mesh is going to be solved, the more there is to gain from building a more efficient model.

Methods for Meshing Solids

There are several ways to build a brick mesh in most FEA preprocessors. You will rarely be forced to pick the eight nodes of a solid brick element. Most efficient methods for building brick element models involve some means of extruding or revolving a surface or surface mesh. The options for performing these tasks are reviewed below. The first task in building a manual mesh is planning: you should be able to visualize a strategy for developing the model before proceeding. Look for patterns and interactions between features. If you cannot visualize the completed model, a manual mesh may be beyond your skills or it may simply be too time consuming to fit into the product design process.

Extruding and Revolving Surface Meshes

One of the first considerations in planning an extrusion or revolution is the quality of the seed surface mesh. If this mesh contains poorly shaped quadrilateral or triangular elements, the resulting mesh will inherit these problems. If the manual mesh was planned for increased accuracy, building the mesh with bad elements somewhat defeats that purpose. Upon completing the extrusion or revolution, you will probably be given the option to delete the seed mesh. There are only two reasons not to go ahead and delete the surface mesh. First, if you are planning to use this surface mesh to extrude or revolve in the opposite direction to complete the mesh, leave it and delete it on the final operation. Second, retain the seed mesh if these shell elements are needed to provide load continuity to the solid mesh from an attached shell portion of the model.

One technique for extruding elements involves converting a 3D surface mesh into a thin-walled solid. This might be done when an initial shell mesh is found to be inadequate for the goals of the study. The nonplanar shell elements are extruded along their element normals to provide a solid mesh of a prescribed thickness. To be used successfully, your preprocessor must provide the capability to adjust the extrusion to maintain element continuity. Fig. 7.13 shows a shell model of a rubber boot (a) that might be better represented as a solid mesh. An important first step in using this technique is to verify that all surface element normals are pointed in the direction you intend to extrude the solids. Fig. 7.13(b) shows the corrected shell model with the normals pointing out. The mesh in (c) is the final solid mesh developed by extruding the boot surface into three elements through the thickness.

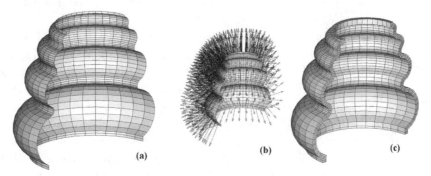

Fig. 7.13. Shell model of rubber boot (a), corrected shell model with normals all pointing out (b), and final solid mesh (c).

One more item to consider when planning revolutions of an existing surface mesh is that an element should not touch the axis of revolution at only one point. Such element will revolve into a degenerate element. Some preprocessors foresee this problem and will break a triangle being revolved into equivalent tetrahedrons. Most of the time, however, it is up to you to control the accuracy and continuity of a manual mesh.

Automeshing Solids

If solid automeshing is expected to be a large part of your design analysis challenges, you should first review the geometry techniques discussed in Chapter 5 before proceeding with the automesh. Developing clean models means taking known problems with the geometry into account. Build your geometry within the guidelines specified in that chapter and you will be ahead of the game. You should always plan to use parabolic tetrahedrons when you expect the results to be meaningful. These will make the model more resource intensive, but can provide accuracy on the same order as a brick mesh. Linear tets can be used to debug boundary conditions and for trend analyses in which the actual loads are unknown. If the magnitude is unknown but kept relatively consistent, a linear tet mesh can provide fast correlation checking.

Element Quality Issues

Three factors control the quality of an automeshed solid model: cleanliness of the geometry, mesh refinement for transitions in density, and your software's ability to correct poorly shaped elements on the fly. The first two are totally under your control, and although the third is not, the meshing algorithm will function more smoothly if the first two are optimal. If your geometry contains many sliver surfaces and short edges, and/or you did not take the time to apply local mesh control to surfaces and curves in order for the mesh to transition gradually around smaller or more important features, distorted tetrahedrons in the model are likely. Tet distortion can take a couple of forms. The most common is a flat element in which the fourth corner node is nearly in the same plane as the other three nodes. In addition to flat tetrahedrons, you may encounter many of the element distortion problems listed in the section on shell modeling. High aspect ratio elements and skewed elements can have a significant impact on the model's local accuracy. Check for such elements prior to and after the solve in the areas of interest to ensure that the results are of the highest quality.

Another factor that affects the accuracy of a second order tet model is the position of the mid-side node. Ideally, the mid-side node is equally spaced on an edge from the two corners. You typically do not have much control over the spacing because it is a function of your mesher's robustness. You may have control over the final positioning of the mid-side nodes with respect to the actual geometric template. Assuming you have the option, projecting the mid-side nodes on the surface of the part will provide better results than mid-side nodes geometrically positioned between the corners. Fig. 7.14 illustrates the difference. In the mesh section of 7.14(a), the mid-side nodes are on the linear edge of the element, creating a chordal representation of the curvature and not fully taking advantage of the second order edge's ability to better map to the part. Because stress results are derived from calculated strain, the closer your mesh is initially representative of the actual part, the more precise the strain calculations. You may also run into problems where the mesher treats the relatively sharp transition between two elements in 7.14(a) as a discontinuity and calculates high stresses at the "corner." In the 7.14(b) mesh section, the mid-side nodes are on the curvature of the geometry. You can see that this mesh almost completely captures the actual shape of the arc and will provide much better stress results if the fillet is subjected to high loads.

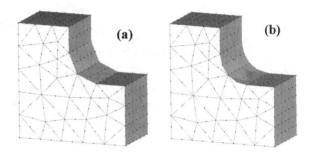

Fig. 7.14. Mesh sections illustrating mid-side nodes on linear edge of element (a) and mid-side nodes on the curvature of geometry (b).

Meshing a Solid Model

As stated previously, one control you have over the quality of an automesh is your ability to force a smoother transition from small features or areas requiring a tighter mesh. The ability to specify local mesh refinement is critical to the automeshing process. Without it, you are forced to continuously reduce the nominal mesh size across the entire part to obtain convergence. If there are regions of localized stress, the model will become prohibitively large very quickly. However, with the use of local mesh refinement, you can pack as many nodes as you need in a critical area and transition to an element size that provides the correct overall model stiffness without burdening the mesh with excessive degrees of freedom.

Local mesh refinement can be achieved by specifying a mesh size on surfaces or curves, or at points. In addition, in most preprocessors hard mesh points can be projected onto the surface of the solid to provide highly localized control. If curves or surfaces are not appropriately placed on the outer faces of the solid, build surface patches as described in Chapter 5 to provide the necessary mesh control.

The common practice of starting the solid automeshing process by meshing the part's outer surfaces with triangles is beneficial for two reasons. First, most automeshers start with a surface mesh in any event, and executing this process in advance will give you confidence that the final solid mesh will succeed. Second, the initial surface mesh will provide a sneak preview of what the solid mesh is going to look like, at least on the outside. You can adjust the mesh control on individual solid faces until the mesh meets your requirements, and then kick off the solid tets. This process is highly recommended when you are trying to calculate precise stresses in a model.

Debugging a Failed Automesh

Despite diligence in geometry building and mesh control, you can expect that an occasional part will not mesh on the first pass. Hopefully, these are few and far between. Most preprocessors provide information on the surface or curve responsible for a failed automesh. If you have no way to locate the offending area and automeshing is your only option on a part, you are dead in the water. That is one of the fundamental problems with the latest batch of CAD embedded FEA tools that hide the meshing from the user. If every part always meshed perfectly, there would be no problem. But, it is a near certainty that if you are actively building models, you will encounter a part that will not mesh. In a more full-featured preprocessor, the steps for debugging a failed run involve investigating the reported problem area, making geometry corrections, and then meshing the surrounding surfaces to observe how the model is constructed. Once the surface mesh completes to your satisfaction, try the solid mesh again.

If it still fails, there may be a geometry definition problem that causes the automesher to get confused, such as a surface that folds back on itself or a high order NURBS surface with highly localized, yet imperceptible, fluctuations in tangency. When you encounter these situations, your most expedient solution may be to simply mesh around the offending surface and manually stitch triangular elements across the areas that cannot be resolved by the automesher. Most full-featured preprocessors provide the capability to fill the volume inside a surface mesh of triangles with tetrahedrons. In fact, the surface meshing (paving) is the most time-consuming part of the automesh process. You will find that the volume enclosed by the triangular mesh fills extremely quickly in most codes. If your preprocessor does not support these advanced meshing tools, you may want to consider making small adjustments to the features local to the problem area in hopes that the model will correct itself with the right combination of geometric factors. After going through this once, you may wish to purchase a better preprocessor. Some of the open pre/post processors described in Chapter 19 are relatively inexpensive and will pay for themselves in a single project if you are not able to mesh a critical part.

Transition Elements

Many problems will involve more than one element type and the transition between the dissimilar elements must be handled carefully to ensure proper load transmission. In addition, you may have occasion to tie meshes of the same element type together when they do not have a common interface mesh. To make these connections, transition elements are often used to simplify the process. Some codes have developed transition elements specific to respective solvers and cannot be used in other systems, such as "links" in Pro/MECHANICA and wedges with rotational degrees of freedom in ANSYS. If your solver supports additional types of transition elements, you should study their capabilities in your code's documentation and practice using them with test models. However, the work-horse transition element available in most codes is the *rigid element*.

Rigid Elements

Rigid elements come in different configurations with different names. The second most commonly used name for these elements is *multi-point constraint* (MPC). MPC is probably a better term because a rigid element is not really an element. The best way to think of MPCs is that they are a mathematical means for coupling one or more degrees of freedom between two or more nodes in a model. The degree of freedom of one node is constrained to a degree of freedom of another; hence the term multi-point constraint. An MPC must have one master (independent) node, and one or more slave (dependent) nodes. A simple MPC configuration might attach all the nodes on one flat surface to a parallel flat surface. One of the surfaces must be designated as the master surface and the other the slave. If the only dependent degree of freedom on these MPCs is the translation normal to the two surfaces, the slave surface will be constrained to maintain its orientation and distance from the master, whatever happens to the master. Although the slave surface will be allowed to slide parallel to the master, it will behave as if it is attached, much like with a magnet. The simple model in Fig. 7.15 illustrates this concept. The member on the left was designated as the master and loaded per the vector shown. MPCs were inserted between the adjacent nodes at the top of each member and actually bridged the small gap. The only dependent degree of freedom was normal to the interface edge. You can see how the slave member follows the master by maintaining the parallel orientation of the two edges and a fixed distance.

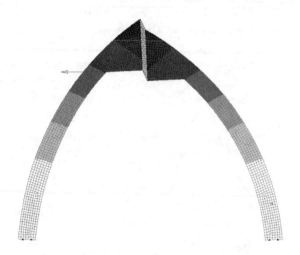

Fig. 7.15. Model illustrating slave surface allowed to slide parallel to master.

Another use for MPCs is in the transition between dissimilar elements. As stated in previous chapters, attaching a shell edge to a solid edge results in a hinged interface unless something is done to couple the rotations at the interface. MPCs serve this purpose well. The simple test model in Fig. 7.16 shows the proper way to tie a shell mesh to a solid mesh with MPCs. The nodal connectivity couples the relative translations of the two meshes. However, the rotations at these nodes must be tied to the relative translation of at least two other nodes on the solid with an MPC to prevent accidental alignment of the dependent nodes which might provide hinging.

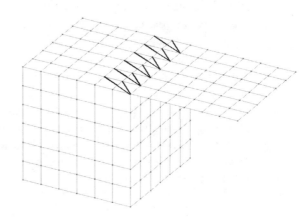

Fig. 7.16. Illustration of proper way to tie shell mesh to solid mesh with MPCs.

Constraint Equations

Constraint equations are similar to MPCs in that they define the motion of one node with respect to another. The difference lies in the constraint equation's ability to make the relationship more complex. Instead of simply a "where I go, you go" relationship between two nodes, the slave node can be instructed, for example, to move two units for every one unit the master node moves. Most constraint equations in design analysis can be accomplished with MPCs and well-positioned coordinate systems. However, constraint equations have proven to be handy tools in some industries. You should at least be familiar with their existence in the event you need a complex interaction defined between two or more nodes.

Summary

This chapter only briefly touched on the many capabilities and variations available in finite element modeling. The choices are endless and many new techniques are developed on the fly as new modeling types and challenges are explored. Whether you primarily automesh or spend most of your time manually meshing models, there are still many choices you must make that will affect the ultimate accuracy of the solution. Remember to check and update convergence by using local mesh refinement and making intelligent decisions on element usage. Plan a mesh before beginning your work on same. With proper planning, you can avoid many obstacles and minimize the shortcuts toward the end when things just do not quite line up. The importance of understanding the accuracy of your tools cannot be overemphasized. An automesh supported by validation data can withstand the scrutiny of even the pickiest specialist if correctly executed. Choose the most efficient and effective way to model based on the goals of the problem. Take each case as it comes and do not fall into the automesh rut. Learn to recognize which situations require more complex meshing and do not be afraid to go there. You have control of the accuracy of your solution and must stand behind it. Do not let procrastination or neglect be the reason for incorrect data. In short, make every mesh count.

8

Boundary Conditions

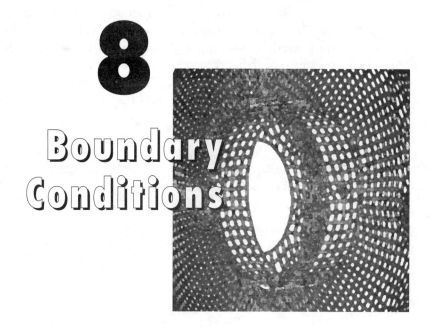

In FEA, the name of the game is "boundary conditions," that is, calculating the loads and constraints that each component or system of components experiences in its working environment. In other words, boundary conditions are those conditions that you must place on a model to represent everything about the system that you have not modeled.

For example, consider a motorcycle wheel. You may have the most sophisticated FEM of this component. You can shade it, rotate it on the screen, and impress the world with the images your workstation provides. Then you can grab an area at the bottom of the rim and apply some "intuitive" loads while constraining the hub. Press a button and show your audience the beautiful stress-contour images that result from the FEA that just ran in the background. You are now dubbed an engineer. What is the problem with this scenario? The flashiness of today's technology can be mesmerizing. It can be very useful as much as it can be misleading. How did you obtain the loads? Did you consider the tire/wheel interface? Did you treat the tire as a simple constraint on vehicle motion rather than a force and moment producer at its contact patch? Did you fully constrain the hub or accommodate for the load transfer characteristics of the bearings? What about braking loads transferred by the attached rotor? Driving loads by the sprocket? The interface between these last two components and the hub? As in any system, the old adage still applies: "Garbage in, garbage out." The difference is that FEA makes garbage look very beautiful and convincing. But if one does not understand and represent boundary conditions accurately, these attractive results have no real meaning.

Fig. 8.1. FEA of motorcycle wheel. (Image courtesy of Buell Motorcycle Company.)

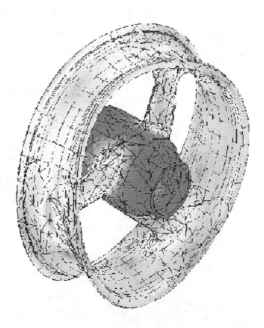

Hence, you must always make the time to justify the boundary conditions used in your model as thoroughly and convincingly as possible. If you cannot persuade your peers of the appropriateness of your chosen loads and constraints, the results of your model do not have a chance of being accepted by them. Of course, you can always opt to simply present your findings without a preface. Yet, the practice of displaying results with no discussion of the boundary conditions used in their development is a risky business, and is definitely not recommended. An elegant derivation of boundary conditions is often more impressive than the results of the corresponding analysis. Challenge yourself to come up with smart and efficient ways to capture the manner in which your models are subject to their respective environments.

What Are Boundary Conditions?

As mentioned above, whether analyzing a component or system of components, the boundary conditions in a model must represent everything in the operating environment that is not explicitly modeled. Moreover, boundary conditions must never restrict or allow deformations that would not be restricted or allowed by the unmodeled parts they represent. To help you digest these extremely important concepts, this section of the chapter will provide an overview of boundary conditions–the big picture, if you will. The remaining sections will then take each concept and present it in more detail. You are strongly encouraged to peruse all sections, because you can never really know too much about boundary conditions.

A Simple Example...

Consider a simple analysis of a chair subject to a load centrally located on its seat. This load might be a rigid crate of auto parts or the intended user of the chair. Assume that the width of the crate and the person are similar so that the load may be applied over the same area. In the first pass at this study, a uniformly distributed load is placed on a 3D shell model of the seat, and the regions where the legs meet the underside of the chair are rigidly fixed. For this model, both input and restraints on the system are modeled with boundary conditions: the weight of the crate or person is input as a load, and the seat's interface regions with the legs are restrained with constraints. The results of this simple model are shown in Fig. 8.2. Some of the assumptions inherent in this boundary condition scheme are summarized below.

- The legs are rigid in compression and do not add any substantial component to the vertical deformation.

- The legs are rigid in bending and force the chair bottom, local to the interface, to remain perfectly horizontal.

- Any sliding of the legs on the floor due to side load components resulting from the seat bending will be neglected; legs are bolted to the floor or friction is sufficient to resist side loading.

- The load can be modeled as being uniformly distributed both at the instant of its application and any time thereafter.

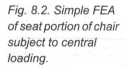

*Fig. 8.2. Simple FEA
of seat portion of chair
subject to central
loading.*

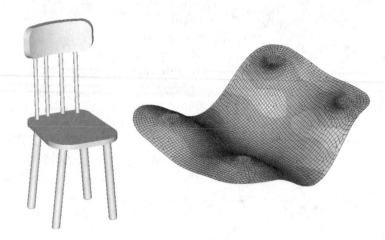

General engineering judgment as well as personal experience with
chairs should suggest that some of these assumptions are difficult to justify. First, a constant curvature of the chair seat under load is intuitively
more realistic than the one shown in the figure, where no rotation at
the constraint regions is allowed. Second, horizontal displacements at
the top of each leg can be expected simply due to shear effects. Third, a
horizontal displacement due to the bending of the seat should be significant. And last, the stiffness of the legs, or lack thereof, both in compression and in bending, may contribute substantially to the final
deformed shape.

For these reasons, the next pass at the analysis might warrant the actual
modeling of the legs and moving the constraints to their "feet." Note
that in this second study boundary conditions model the same input as
before, but now the constraints model the floor. In the case that stresses
or frictional effects at the floor must be considered, or the load distribution due to the actual input geometry calls for a more detailed definition, a third study could be developed. This study would utilize contact
interfaces between the seat and the modeled input geometry and
between the legs' feet and the modeled floor. The new, more complex
boundary conditions of this FEM would be the foundation under the
floor and an acceleration force due to gravity. In either the second or
third studies, the deformed shape resulting from the more relaxed constraints may be an order of magnitude greater than that exhibited by
the first study. This can be attributed to the fictitious rigidity that is
caused by too much constraint. Make a note of this important concept.

As you can see in this example as well as in the study of the motorcycle wheel mentioned earlier, the choice of where to stop modeling is related to the goals of the analysis and your ability to define boundary conditions that accurately represent the world beyond that stopping point. Most finite element codes are displacement based. Therefore, all results of interest are extracted from the calculated displacement or strain. Your best tool for evaluating the accuracy of an FEA is to review the resulting displacements in light of the environment you represent with boundary conditions. If the model does not look quite right using this measure, it is fair to say that all other results data will be suspect as well.

Types of Boundary Conditions

Boundary conditions are applied as *constraints* and *loads.* Typically, loads are used to represent inputs to the system of interest. These can be in the form of forces, moments, pressures, temperatures, or accelerations. Constraints, on the other hand, are typically used as reactions to the applied loads. Constraints can resist translational or rotational deformation induced by these loads, although this will be dependent upon the choice of elements in the model. Sometimes, however, using constraints to impart the load input to the model is convenient, such as in the case of applying known deflections. On other occasions, upon inputting actual reaction forces, the loads will be used to place the system in equilibrium to minimize local rigidity that might be imparted by constraints. A hand-held device has no real ground point or fixed area. Consequently, it may be more realistic to calculate a balanced force distribution, perhaps by using a free body diagram, and employing constraints only to stabilize the model, rather than to react any input.

Boundary conditions can reference a coordinate system other than the system-defined world or global coordinate system, which is usually Cartesian. They can be defined relative to cylindrical or spherical coordinate systems if either frame of reference is more accurate or if it allows for simpler construction of the desired loads and constraints. In a linear static analysis, the boundary conditions must be assumed constant from application to final deformation of the system. In a dynamic analysis, the boundary conditions can vary with time and, in a nonlinear analysis, the orientation and distribution of the boundary conditions can vary as the displacement of the structure is calculated.

Spatial versus Elemental Degrees of Freedom

An essential factor in the evaluation of a boundary condition scheme is the choice of loads or constraints in light of the elements used in the model or those elements used locally at the application regions themselves. This choice is always governed by the available degrees of freedom (DOFs) in these elements. The concept of DOFs is used in two ways in finite element modeling. First, spatial DOFs refer to the three translational and three rotational modes of displacement that are possible for any part or system in 3D space. While this will be discussed in more detail later, a constraint scheme must remove all six DOFs for the analysis to run. These may be removed with a variety of combinations of translational or rotational constraints. Any spatial DOF left unconstrained will cause the solution to fail due to what is best described as an unsolvable system. Refer to Chapter 3 and the simple two-element finite element model shown in Fig. 3.2. If the constraint of *U0* is not included in the model, the spring would simply translate axially indefinitely.

The second meaning of degrees of freedom refers to elemental DOFs, or the ability of each element to transmit or react to a load. This was described in the previous two chapters in great detail. What is pertinent here is that a boundary condition cannot load or constrain a DOF that is not supported by the element to which it is applied. While your particular preprocessor might make it possible to apply unsupported boundary conditions, the solver will not recognize them; the solution will proceed as if they were not present at all. To avoid possible confusion and miscommunication, it is strongly recommended that you do not declare unsupported boundary conditions in your model.

Boundary Conditions and Accuracy

As stated in the previous example, the choice of boundary conditions has a direct impact on the overall accuracy of the model. The global, or overall, stiffness of the system must be modeled correctly for any of the more local behaviors to be correct. Two of the most unwanted FEA effects to watch out for are the overstiffening and/or understiffening of a model by your choice of boundary conditions. The first of these two is usually the most encountered, because, unlike their real world equivalents, constraints in FEA land are perfect. Hence, avoiding the introduction of an additional level of rigidity around constraint regions is tricky. An overly stiff model due to poorly applied constraints is typically called *overconstrained*. Overconstrained models can be created by using redundant constraints, using excessive constraints, or by failing to observe the strain coupling in a 2D or 3D model. An understiffened model can be the result of two boundary condition scenarios. The first is that of an *underconstrained* model, which simply has too few constraints to prevent rigid body motion. The second is actually caused by the use of loads, which impart no stiffness in FEA, to represent a structure that in actuality does impart stiffness to the system under evaluation. Both overstiffening and understiffening can have unpredictable effects in the final results of your analysis.

Overconstrained Models

The accuracy of the previous chair example was shown to be dependent on the level of rigidity assumed for the unmodeled components—the legs. While the legs might be somewhat more rigid than the chair seat, very small horizontal displacements may have a major impact on the total deformation of the chair. The relative rigidity of the parts being modeled using boundary conditions will determine if a particular degree of freedom should be constrained. Failure to consider the validity of constraints will cause an overconstrained model for one or more reasons. The following discussion describes general conditions to be aware of, although more complex conditions in real world parts will require an even more detailed study of the local and global behavior caused by constraints.

Redundant Constraints

Due to the presence of both spatial and elemental DOFs in an FEM, you must always be aware of redundant constraints. In many cases, redundant constraints will have no effect on the overall behavior of the model. For example, if all nodes of a shell element are constrained in its normal direction, constraining their rotations about either of its parallel axes would be redundant. In general, however, the application of redundant constraints suggests a poorly constructed constraint scheme. It is likely that redundant constraints will also be excessive and thus have a noticeable effect on the local and/or global stiffness of the structure.

Excessive Constraints

Excessive constraints result both from a poor understanding of the actual supporting structure being represented by them and insufficient planning of the total boundary condition scheme. Constraining the horizontal displacement of the chair seat at its interface with the legs is a classic example of excessive constraints. In many cases, the application of excessive constraints stems from the geometry easily accessed by the analyst. This is yet another reason why the analysis model should often be different from the design model. Resolving the excessive constraints of the simple chair seat model serves to illustrate these concepts.

Fig. 8.3. Revised FEM boundary condition scheme for chair seat.

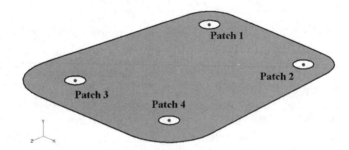

Fig. 8.3 shows a shell model of the chair seat with patches representing the tops of the legs. It may appear to be acceptable to constrain each circular patch in vertical translation while leaving the rotational DOFs unconstrained. However, a normal constraint on three points of a planar surface will eliminate any chance of rotation. Therefore, if you chose this constraint scheme, you unwittingly eliminated the rotational DOFs as well, and the seat will behave as if the leg-to-seat interfaces were fixed completely. A more realistic constraint scheme would be to pin the center point of each circular patch. If the points are fixed only with translational constraints, the entire patch will be allowed to rotate fully, as if on a ball joint. Each point should be fixed vertically, and horizontal constraints should be selectively applied so that in-plane spatial rotation and rigid body translation is removed without causing excessive constraints. Note that symmetry should suggest itself to you in this model, but that discussion is left for later in the chapter. A constraint scheme that satisfies the needs of this model consists of the following items.

- Constraining the center point of Patch 1 in all three translational DOFs.

- Constraining x and y translations of the center point of Patch 2.

- Constraining z and y translations of the center point of Patch 3.

- Constraining just the y translations of the center point of Patch 4.

Prove to yourself that this scheme succeeds in allowing in-plane translation induced by the bending of the seat without rigid body translation or rotation.

Coupled Strain Effects

Strain or material deformation in one direction is dependent on deformation, or the freedom to deform, in other directions. This coupled effect is governed by the Poisson's ratio of the material and must be considered in the application of constraints in shell, solid, or planar models. Beam elements cannot represent two- or three-dimensional coupling due to the one-dimensional nature of the element. The simple plate in tension shown in Fig. 8.4 illustrates the coupling of the in-plane strain very clearly. As the loaded edge is pulled to the right, the top edge wants to contract so that the plate thins as it elongates. The constraint scheme used in Fig. 8.4(a) fixes only the x DOF on the left vertical edge and the y DOF on the bottom edge. The resulting deformation is the natural deformation of the system and will result in a uniform stress distribution equal to the applied load divided by the cross-sectional area of the plate. For comparison, Fig. 8.4(b) has both the x and y DOFs constrained on all nodes of the left vertical edge. Note how this constraint scheme restricts the coupled strain in the plate, a phenomenon known as *Poisson's effect*. It is important to note that this vertical restriction on the left edge actually reduces the horizontal deformation by 5%. In a more complex model, the effects of overconstraining the coupled strain can be even more significant.

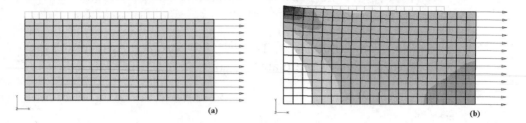

Fig. 8.4. Plane stress model of plate with coupled strain unrestricted (a) and with coupled strain restricted (b).

As another example, consider the analysis of a joint member at the end of a long shaft, which is in pure tension, Fig. 8.5. Due to the uniaxial loading in this part, the shaft can be cut close to this end to reduce the overall model size. The behavior at the cut surface can then be modeled by applying a properly distributed reaction load or by allowing the reaction to develop with a uniaxial constraint on the cut surface. This example will take the second approach, as shown in Fig. 8.5(a).

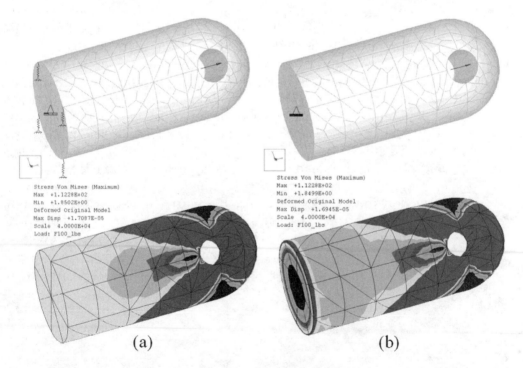

Fig. 8.5. FEM and results of joint member at end of long shaft in tension, using uniaxial constraint (a) and full constraint at the modeling cut (b).

Note that while there are no loads parallel to the cut plane, any small inaccuracies in the run due to numerical round-off will send the model translating and/or rotating into space along the cut plane (in the unconstrained directions). To deal with this in the previous example, the bottom edge of the plate was constrained, which caused the deformation to contract toward it. There are many ways to numerically stabilize a model without introducing Poisson effects. In the current example, a constraint scheme is to be chosen such that the resulting deformations resemble reality; that is, the shaft contracts toward its center axis. One way of achieving this consists of constraining a node precisely located at the center of the cross section in both translational directions parallel to the cut plane. A second node must then be constrained to prevent rotation about the axis of the part. This node should be on one of the principal axes of the cross section, and the constraint should be normal to the plane defined by the chosen principal axis and the part's longitudinal axis. If nodes are not located in these precise locations, an alternative is to use two uniaxial soft springs to anchor the cut to ground. Their location and stiffnesses are chosen so that they weakly limit tangential translations that are orthogonal to one another. Because the expected value of these translations is zero, the springs should only serve to stabilize the run.

For this example, if you did not necessarily care about the resulting mode of deformation of the shaft material (contracting toward its center axis), you would only need two appropriately located, orthogonal springs along the perimeter of the cut to stabilize the run. However, since the requirement was made above as to the desired deformation mode, you need four springs to keep the shaft centered. This is the technique used in Fig. 8.5(a). Note that as expected, the results of this FEA show a constant stress level at and near the cut surface. Had you applied a full constraint instead, Fig. 8.5(b), a Poisson effect would cause a fictitious stress gradient at the cut, where the material is now unable to "suck in." The absence of this Poisson effect stress in an over-constrained FEA model should not be attributed to "beginner's luck." It is more likely an indicator of poor convergence in that area. Chapter 10 will discuss this in more detail.

Understiffened Models

Understiffened models can result in levels of inaccuracy of equal magnitude to those of overly stiff models. However, the cause and symptoms are quite different. While diligently avoiding overconstraint of your model, keep these points about understiffened conditions in mind.

Underconstrained Models

The most common underconstrained modeling errors stem from neglecting one or more spatial degrees of freedom. If rigid body translation or rotation is allowed in your model, the solution will most likely fail. If this occurs, you will have to interpret the error messages to determine which DOF was unconstrained and then develop a solution that does not overconstrain the system. While FEA run errors of the "insufficiently constrained" type usually point to missing or incorrectly applied constraints, there are times when such errors refer to free nodes in the model, cracks in the model, excessive displacements, or abrupt stiffness changes that make the model behave like a mechanism. If a model is truly insufficiently constrained, a failed run is actually preferable: if the model is stable but poorly constrained, the error may not be noticeable in the final results. Again, careful results interpretation is critical.

Insufficient Part Stiffness

In the effort to develop a more elegant model, it is common to try to replace as many assembly components as possible with boundary conditions. In fact, this process is even recommended in Chapter 4 in the discussion of component contribution analyses. It is possible, however, to overdo it. Many parts are stiffened considerably by attached components, even if they are not rigidly attached in all directions. Because loads impart no stiffness in a linear analysis, replacing an attached component with an equivalent load could allow the modeled part to have much greater flexibility than it should. There will probably be no error message indicating that the part has insufficient stiffness. The FE solver has no way of knowing that you did not mean to analyze the model in the way you did so. The only sure way to catch these errors is to examine displacements of the model local to any boundary condition that replaces a system component. Ask yourself if the part that was not modeled could really allow the deformation that you are seeing. If it cannot, it may be necessary to adjust the boundary conditions or to model all or a portion of the component in question to ensure that the stiffening effects are accounted for.

Local versus Global Accuracy

In the chair example described earlier, the intermediate constraint scheme of Fig. 8.3—used to eliminate the excessive constraints of the first study—would have certainly improved the accuracy of the chair seat deflection. Yet, the stress and deformation local to the constraints would be suspect. The interface between the top of the leg and mounting features on the chair seat may be too complex to capture with the simple constraint scheme offered. Many times, you will not be interested in results in the close vicinity of these regions, and general constraints will prove adequate. However, if you are interested in local results, you will likely need to model more of the system around these regions. Moreover, you may have to use more involved modeling techniques, such as contact.

One of the fortunate characteristics of finite element modeling is that local effects will most likely have little effect on global behavior. Consequently, in the interest of speed, it is often advantageous to use coarse constraints or loads far from the area of interest. While convenient and acceptable in many cases, this technique should not be used casually. It is your responsibility to determine how far is far enough when relating the boundary condition application to the area of interest. In many cases, the exact location and size of this area is not known until the first few models have been solved. Therefore, it is not uncommon to spend more time on boundary condition creation in the initial models and then back off on the complexity when it is determined that effects local to the boundary conditions do not affect the critical behavior elsewhere in the model.

The opposite is also true. You may learn that high stresses are present near boundary conditions, and you must determine whether these are fictitious effects due to the modeling techniques, such as point loads, point constraints, or Poisson effects. Once again, the only way to make this determination may be to use more complex modeling methods that better represent the real interaction of the modeled part with its surroundings.

Singularities

While it is common practice to utilize point constraints or loads, the fact of the matter is that these still represent unrealistic conditions. Any two elastic bodies in contact, even a needle on a steel plate, will interact over a finite area. Local deformation on the steel plate will cause spreading of the concentrated load. Even a needle point, when observed under a microscope, shows a finite tip radius. However, in FEA, point loads and point constraints are just that, point phenomena. Consequently, the applied load or reaction force is interpreted as a finite force over an infinitely small area. A point or node in a FEM has no area. The local stress, as calculated by *force/area*, is, for all practical purposes, infinity. A "divide-by-zero" condition such as this is often called a *singularity*. In nonadaptive technologies using a linear solution, such as the more common h-element codes, singularities will dissipate quickly over several elements and the global response will not be affected. Meaningless, high local stress values will be calculated based on the mesh density and material stiffness in the surrounding area. These can be ignored as fictitious, or as known inaccuracies. Consider the example shown in Fig. 8.6. The plate is loaded with a distributed load on the right side and has a point load of equal magnitude on the left. The stress field around the point load distributes itself well before the center of the plate is reached.

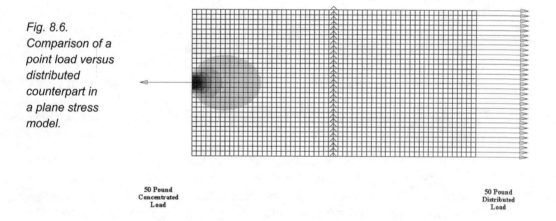

*Fig. 8.6.
Comparison of a
point load versus
distributed
counterpart in
a plane stress
model.*

50 Pound
Concentrated
Load

50 Pound
Distributed
Load

The definition of complex loads should take place only when you are interested in results at or in close proximity to the load application region. If your analysis is more concerned with locations far removed from these regions, simpler loads, including those causing singularities, may be utilized. The ability to use singular boundary conditions stems from St. Venant's Principle, which states that stress and deflection far from an applied load can be represented by a statically equivalent loading scenario. While this greatly simplifies the boundary condition application of some systems, the nagging question of "how far is far enough?" must still be resolved. Because there are no hard and fast rules for determining this distance, singularities should be avoided if there is any doubt about their validity. In addition, adaptive technologies such as p-element solvers will try to converge, or adjust the edge definition, local to a singularity in an attempt to determine the actual stress at the point. Due to the infinite nature of the singularity, the p-element solver will never converge in this area but may solve many more iterations than necessary in such a futile attempt. It is nearly always worthwhile in p-element solvers to put more thought into boundary conditions so that singularities are avoided. In p-element models, the balance must always be made between the user time spent setting up the model with more complex load definition and the computer time spent solving what might be the less efficient but simpler model. Again, the use of test models proves invaluable. You can also use Table 8.1 to assist in identifying possible singular boundary conditions:

Table 8.1. Boundary condition singularities for each element type (Yes entry indicates singular condition)

Load or constraint on	Beam	Shell	Solid
Point	No	Yes	Yes
Curve or edge	No	No	Yes
Surface or face	N/A	No	No

A few additional comments on singularities are warranted. First of all, singular conditions will cause high local stresses that may obscure or confuse results interpretation. Most postprocessors report the maximum and minimum values of each output type in a summary report. If the maximum stress stems from a singularity, the initial perception of the stress state may be unrealistically severe. In addition, most postprocessors will scale the color results contours based on the maximum and minimum values in the model. When singular stresses are present, the color legend can be so skewed to these magnitudes that the real stress distribution cannot be seen. Fortunately, most postprocessors also allow the color legend to be scaled to some user-defined minimum and maximum so the results of interest can be viewed more easily. However, this forces another task into the FEA process that might have been avoided by more thoughtful boundary conditions. The up side to this is that a skewed results legend may be the first indicator of an unexpected singular condition. The high stresses should be located and qualified quickly. If you feel that the singular condition does not affect the results of interest elsewhere in the model, it can be ignored. If there is some doubt, however, the boundary conditions should be adjusted to remedy the situation.

Another comment regarding singularities relates to the constraints that might be needed to ensure numerical stability in a model. By now, you have probably questioned the recommended boundary condition schemes for the example of Fig. 8.5(a), since all of these involved constraints on points. However, because the expected response in the constraint direction at these points is zero, singularities should not develop. You may actually use this fact to your advantage; if singularities arise, then the point constraints are most likely not accomplishing what you intended.

The last note on this topic involves nonlinear analyses. Note that the use of singular loads and constraints was qualified as being acceptable in linear analyses. A nonlinear material analysis will suffer from a similar problem as the adaptive linear solution. High local stresses result from high local strains. A nonlinear material analysis will try to resolve the stiffness at a singularity by adjusting the material modulus as defined by its stress-strain relationship. Fictitiously high strains may cause additional nonlinear iterations that should not have been required by the true strain state of the model. Due to the length of nonlinear solutions, every attempt should be made to avoid singularities that might unnecessarily extend the solution time.

Bracketing Boundary Conditions

It is often difficult or even impossible to accurately define the boundary conditions on a model because the actual operating conditions are unknown or vary considerably depending on the use or user, or because interactions between parts are difficult to quantify. Frictional effects and modeling joints in assemblies constitute many of these uncertainties. Determining the minimum constraint set that will make the model stable while still allowing realistic displacements is good practice. Once the optimum minimal set is established, additional constraints should be applied sparingly. This process will result in the most flexible system allowed without rigid body motion. However, you may decide that the actual system could easily be more rigid with a different interpretation of the entire operating environment, due to effects that are hard to quantify. Similarly, loads may vary due to operating specifications or tolerances. The actual loads may be difficult to measure in the actual operating environment. When the optimum or correct boundary condition scheme is hard to model or determine, consider *bracketing* the system with conditions that take into account the various options you are considering. If the results from these cases are similar, use the case which is easier to apply or more computationally efficient. If they are different, the differences must be scrutinized to determine which one models the behavior more realistically.

To illustrate the concept of bracketing, consider the now familiar chair model, shown again in Fig. 8.7. Because the legs of the chair are not perpendicular to ground, friction will have an effect in keeping them from spreading out. A full contact model using friction would prove very resource intensive, assuming your software even provides the capability. A different approach would be to run two different analyses using standard constraints. The first assumes infinite friction [Fig. 8.7(a)], while the second assumes no friction [Fig. 8.7(b)]. If the solutions look very similar, use the one that proves more efficient. If the solutions look very different, you must determine which is more appropriate through testing or other analysis methods. Of course, if an expedient solution is required, and the solution varies not so much in the distribution of the stress field but in its magnitude, you could choose to be conservative by using the analysis that provides the highest stresses.

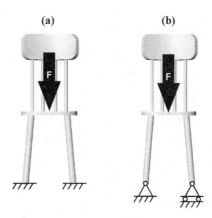

Fig. 8.7. Bracketing the boundary conditions of a chair analysis, using (a) infinite friction and (b) no friction.

Review of General Boundary Condition Concepts

Preparing Geometry for Boundary Conditions

The concept of load or constraint patches was discussed in Chapters 4 and 5. Essentially, these patches constitute boundary condition application areas, that is, the interface with unmodeled parts. For example, dividing the walls of cylindrical holes or protrusions into two or more sections provides logical regions for load or constraint application. Spending time setting up an initial mesh by choosing local element sizes and working around any geometric inconsistencies proves to be discouraging when you find that the boundary conditions require a missing patch or some other feature for proper boundary condition application. Resist the temptation to use the best distribution allowed by the premature mesh to proceed with the analysis. Proper planning at the beginning of the study should allow you to make the necessary adjustments in the geometry stage so that this obstacle can be avoided.

Applying Boundary Conditions to Geometry

Many preprocessors allow you to use the underlying geometry to assign boundary conditions to a mesh. Historically, an analyst had to resort to manual, time- and patience-consuming methods to locate all nodes or elements on surfaces or edges in order to apply the desired loads and constraints. Nodes and elements had to be carefully numbered to allow for sequential selection by ID. Alternatively, multiple screen picks in a graphical interface were required to build a selection set. Adding to and removing from this set involved the manipulation of several model views in order to better access entities. By comparison, in the present state of FEA, boundary conditions should be applied to the underlying geometry whenever possible. First, it is much easier to grab a surface or curve than to pick nodes on the surface or curve. Second, a preprocessor that allows geometric application of boundary conditions should correctly distribute the boundary conditions to the proper nodes. This is extremely important for nodal loading, as will be discussed later.

One of the less tangible benefits of using underlying geometry is simple model manipulation speed. An extremely large model will slow down even high-end workstations with excellent graphics cards. Once the mesh exists, it may take two or three times longer to spin the model around to apply the boundary conditions to the appropriate nodes or elements. If the boundary conditions were applied to the underlying geometry before the mesh was created, this time-consuming manipulation would not be required. Another benefit of applying boundary conditions to geometry prior to meshing is that the need for patches or feature adjustment will be highlighted earlier in the process. Consequently, you could save minutes or hours, depending on the time required to build or clean up a mesh.

Using Load and Constraint Sets

All FEA codes require that boundary conditions be defined in sets. Unique load and constraint sets can be used to differentiate independent scenarios that represent separate tests or operating conditions. If you think that the results of independent conditions–loads or constraints–may be important, it is in your best interest to use multiple sets. A good example is a massive structure on which the effects of gravity are significant. While the worst case scenario is likely to involve the operating loads in addition to the dead weight of the system, differentiating results that are related more to gravity than to the operating environment may make structure improvement more straightforward. If multiple boundary condition sets or cases are submitted to the solver, your postprocessor should provide some means for combining the results linearly in various ratios so that many conditions can be evaluated from only a few actual solutions. Similarly, it is often possible to combine sets of boundary conditions to create a third set within the preprocessor. In these systems, it is much easier to combine component boundary conditions than to recreate or delete boundary conditions from a complex set. If your system does not allow copying and combining boundary condition sets, you may wish to be more judicious about how finely you break up the applied loads and constraints.

If multiple load and constraint sets are used, care must be taken that the proper set be active or accounted for when additional boundary conditions are added. Similarly, if a change in the general boundary condition scheme is suggested by a peer review of preliminary results, be sure to incorporate the change in all sets that will be used in the final studies. Logical numbering and naming of the sets can only help keep the data organized while the project is underway and will make it much easier for someone who must use the data downstream to decipher what was done originally.

Coordinate Systems

A very powerful tool for the application of boundary conditions is the ability to define user coordinate systems (UCSs). It is often difficult, if not impossible, to define the orientation of a load vector or a constraint in terms of the system's global coordinate system. Placing a UCS whose orientation may be aligned with these boundary conditions proves to be much more efficient. Fig. 8.8 shows the most common UCSs available in FEA. Cartesian coordinate systems with various orientations are the most common type of UCS, cylindrical coordinate systems are the next most popular, and spherical coordinate systems have limited use. When employing the various types of UCSs, remember that the six DOFs must be interpreted with respect to the active coordinate system. In general, the three translational DOFs are labeled T1, T2, and T3, and the three rotational DOFs are labeled R1, R2, and R3. In many systems, the DOFs are also labeled simply by numbers as DOF 1 through 6, respectively. The equivalent DOFs in the three common coordinate systems are shown in Table 8.2.

Fig. 8.8. User coordinate system types supported by most analysis codes.

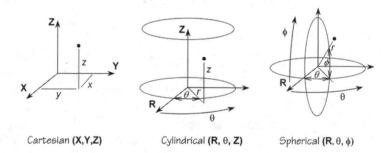

Cartesian **(X,Y,Z)**   Cylindrical **(R, θ, Z)**   Spherical **(R, θ, φ)**

Table 8.2. Equivalent degrees of freedom in common coordinate system types

	Degree of freedom	Cartesian	Cylindrical	Spherical
1	T1	TX	TR	TR
2	T2	TY	Tθ	Tθ
3	T3	TZ	TZ	Tφ
4	R1	RX	RR	RR
5	R2	RY	Rθ	Rθ
6	R3	RZ	RZ	Rφ

Always evaluate your model and determine the most appropriate UCS types and orientations before beginning a study. Review geometry, loading, and constraints, as well as required output data. Remember that UCSs are "free"–they do not add to either file size or run time, yet make setting up the problem significantly easier.

As an example, consider Fig. 8.9, a plane stress model of a lifting lever, which is free to pivot about a pin through its mounting hole. Note that the material around the hole may not displace radially into the pin. Hence, the outline curve of the hole must be properly constrained. Attempting to define this type of constraint in terms of the global Cartesian coordinate system–even if the lever's pin hole was centered at the origin of the system–would prove to be very difficult indeed. Do not fall into the trap of attempting to fix the translations on the outline curve and free rotations with respect to a Cartesian z axis coaxial to the center of the hole. This cannot work because as the lever pivots, each point on the curve wants to translate in both x and y about the hole's center, or in translations that have been constrained. In general, rotational constraints should be reserved for shell edges or beam nodes. Nodal rotational constraints only fix the local rotation in the specified DOF, not the global. Although some special applications of rotational constraints on shell faces will be mentioned later, their general usefulness is limited.

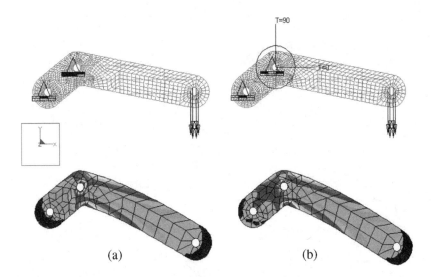

Fig. 8.9. Plane stress FEA results of lifting lever using "incorrect" Cartesian constraint (a) and "correct" cylindrical constraint (b).

Next, assume that by placing a cylindrical UCS at this center point, you simply need to constrain motion of the curve in the radial direction to achieve the desired pivoting effect. Each point on the curve is then free to move tangentially. Of course, as will be discussed in more detail later in the chapter, if you care about results around the constraint region, you must ensure that such action simulates reality. Suffice it to say for now that, unlike a radial constraint, a pin cannot pull the contacting lever material toward itself; in reality, this material is free to pull away. Note in Fig. 8.9 the differences in the results between the "incorrect" Cartesian constraint and its "correct" cylindrical counterpart. You should become very familiar with cylindrical coordinate systems–most rotating components, such as the motorcycle wheel mentioned at the beginning of this chapter, require that boundary conditions be applied both in radial and tangential directions.

One last note on UCSs: they are attached to ground. Although geometry will deform under loading and displace from its original location, the UCSs will stay anchored to the global coordinate system. They will not track geometry that was used to define respective locations at zero deformation. Because these deformations will generally be very small, any errors introduced resulting from the same should be negligible. Nonetheless, it is a good idea to keep this in mind upon reviewing results.

Constraints

Constraints in FEA serve to remove spatial degrees of freedom (DOFs) from the model. In general, FEA models must somehow be constrained in all six spatial DOFs to remove the possibility of rigid body motion (displacement without structural strain). As mentioned previously, a constraint on a node of a DOF that is not supported by an element attached to that node will have no effect. Consequently, you must be aware of the element degrees of freedom to correctly constrain the spatial degrees of freedom.

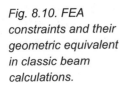

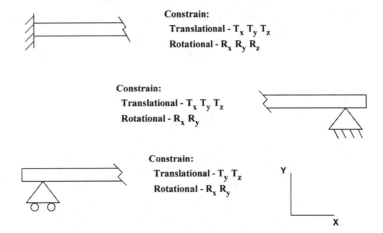

Fig. 8.10. FEA constraints and their geometric equivalent in classic beam calculations.

Constrain:
Translational - T_x T_y T_z
Rotational - R_x R_y R_z

Constrain:
Translational - T_x T_y T_z
Rotational - R_x R_y

Constrain:
Translational - T_y T_z
Rotational - R_x R_y

Constraints on Different Element Types

When trying to understand the DOF requirements of different element types, it helps to visualize the actual and understood geometry of the element and the terms required to fully stabilize it. For example, consider that the single end point of a beam element describes an area only in theory. Because a couple is produced at the end of an actual beam by opposing loads on two noncoincident points, a rotational term, or DOF, is required in its elemental counterpart to allow the generation of equivalent moments at its end point. Therefore, when fixing the end of a beam completely, all six translational and rotational DOFs must be constrained. Similarly, a shell edge has no physical area and can only resist or impart moments tangent to it with a rotational term. Hence, to completely constrain a shell edge, all DOFs except for the two rotations normal to it must be constrained. When dealing with more than a single shell element, it is impractical, if not impossible, to exclude these normal rotational constraints using a consistent coordinate system orientation. Hence, although redundant for some of the elements, these constraints should generally be applied as well.

Unlike beams and shells, a solid face should always have at least three points in contact with the rest of the structure, and therefore, is stable simply with translational terms. A solid element should never be constrained by less than three points and only translational DOFs must be fixed. Most solvers will simply ignore rotational constraints on a solid so that you can choose to leave them on or off, depending on the default settings of your preprocessor. However, as mentioned previously, you are strongly advised not to define constraints unsupported by the corresponding elements, because they may confuse your boundary condition intent upon review of the model. For additional information on element properties, refer to Chapters 4 and 6.

Conditional Equilibrium

A mention must be made of the analysis of components that exist in a state of conditionally stable equilibrium. These components are placed in equilibrium not by the system constraints, but by the magnitudes of the loads present. A numerically stable model must not allow rigid body motion for any magnitude of the applied loads. Although there are many systems that require a force balanced system, numerical approximations in even the most diligently balanced solution will generally render the analysis unstable and result in an "insufficiently constrained" error. To get around this, properly positioned point constraints, symmetry conditions, or "soft" springs to ground may be put in place to mathematically stabilize the analysis. You should become familiar with these techniques. Tie your model down to ground at points where you expect corresponding directional displacements to be zero. Then, if the model is not in the equilibrium state you expected, finite stresses at constraint points or large displacements of the stabilizing springs will flag this in the results. This technique was briefly introduced and utilized in the example of Fig. 8.5(a). Throughout the rest of the chapter, additional examples will further highlight its use.

> ⟿ **NOTE:** *MSC/NASTRAN allows the definition of an "inertia relief" constraint, which eliminates the need for utilizing modeling tricks on models that are in conditional equilibrium. This constraint provides stability by internally calculating and applying an acceleration based on system mass to counteract any unconstrained DOFs as specified.*

Symmetry

The use of symmetry is a very efficient means of simplifying a model, both in terms of size and run time. If geometry and boundary conditions are, or can be approximated as being, identical across one, two, or three axes or planes, you should use symmetry to simplify the model.

Refer to Chapter 4 for background on symmetry modeling types. To create symmetry using boundary conditions, you must use loads and constraints to account for the symmetric portion that you are omitting from the model. When using this type of modeling, always watch out for the Poisson effects as discussed in earlier in this chapter. Note that due to the possibility of asymmetric vibration modes in a symmetric model, symmetry boundary conditions should never be used in a modal analysis. For the same reason as well as other less obvious concerns regarding the effective size of the structure in the model, symmetry should also be avoided in a buckling analysis.

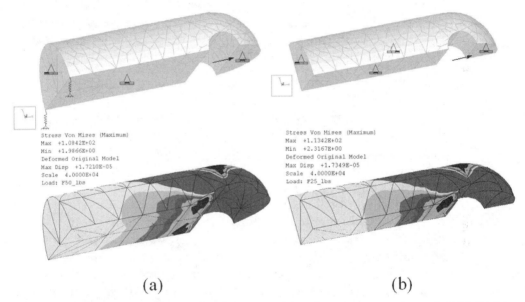

(a)   (b)

Fig. 8.11. FEMs and results of half-symmetry (a) and quarter-symmetry models (b) of joint member presented in Fig. 8.5.

As an example of symmetry, revisit the model of Fig. 8.5. This model could be simplified as a half- or even a quarter-symmetry model. Both are shown in Fig. 8.11. In fact, if the original model were that of a short, symmetric shaft in tension, and the cut plane of example 8.5 were a bisector, these would be quarter and eighth-symmetry models instead. Again, be careful to prevent the constraints from adding a Poisson effect at the cut planes. The material on these planes is simply straining in equal and opposite directions normal to them. There should be no strain along or through these planes; if there is, the model's boundary conditions cannot be interpreted by symmetry. Therefore, uniaxial constraints applied to the normal direction of the cut plane should be sufficient. This holds true for solid models and planar approximations such as plane stress, plane strain, and axisymmetric models. Observe how results of the two symmetry models are almost identical to one another and to those of Fig. 8.5(a). The quarter symmetry run was fastest and converged to the tightest level, so that its results should be considered the most accurate. Note also that only one uniaxial soft spring is needed for the half-symmetry model, and no springs are required for its quarter-symmetry equivalent. (Two springs were used in this half-symmetry example to let the model contract toward the center axis.) For each symmetry plane you create, you effectively eliminate an undesired DOF. You should now be able to recognize that there are many advantages to symmetry modeling.

Symmetry models of elements that use mathematical representations (or DOFs) for their missing spatial dimensions, such as shell and beam models, must be treated a little differently. The boundary conditions at the cut planes of these models must account for their rotational DOFs in addition to translational. For example, since the geometry of a 3D shell element is missing a thickness dimension, the shell is able to hinge about its edges unless properly constrained. Fig. 8.12 shows how an additional rotational constraint must be placed on the shell edge to define a proper symmetry model. In general, rotations about both axes that lie on the symmetry plane must be constrained in shell models in addition to the translations normal to the symmetry plane. Following a similar line of thinking, it is easily demonstrated that all rotational DOFs of a beam model must also be eliminated at the symmetry plane.

Even if a model is not exactly symmetric, symmetry can be used to develop a better understanding of the system. Choosing the weaker half of a nearly symmetric system constitutes a worst case model and can provide design data on an initial concept. Remember that because of its ability to reduce size, complexity, and spatial DOFs, symmetry modeling is often the best way to minimize run time and avoid errors such as false constraints.

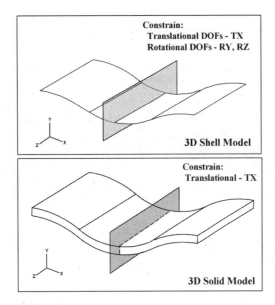

Fig. 8.12. Equivalent symmetry constraints for solid and shell models.

Enforced Displacement

A very useful constraint type available in most FEA packages is known as an *enforced displacement*. This constraint allows the definition of the location where nodes or elements will end up at the end of the analysis. Remember that an enforced displacement constraint moves a node *and* constrains the node at that adjusted position. Recall also that an enforced displacement always overcomes material stiffnesses and applied loads. The reaction force at an enforced constraint should be checked after an analysis to ensure that the forces required to effect this displacement are realistic. A basic example of using enforced displacements involves a fastener interface with an aluminum block. If you knew that the washer under the head of the fastener displaced 0.005" at its torque specification, you could define an enforced displacement constraint on the elements corresponding to the underside of the washer surface. Fig. 8.13 shows the FEM and corresponding results of such an analysis.

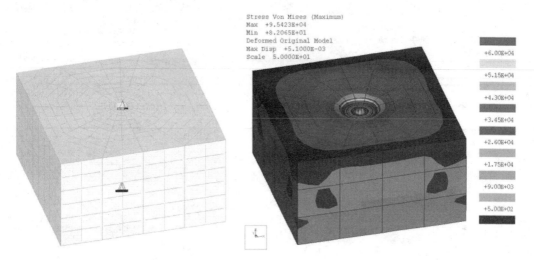

Stress Von Mises (Maximum)
Max +9.5423E+04
Min +8.2065E+01
Deformed Original Model
Max Disp +5.1000E-03
Scale 5.0000E+01

+6.00E+04

+5.15E+04

+4.30E+04

+3.45E+04

+2.60E+04

+1.75E+04

+9.00E+03

+5.00E+02

Fig. 8.13. Enforced displacement fastener interface.

Although not all codes permit it, an enforced displacement in terms of a coaxial cylindrical UCS is a very simple way to model the effects of press fits or internal pressure in a cylinder. However, you should be careful with how you define press fits. For the FEA to be correct, the enforced displacement must use the transition radius in its definition. This radius is the final assembly dimension that takes into account the deformation of both members in the press fit. Generally, you will not know what this transition radius will be ahead of time. If this is the case, you will be better off using thermal expansion or contact to carry out this type of analysis.

An additional note of caution: enforced displacements always make every single node or element involved in the constraint go where it is told. For example, consider the solid, quarter-symmetry model of a shaft clamp shown in Fig. 8.14. To simulate the rigid shaft within the clamp, the inner surface of the clamp is constrained radially using a cylindrical UCS. A surface patch equivalent to the washer diameter at the bolt through-hole location is then loaded with an enforced displacement parallel to the axis of the bolt. While one would expect that the top surface of the clamp would deform continuously, the results show that, as requested, the entire washer patch deforms parallel to its initial orientation. This creates a discontinuity in the top surface that produces unrealistic stresses local to the constraint and has a significant impact on the accuracy of the overall stress behavior. A better solution would be to use rigid links to tie a centrally located point to the washer patch and then to displace the point. This will allow the top surface to deform in a more natural way. An even more accurate solution would be to model the actual bolt head/washer geometry and use a contact interface between the bolt head and the clamp. Try these solutions and compare the results.

Be careful to distinguish between small versus large displacements. The latter fall in the nonlinear regime and the FEA code should be able to account for such boundary conditions. Nonlinearities in general require the use of detailed material properties, such as a stress-strain curve, tangent modulus, and so on. For models using shell elements, the anticipated or prescribed displacement should be much less than the thickness of the element.

Fig. 8.14. Symmetry FEM of shaft clamp with corresponding results.

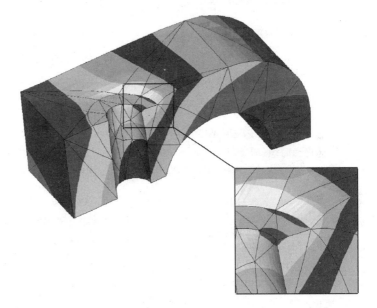

As always, simple test models go a long way toward ensuring that you understand techniques new to you. Test models should also be used to better understand new implementations of enforced displacement. Note also that an enforced displacement allows a static analysis to be solved without a load case. In fact, some codes require that an enforced displacement be input as a load.

Loads

Developing the loading for an FEA model is another critical aspect of building an accurate simulation. Any load application must consist of four components: (1) magnitude, (2) orientation, (3) distribution, and (4) time dependence. The orientation and distribution of the load are often easily understood, but its magnitude is hard to quantify. Sometimes the opposite is true. Determining the magnitude of the load in complex assemblies may involve iterative testing or the development of a dynamic or kinematic simulation model. A free body diagram is often the best means to determine the initial loading on a component or assembly. The FE analyst should feel very comfortable using free body diagrams; refer to Chapter 2 for a detailed review of this technique.

Units

In defining loads, you must always verify use of a set of units consistent with the rest of your model. It is not uncommon to encounter a model with metric geometry that has been loaded with forces in pounds. A mistakenly mixed set of units is very difficult to debug because nothing seems to be wrong with the model itself. Until the entire world finally agrees on a single standard, you must be diligent with the units you use. Even within a given system, be careful to avoid inconsistencies such as mixing inches with ft-lbs. Tables 8.3 and 8.4 should help you define and convert units between the two most popular unit schemes, inch-pound-seconds (I-P-S) and millimeter-Newton-seconds (mm-N-S).

Table 8.3. Consistent set of units for standard steel in I-P-S and mm-N-S unit schemes

Quantity	inch-pound-seconds	millimeter-Newton-seconds
Young's modulus	30×10^6 pounds per square inch (PSI)	2.07×10^5 megapascals (MPa)
Mass density	7.324×10^{-4} pounds mass per inch cubed (lb_m/in^3)	7.83×10^{-9} metric tonnes per millimeter cubed (mt/mm^3)
Poisson's ratio	0.30	0.30

Table 8.3. Consistent set of units for standard steel in I-P-S and mm-N-S unit schemes

Coefficient of thermal expansion	6.5×10^{-6} inch per inch per degrees Fahrenheit (in/in/°F)	11.7×10^{-6} millimeter/millimeter/ degrees Celsius (mm/mm/°C)
Thermal conductivity	2.071 British thermal units per hour-inch-degrees Fahrenheit [Btu/(hr-in-°F)]	0.043 watts per millimeter-degrees Celsius (W/mm-°C)

Table 8.4. Commonly used conversion factors

	Unit	Multiply by	To get
Young's modulus or stress	PSI	6.895×10^{-3}	MPa
Length	in	25.40	mm
Force or weight	lb_f	4.4484	N
Weight to mass	lb_f	1/386.4	lb_m
Weight to mass	N	1.02×10^{-4}	mt
Mass	lb_m	0.17524	mt
Mass density	lb_m/in^3	1.0694×10^{-5}	mt/mm^3
Density (SG)	specific gravity	9.344×10^{-5}	lb_m/in^3
Power	Btu/hr	0.2931	W
Temperature	°F	(°F-32)/1.80	°C
Coefficient of thermal expansion	in./in./°F	1.80	mm/mm/°C
Thermal conductivity	Btu/(hr-in.-°F)	6.411×10^{-3}	W/mm-°C
Convection coefficient	Btu/(hr.-in²-°F)	0.02077	W/(mm²-°C)

Consider bookmarking these tables. You will need to refer to this data on a regular basis if your analyses require conversion between units. Even if your company always uses consistent units, material properties may not always be supplied in the same units. Hence, conversion between units sets is generally inevitable.

Load Distribution

The distribution of an applied load can be defined in several ways: uniform, per unit length or area, interpolated, or functionally defined.

Interpolated loads vary according to specific points selected over a surface or edge. The interpolation fit used on these points may be linear or quadratic depending on the number of points defined and the capabilities of your preprocessor. Functional loads are based on user-defined equations and are defined in terms of system or user coordinates. Complex functions are able to simulate most measurable load distributions. The pressure on the walls of a water column is a good example of a coordinate system dependent load.

When considering distribution options, remember one of the initial guidelines for the application of boundary conditions. Boundary conditions should not cause or restrict deformations that are not caused or restricted by the parts they are representing. This is of great importance when the load is being applied near an area of interest, or when you do not know up front where the area(s) of interest will be.

The difference between point loads and distributed loads was discussed earlier. Another important concept in load distribution can be explained using the previous chair example. The assumption was made that the load would be distributed uniformly over the chair seat, regardless of whether the input was a person or a rigid crate. At the time, this assumption was not challenged but some additional thought is warranted now. If the input was indeed due to a person seated on the chair, it is fair to say that the flexibility of the interface between the person and the chair seat would follow the deformation of the seat and continue to apply the load for large amounts of seat bending. Would the same hold true if the input came from a rigid crate? More specifically, if the bottom face of the crate was at least as large as the rectangle defined by the interface of the four legs with the seat, would the applied load follow the bending of the seat and continue to impart a uniformly distributed force?

In reality, as soon as the seat began to deflect under the distributed load of the crate, the force distribution would redistribute to the supported portions of the seat. This phenomenon is called *bridging*. The load, just after its initial application, would be concentrated around the regions surrounding the legs. Because a static analysis assumes steady state conditions, a more realistic load distribution would be an interpolated load with a unity factor at the centers of the legs and factors tapering off to zero somewhere toward the center of the seat. The actual locations of these reduced interpolation factors may need to be developed using an iterative process, but the difference between this distribution and the first, uniformly distributed assumption may be significant enough that the extra work is usually warranted.

Load Orientation

In most cases, the orientation of an applied load will be defined by specifying the load components in the directions of the active coordinate system. As discussed in the section on user-defined coordinate systems, it is usually in your best interest to define a UCS with an axis aligned to the load orientation so that you are not forced to break it into components. If UCSs are not used, these vector components must be developed manually using trigonometric techniques. Loads can also be defined normal to an element's coordinate system in most preprocessors. Pressure loading is a special case of this practice.

Nonlinear Forces

In a linear analysis, the load magnitude, orientation, and distribution must remain constant throughout the deformation of the system. In most cases, this assumption is valid. If the surface or edge on which a load is applied deforms so much that an update to the load orientation is required, a nonlinear, large displacement analysis is probably warranted. Contact, discussed to varying extents earlier, is a nonlinear technique. This is due to the fact that contact reactions are not accounted for as the load is initially applied, but develop as the specified entities approach and deform into one another. Consequently, the distribution of forces in the model changes throughout the contact event.

Another type of nonlinear force is called a *follower force*. Follower forces are loads defined with respect to local nodes or elements, not a fixed coordinate system. As the part deforms locally, the load orientation changes. A simple example of this type of load is shown in Fig. 8.15. The beam in 8.15(a) has a fixed weight hanging from the end. The weight imparts a load of magnitude, W, to the beam end. The load is always oriented vertically downward. This is a linear load. The scenario depicted in 8.15(b) has a rocket engine welded to the end of the beam that imparts a thrust of W. As the beam deforms under the thrust of the rocket, the orientation of this thrust changes. Hence, this is a nonlinear event. Note the difference in the resultant deformations of the two loading scenarios. If follower forces are required, large displacement effects should also be solved for.

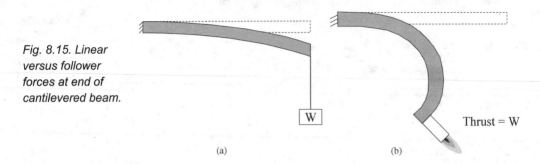

Fig. 8.15. Linear versus follower forces at end of cantilevered beam.

W

Thrust = W

(a)   (b)

Types of Loads

There is a greater variety of loads than constraints in FEA. They can be applied in the form of forces, moments, accelerations, temperatures, pressures, or periodic excitations. In most cases, the load type will be chosen to best match the actual applied load. Internal pressure in a manifold would be modeled with a pressure load. Reaction or spring forces would be modeled with force loads. An exception might be the use of a temperature load to induce thermal contraction or expansion of a component for the purpose of preloading a spring or bolt. Another exception might be the application of a pressure load to a circular patch with a variable diameter. While codes with built-in optimization or sensitivity routines will usually allow variable control of geometry, few, if any, will allow variable control of an applied load. Using the patch diameter as the variable with a pressure, or area dependent force, will effectively vary the applied load. This technique may also be used with moments, but beware of the elemental DOFs.

Forces and Moments

Forces, the most basic load entities, are defined in terms of the active coordinate system directions and may be applied to any element or geometric entity. Moments, on the other hand, require a bit more discussion. While their application in your FEM is very straightforward on shell elements and beams, applying a moment to a solid face is somewhat more involved. Remember that these entities do not have elemental rotational DOFs; that is, they cannot carry a moment. There are a few ways to get around this. The simplest is to place shells on those faces and apply the moment to the shells. A drawback to this approach is that you are overlapping material in your FEM. To avoid inducing a rigidity variance in the model, verify that the material properties assigned to the shells are the same as those of the solid, and that the shell thickness is small compared to the solid model size. Another way to approach this consists of defining two opposing levers of rigid elements or stiff beams that attach to the solid face. They may then be loaded with equal and opposite forces at each end so that the magnitude of the forces cancels out while a couple of the desired magnitude is applied. This technique is a little more involved in terms of the necessary addition of extraneous lever geometry, and reconciling the interface between this geometry and the face to avoid a false variance in model rigidity and/or stress risers.

> ◆ **NOTE:** *One additional technique for applying moments and forces to solid faces is unique to Pro/MECHANICA and involves the definition of a "total load applied at point" (TLAP). This load type allows the user to apply a load to a group of entities of the same type (solid faces or otherwise), and it statistically distributes the load according to the relative location of each entity with respect to the application point. To apply just a moment to a solid face, two imaginary levers with opposing forces at its ends can be simulated with the simple definition of two opposing TLAPs, each at one of the lever end point locations. It must be noted that a TLAP does not contribute rigidity to the model.*

Pressure Loads

Pressure loading is typically interpreted as normal to the surface or element face to which it is applied. Surface elements, or shells, have a defined inside and outside surface normal. A pressure load is set by default from the inside to the outside surface. Therefore, a positive pressure on a shell model may actually be interpreted as a vacuum if the orientation of the element normals is opposite of the expected. The orientation vectors displayed in a graphical preprocessor should always be checked when using pressure loads. In addition to orienting these loads, surface normals determine which side will carry the "top surface stresses" and "bottom surface stresses." Hence, when pressure is to be applied to a shelled surface, the surface or element normals must be consistently oriented for both load orientation and results interpretation. In the case of a pressure load applied to a solid face, the orientation is automatically correct because the normal vectors of all solid element faces are always directed away from its centroid, and a positive pressure will be oriented opposite this vector.

Pressure can also be applied to the edges of planar elements for a plane stress, plane strain, or an axisymmetric model. The orientation of a pressure load on these elements is similar to that of a solid. A positive pressure will be oriented toward the center of the planar element.

Acceleration Loads

Simulating a load resulting from the inertia of the component or system of components is common. Consider a model of a car seat structure and how you would simulate its response to the car's acceleration force. Of course, this acceleration affects the person sitting on the seat, "pushing" him or her back into it. You may be able to figure this load distribution and apply it directly to the seat model. Yet, what about the inertia of the seat itself? If the seat's weight magnitude and distribution are comparable to those of the person sitting on it, it would be incorrect to ignore its inertia effects.

Two ways of correctly simulating this scenario with FEA involve turning on an acceleration load of the same magnitude and direction as that experienced in the car. The first way consists of including the inertia of the person in your model. This could be achieved with either appropriate geometry and material properties or with appropriately located inertia elements. In either case, the connections between these inertia components and the structure must be well thought out and justified. The second way retains the load distribution of the person's body on the seat, which you might have already obtained. This load is then combined with the acceleration load.

Keep in mind that acceleration loads can be applied in any direction. You may usually ignore gravity or any other type of acceleration if the resulting inertia loads are small compared to the applied loads. Whenever in doubt, however, apply acceleration loading. Accelerations may be obtained from testing or a kinematic analysis. Note that a centrifugal acceleration load allows rotating systems to be evaluated statically. Note also that a dynamic base excitation system may be simulated by changing frames of reference and constraining the base of the system while modulating a gravity load.

Temperature Loads

Temperature loads may be used to evaluate the effects of thermal expansion. They are applied as a temperature rise over a reference temperature, usually ambient. Some common examples of phenomena that may be analyzed using this tool include the prestressing of material due to welding, press, or shrink fits, and cooling of injection molded parts. For a thorough review of this type of loading, refer to Chapter 14.

Checking Applied Loads

Once you have applied all loads to the model, check its total load resultant. Is it what you expected? This final check is quick and can flag some obvious errors before you actually run the model. A visual check of the load symbols in your preprocessor can also be helpful in identifying obvious mistakes. Recall as well that loading in FEA assumes a gradual application with normal strain rates. If you are attempting to simulate a high strain-rate event, such as impact, you should back up your analysis with a thorough testing program and consider a transient dynamic analysis.

Comparison of Boundary Condition Schemes

The following example will serve to illustrate many of the boundary condition concepts discussed in this chapter.

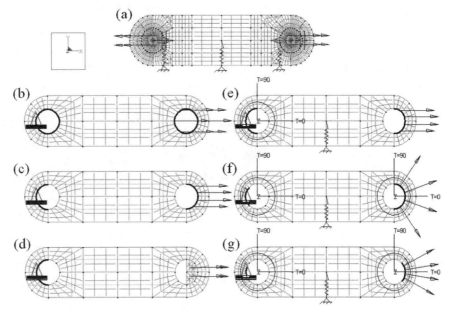

Fig. 8.16. Plane stress FEMs of suspension coupler link using different boundary

conditions.

Figs. 8.16 through 8.20 show a Pro/MECHANICA plane stress boundary condition study of a steel coupler link used in a rising rate suspension mechanism. This link has needle bearings at both ends and is always in tension. For simplicity, it is assumed that these bearings constitute a frictionless, fully contacting interface between each pin and the link. Fig. 8.16 presents each of the boundary condition sets used in the study. The design load is 500 lbs. Note that a half-symmetry or even a quarter-symmetry model could be used for this analysis. However, by modeling the full link, a discussion of other possible constraints can take place.

Fig. 8.16(a) consists of a contact model where the load is applied directly to the pins, which have a 0.001" clearance fit with the link at each end. This model can be used as the baseline, because it very closely represents the link system as assumed above. Its stress results are presented in Fig. 8.17. Note that the model has no constraints, since the opposing loads place the pins and link in equilibrium. However, "soft" springs to ground are put in place at the mid-section of the link and at points on the pins to mathematically stabilize the analysis. The spring on the link has low k_x and k_y spring stiffness values, while the springs on the pins only have a low k_y (zero k_x). For the remaining discussion of this example, the pins themselves will not be displayed in the contact results so that magnified deflections of the link may be inspected more clearly.

Fig. 8.17. Baseline: FEA von Mises stress results of Fig. 8.16(a). Deformation display scale factor = 1.

Fig. 8.16(b) probably constitutes the scenario with the least forethought. In it, the "bearing curve" at one end is fully constrained, while a uniform 500 lb load is placed on the curve at the other end. Observe in Fig. 8.18(b) how different the results are to the baseline, Fig. 8.18(a). The loaded end does not "egg" out as much as in the contact model. This is due to the fact that the load is perfectly distributed over the entire perimeter of the curve, instead of being focused on a small contact region. The constraint end shows a similar and even more pro-

nounced effect, as the curve has been made infinitely rigid, with infinite friction all around what would be the pin interface. Note that, unlike this load and constraint, in reality the pins cannot pull on the inside portion of the curves.

Acting on this last observation, Fig 8.16(c) shows a link where boundary conditions similar to (b) are placed just on the outside halves of the curves. Note in Fig. 8.18(c) how the stress results at each end more closely resemble those of the contact model. When attempting to use this technique, you should iteratively determine where the point of lost contact actually occurs so that the appropriate portion of the curve may be selected. This may be achieved by utilizing less or more of the curve until its displaced ends remain tangent to the rest of the curve. Do not forget that friction and rigidity issues must still be addressed.

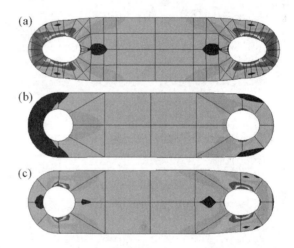

Fig. 8.18. FEA von Mises stress results of Figs. 8.16 (a), (b), and (c). Deformation display scale factor = 500.

(a)

(b)

(c)

The model in Fig. 8.16(d) demonstrates a technique often used in an attempt to simulate contact. Half of the bearing is modeled using a shared boundary curve with the link. The problem with this technique is that now both ends have infinite friction. Note that the stress results and deformations shown in Fig. 8.19(b) are actually farther from the baseline. Hence, Fig 8.16(e) tries to eliminate friction in the constraint end of 16(c) instead. It places a cylindrical coordinate system at the center of the missing pin and radially constrains the outer half-circle. Note the soft k_y spring for mathematical stability. As observed in Fig. 8.19(c), results at the constraint seem to be moving in the right direction.

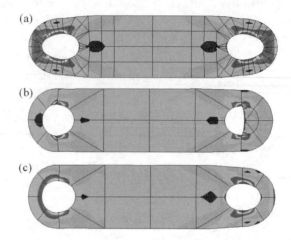

Fig. 8.19. FEA von Mises stress results of Figs. 8.16 (a), (d) and (e). Deformation display scale factor = 500.

Fig. 8.16(f) and (g) show models whose boundary conditions require the most forethought. First, the application of the constraint in (e) is modified in (f) to only include the outer quarter segment of the curve. This is done in attempt to satisfy more accurately the tangency condition as described above. At the other end, the definition of the load is a bit more involved. Because of the lack of friction in the contact model, it should be clear that an equivalent load must always be normal to the application surface (curve). Hence, another cylindrical coordinate system is placed at the load end of the link so that a radial load can be specified. This load's definition follows a textbook derivation used by some FEA codes for bearing simulation; an angle dependent sinusoidal function applied radially to the outer half segment of the hole region (in this case, the curve). It accounts for the reaction forces at the interface between each pin and the link, which must taper off to zero at the points where contact is lost, and be at a maximum at the mid-point of the contact region.

For the cylindrical coordinate system orientation shown in Fig. 8.16(f), the function in this model is equal to $636.6cos(\theta)$. In this definition, a load magnitude higher than 500 lbs is used to make up for the cancellation of the y components of the radial force. Its value is quickly determined by applying a unit load, checking the resultant due to it in the x direction, and using its reciprocal as a scale factor for the desired load. Note that stress and deformation results for this model, shown in Fig. 8.20(b), are thus far the closest to the baseline at both ends.

Fig.8.20. FEA von Mises stress results of Figs. 8.16 (a), (f), and (g). Deformation display scale factor = 500.

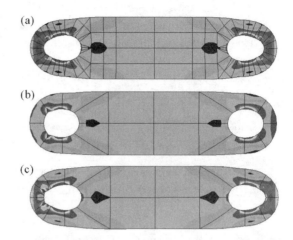

The final model, Fig. 8.16(g), simply demonstrates some fine tuning of (f). First, by inspecting the results of Fig. 8.20(b) at the constraint end, you should note that even with the use of a quarter curve segment, tangency has not been achieved. Again, the infinite rigidity of the constraint is at fault. Hence, an even smaller portion (an eighth) of the curve is constrained. Similarly, at the load end, the application region is made smaller. The size of this region is determined by a closer examination of the contact results, in which it appears that only a quarter segment of the pin's boundary curve is transmitting the load to the link. A more aggressive load is applied to this region as well in the form of a steep, quadratically distributed radial load, with zero values at the end points of the quarter segment and a maximum at its mid-point. In this case, using the same load factor calculation as outlined above, the magnitude necessary to achieve the 500 lb resultant in the x direction is 532.1 lbs. Note in Fig. 8.20(c) how similar the deformation and stress results at both ends of the link have become to the contact solution. Although there are still high stresses at the discontinuity caused by the rigidity of the constraint, these are very localized.

It is important to note that all assumptions made to achieve similarity with the contact results are unique to this model. A difference in material pairs, bearing fit dimensions, and/or link geometry might suggest a different boundary condition scheme or that more or less of the curve be utilized as an application region at either end of the link. Of course, the question to ask at this point is: "Why bother?" Why not simply run a contact analysis and get it over with? Well, for simple models this might be the solution–assuming that your FEA package does indeed provide the capability. Yet, for larger, more complex models, contact runs are very resource intensive, require more elements, and take longer to set up than their noncontact counterparts. Note the relatively large number of elements needed with contact in this simple link example to obtain good results. Hence, the use of a contact run and the additional setup and run times required should always be balanced against the goals of the analysis, the expected area of interest, and the time available to achieve a solution so that design decisions can be made.

It is quite enlightening to see the vast differences in results due to different modeling techniques in a simple two-force member. Do not be discouraged or intimidated. This example simply serves to highlight the importance of selecting a proper and efficient set of boundary conditions. For this, you will find test models to be very good friends. Try out your ideas in simple, quick running models before attempting to tackle the full model. If it is not effective or plainly impossible to model the actual boundary conditions, try to bracket them with a couple of model runs. For example, if a significant amount of friction exists in the coupler link's bearings, and your FEA package does not support friction, run the extremes: a frictionless contact model and a full friction general constraint model. The two will encapsulate the actual conditions you are trying to model and may be adequate to make design decisions about the structural soundness of the design.

One final comment on the above example: had you been interested only in the stresses at the mid-section of the link, note that any of the models would have been appropriate. Remember St. Venant's principle: Do not lose sight of the study's real goals.

Summary

At this juncture, you should be aware of the fact that boundary conditions are arguably the toughest aspect of FEA. Although creating geometry and developing a good mesh from the same are quite challenging, the real engineering comes into play as you decide how much of the system you are going to model and how you are going to represent what is left out in the model. Remember that the safety factor used in an analysis must directly correspond to how well you have modeled its boundary conditions. Do not fall into the trap of using small safety factors with boundary conditions that are not well understood and have not been properly verified. This admonition is especially relevant when you begin working on systems with which you have not had a lot of experience. Make it a habit to discuss boundary conditions with someone else in your company, preferably someone who has had experience, analytical or otherwise, with the system you are about to analyze. If nothing else, listening to yourself having to explain your choices of boundary conditions gives you a chance to review them outside of your own mind–it is always good to take a step back and verify that you have not lost perspective.

Once you have decided how much of the system you are going to model and the boundary conditions you will utilize, plan both the geometry and mesh of the model accordingly. Do not leave the definition of boundary conditions until the end. Trying to accommodate loads and constraints to geometry and elements that were not set up for them is very frustrating. If necessary, create load and constraint patches ahead of time and do not begin the mesh until you are somewhat certain that you have everything you need for boundary condition definition. It is even recommended, if your software allows, that you apply the boundary conditions to the geometry itself before you even begin the mesh. Being forced to redefine geometry as you try to place loads and constraints in a model is difficult enough without having to delete and recreate elements as well.

Of course, no discussion of any aspect of FEA is complete without mentioning correlation testing. As boundary conditions become more involved and you try to simulate more and more features of reality in a model, correlation with empirical work becomes increasingly important. Examine your results to ensure that they correspond to reality and pick points of interest that may be accurately measured in the lab. Proper boundary conditions can often be achieved only through an iterative process, whereby fine tuning of your FEM takes place until an adequate level of correlation is achieved. One notable aspect about achieving a good correlation with reality is that correlated sets of boundary conditions may then be more comfortably used in future analyses. Therefore, with every successful analysis you build an arsenal for yourself that will make you more efficient the next time around. In addition, by storing this information in the project or company "best practices" notebook, your peers will be more efficient as well.

In most full-featured FEA systems, the tools exist to define boundary conditions that truly mimic the operating environment. Your choice of boundary condition complexity will depend on the goals of the analysis, available time and computing resources, and the need for local versus global results. However, it cannot be overemphasized that these issues must be consciously considered. The definition of boundary conditions is arguably the most essential FEA assumption that is within your control, while it is also the most responsible for inaccurate FE results reported by both beginning and experienced users.

9

Solving the Model

Once the mesh is complete, and the properties and boundary conditions have been applied, it is time to solve the model for the results which will help you make design decisions. In most cases, this will be the point where you can take a deep breath, push a button and relax while the computer does the work for a change. Historically, this is also the point where you would walk away, go home for the evening, or get a cup of coffee. However, with the speed of today's solvers and computers, any break based on run time will be short. Solutions which would take a weekend just a few years ago may now solve in under an hour. This speed increase is one of the key reasons that FEA can play an active role in the design process.

What has not changed significantly are the steps to initiate a solution and to ensure that it runs to completion. While the specific tasks for each software package may be unique, some underlying principles are consistent and will be reviewed in this chapter. It is highly recommended that you do not take your solver for granted. Learn what options are available for improved accuracy and speed. Sometimes settings for a particular solution can be saved and reused, thereby spreading the time spent compiling them over many runs. Once you are comfortable with the basic steps to run an analysis, plan some time with your support provider to learn about additional available options.

Multiple Load and Constraint Cases

In Chapter 8, load and constraint cases are discussed as means to separate independent load inputs which can then be combined later in various ratios. Upon reaching the solution phase of your study, the decision must be made to solve for multiple cases in a single run or multiple runs. The decision must be based on two factors: system resources and your software's ability to work with output from different runs.

In most cases, submitting a run with multiple load cases will be faster than running sequential, complete solutions for each load case. This outcome results from the fact that the stiffness matrix must be built and conditioned only once if load cases are submitted in a single run. For a large model, this can be one of the most time-consuming portions of the analysis. Upon completion of the task, the various load cases are applied to it and separate solutions are calculated. The drawback here is that the scratch directories must be large enough to simultaneously hold temporary data from multiple solutions. Scratch directory size will not be an issue for smaller models, but if you are concerned about disk space for a single run of a large model, chances are your system cannot handle two or more load cases in a single run.

The alternative is to run each load case with a separate solution. This procedure allows the scratch data to be deleted and rebuilt for each run. System requirements should only be slightly greater than a single load case due to the storage needs of the final results. While total elapsed time will be greater when running sequentially, sequential runs may be your only option. Remember that the runs can be scheduled for overnight or an off-peak time. Most codes provide some sort of batching capability to allow unattended completion of multiple solutions. If not, a simple DOS or UNIX batch file will usually be sufficient.

The second consideration for the decision between a single run or sequential runs relates to your postprocessor's ability to work with results from multiple output sets. If your load cases are truly independent and mutually exclusive, or will never need to be combined, the choice will be based solely on resource requirements. However, in many cases, combinations of independent loads may be desired. This is the power of linear superposition. When this capability is required, you must know how your system works with the results data. The three primary options are summarized below.

- Option 1. Results from multiple load cases in a single run may be combined.

- Option 2. Results from multiple runs in the same database may be combined.

- Option 3. Results from the above two scenarios and from multiple databases may be combined.

If your postprocessor can only work with Option 1, your choice is made for you. If you need to combine multiple load cases, they must be in the same run. If you do not have the system resources to manage the combination of multiple load cases, you must resort to sequential runs with the loads combined in fixed ratios in single load cases. This procedure can be limiting if many combinations of independent loads are required.

While most systems provide the means to solve a single model with multiple load and/or constraint cases, results from solutions with multiple constraints are rarely combined. When they are combined, care must be taken to ensure that the combination is valid. While learning about your software and hardware capabilities, solve different constraint cases in separate runs.

Final Model Checks

Before initiating a solve, making a couple of quick checks of your model is recommended. The importance of these checks increases proportionally with the size of the model. On small, quick models, a mistake might not be too costly in terms of time. However, for a solve requiring several hours of run time, a poorly attached shell model or free node may run for half that time before an error surfaces.

Free Node Check

Most preprocessors give you the option to delete all nodes or all unattached nodes. Because you are typically prohibited from deleting a node used by an element due to database hierarchy issues, deleting *ALL NODES* is usually the fastest way to eliminate unwanted nodes. While some solvers have the ability to ignore or constrain out free nodes (e.g., *AUTOSPC* in MSC/NASTRAN), it is best in this and all cases to take control of your model and leave as little as possible up to chance. In a structural analysis, a free node will typically result in an insufficiently constrained error. In a modal analysis, free nodes will manifest themselves in a zero frequency mode. Regardless of analysis type, free nodes should be avoided.

Model Continuity Check

While free nodes will usually result in a failed solve or an obviously incorrect mode, some model discontinuities can result in a completed solution with erroneous results. This scenario is covered in more detail in Chapter 7 for shell models. Discontinuity is least likely in single volume, automeshed solids, and the probability of discontinuity increases as transitional and manual meshing are used with greater frequency. If you are lucky, a discontinuous mesh will result in a failed, insufficiently constrained solve. However, if the model is geometrically stable, it may solve but the load path will be incorrect, thereby yielding incorrect results.

First, merge or join all coincident nodes within a tolerance small enough not to affect correct elements. Use your judgment based on the smallest element edge or length. An exception is intentionally arranged, nearly coincident nodes across a contact pair. If such nodes are present, use your preprocessor's tools to pick around these areas.

Second, check boundary edges of shell models. Figs. 9.1 and 9.2 show a complete shell model of a motor end cap, and correct and incorrect boundary edge plots.

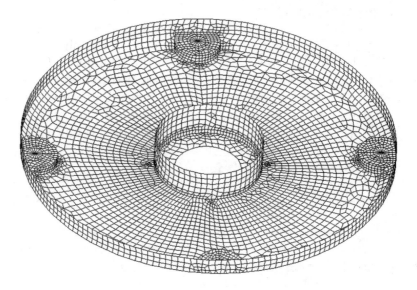

Fig. 9.1. Shell mesh of motor end cap.

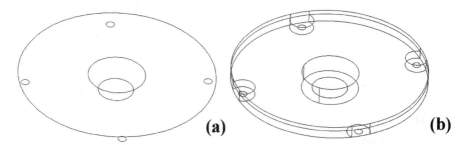

(a) **(b)**

Fig. 9.2. Correct free edge plot of motor end cap (a), and incorrect free edge plot of motor end cap showing cracks between sections of the mesh (b).

A correct boundary edge display should show the outline of the part and any holes intentionally added. Discrepancies in this check should be resolved before initiating a solve. A boundary face plot in a solid model may provide information on cracks. However, automeshed parts may be too complex to see anything meaningful in this view. A hex mesh on a p-element model is more conducive to a free face check.

Sanity Checks

Material Properties

A quick run through of the material properties assigned to a model is always helpful. Because the number of materials is typically small, this run should take little time. The most common mistake is the assignment of metric properties to an English model or vice versa. The resultant error in the solution may not be obvious. The best time to double check such properties is before the run.

Boundary Conditions

Check boundary conditions for three things. First, if multiple load and constraint sets have been defined, verify that the proper boundary conditions are in the proper set. Second, visually check the boundary condition icons. If possible, turn on label display of constrained DOFs and load magnitudes. A quick scan may point out errors made in haste. Finally, perform a load resultant check to provide the sum total of external loads. Compare this result to your free body diagram or initial loading estimates. If loads should zero out, as in the case of a pressurized vessel, the resultant should reflect the same. Most preprocessors will ask for a point about which to establish resultant moments. Unless moments are important, this point may be chosen arbitrarily. If they are important, the location of the point should be obvious.

Mass

In cases where gravity or acceleration loads are used, model mass is critical. If a dynamic solution is chosen, the total mass may be equally important. Most preprocessors can calculate the total mass of the mesh which can easily be converted into weight. A solid, automeshed model should reflect the actual mass of the system. A shell model may be off by 10 to 20% due to approximations inherent in the modeling method. Review the mass reported by the system with respect to your expectations. If there is a discrepancy, resolve it before starting the solution.

System Resources

RAM and disk requirements for a specific model size and type differ according to the solver and sometimes according to the operating system used. Your experience with previous models should provide guidance as to current needs. Documenting models' disk space and RAM requirements in a notebook or on-line scratch pad is always worthwhile, but especially in a multi-user environment. This database will help minimize the chance that a model will be attempted with insufficient resources. It may also provide useful data when a system upgrade is requested. The database should include the following items.

- Number of elements and type

- Number of nodes

- Number of degrees of freedom

- Type of solution

- Run time

- Disk usage

- RAM usage

- Machine name and type

- Machine RAM

Most of the above data can be obtained from the log or summary files created during a solution.

If you suspect that resources are marginal, you may have a couple of options. At least once in their careers, most experienced users have monitored disk space availability during a run, and scrambled to move, delete, or compress files to stay ahead of the solver. While exciting, such activity is not recommended. If your solver requires that data required for a restart be specified prior to the solve, make these adjustments. A restart will typically reuse much of the stiffness and other matrix formulations to allow for a faster second solution. Finally, if the disk space scramble mentioned earlier is even an option, do it before starting the solution. The likelihood of deleting something you will regret will decrease. Keeping your scratch disk uncluttered and defragmented is good practice because it optimizes disk space availability.

Element Check

While you should be checking element quality in the meshing process, a final check may reinforce the need to monitor areas of lower quality. This topic is covered in depth in Chapter 7.

When the Solution Fails

If everything went according to plan, you should be ready to jot down run statistics and move on to results interpretation, or the engineering part of the process. However, in many cases, despite diligence in model building, a run may fail. Unfortunately, most solvers still provide cryptic error messages which provide little help in solving the problem. Understanding the error messages that reflect insufficient disk space or RAM and insufficiently constrained models is recommended. These are the most common errors and there may be more than one error message to indicate each. While it is unrealistic to assume that you can learn all the diagnostic codes of complex solvers, noting run failure messages, causes of failures, and workarounds in the notebook or on-line scratch pad mentioned earlier in this chapter is recommended. A good relationship with a hotline support technician (get his/her direct dial number!) who seems to have all the answers is the next best defense. Selected workarounds and diagnostic techniques are described below.

Insufficient System Resources

Unfortunately, when a solution fails due to insufficient disk space or RAM, finding more resources is the best solution. When this is not an option, your only solution may be to reduce the model size. If this happens frequently, document the incidents and the time required to work around the deficiency and use such documentation as ammunition to get more RAM or disk space. Use the following techniques with caution to reduce model size.

- *Symmetry.* If symmetry was available but not used for various reasons, use it now. This is the safest way to significantly reduce model size.

- *Clean up questionable features.* Spending more time on geometry cleanup and remeshing may provide significant improvements in model size.

- *Use local refinement.* If this technique was not previously used, remesh the model with a coarser global mesh size with local refinement assigned to areas of interest. For many features that are outside the load path but are required for mass or general stiffness, you can use a coarser mesh with negligible effect on results. A solution for displacement may have a coarser mesh than a solution for stress.

- *Manual mesh.* This workaround should not be entered into casually, but a manual mesh can often be 30% smaller than an automesh. As mentioned in Chapter 7, one eight-noded brick element can provide the accuracy of three to five ten-noded tetrahedrons. However, the time and difficulty associated with manual meshes can make them prohibitive in a design engineering environment.

Whichever method you use to reduce model size, remember to evaluate the effect on solution accuracy. A simplification which compromises the desired result is rarely warranted. Document simplifications required as a result of resource deficiencies with appropriate notes on the expected effect on the results.

Insufficiently Constrained Models

Most causes of insufficiently constrained models should be caught in the pre-run checks described earlier in this chapter. However, if enthusiasm precluded diligence, perform such checks upon notification of solution failure due to poor constraints. If your solver indicates the nodes and degrees of freedom responsible for the error, focus your evaluation on that part of the model. If the quick checks do not work, the following workaround should point you toward a solution.

Solve Model for First Mode

The first mode of a structure is the shape of minimum stiffness. It is fair to say that if a portion of the model is unconstrained or poorly constrained, it will result in a zero or nearly zero mode. While this may be a time-consuming solve, it is usually sufficient to identify unattached portions of the mesh. After the modal solution, note the frequency calculated. If it is near zero, you have found the culprit. An exaggerated deformed animation will make the offending portion of the model jump out at you. When said portion has been identified and fixed, scan the model for similar occurrences of the modeling method responsible for the error.

A follow-up modal run is recommended after fixing the mistake. First, the follow-up run will verify that the fix was adequate. Second, when one discontinuity is found, multiple discontinuities are not unusual. Continue this process until the first mode is significant *and* as expected.

Abrupt Change in Stiffness

Another cause of an insufficiently constrained error is an abrupt change in stiffness. This can be illustrated by a thin plastic plate attached to a thick steel plate. Large local rotations may trigger a poorly constrained failure. This error is typically created in transitional meshing of dissimilar elements. While the large rotations may be caught by the modal diagnostic mentioned previously, examination of the model for areas of dissimilar materials or elements may be necessary.

Additional sources of failure may be the wrong element choice for a particular solution or poorly defined material models (e.g., omission of density when an acceleration load or dynamic solution is used). Beyond the failure sources mentioned here, a good relationship with your manuals, on-line help, and support organization is the best way to work around these problems.

Summary

An efficient solution of a finite element model is based on your solver's speed and configurability and your diligence in presenting the model in the best possible condition to the solver. Common sources of solution failure are free nodes, insufficient constraints, or insufficient system resources. Experience and patience will make recovering from these failures a minor task. Although tempting to believe, your responsibility does not lapse when the "run" button is pressed. Learn the intricacies of your solver to ensure that you are working productively and efficiently. Understanding your solver will allow you to progress to results interpretation more quickly.

10 Convergence

Results must always be reviewed in the proper context. As discussed in previous chapters, the assumption set the model was based on must be critiqued in light of the calculated data and the two must be reconciled. One significant assumption that can be addressed only after a run is the assumption that the mesh was sufficient to capture the behavior of interest. An initial mesh must typically be refined multiple times to accurately capture stress behavior. The process of successive mesh refinement to produce the optimal results is called *convergence*.

Importance of Convergence

How important is convergence? This question is best answered with the question, "How important are accurate results?" While sounding facetious, this is a very pertinent question. Before proceeding with this discussion, the working definition of accuracy developed in Chapter 3 should be reviewed. An accurate solution in FEA is the best solution to the geometry, properties, and boundary conditions presented. For a given set of properties, geometry, and boundary conditions, accuracy is controlled by the mesh. The accurate solution may be different from the correct solution, which represents the actual behavior of the physical part in testing or in the field. Qualifying the validity of the geometry, properties, and boundary conditions has been discussed in previous chapters. Thus, when considering convergence, the importance of accurate results is dependent on both the goals set at the beginning of the study and the degree of confidence you have in other assumptions. Converging a mesh to less than 5% change in results is rarely warranted when other key assumptions are potentially off by much more. On the other hand, some indicator of the accuracy of the model should be found in all cases so that the results, even in a gross behavior or trend analysis, can be properly interpreted. The least ambiguous indicator of accuracy is a comparison between results after one or more convergence runs. Single run error estimates are also discussed in this chapter.

Understanding Convergence

To fully appreciate the value of convergence, it is important to understand what is really happening to the model with successive mesh refinements. A trip back to high school calculus will provide an analogy which should clarify the process. Do not panic: there are no integrals to solve. However, the process used to prepare you for the concept of integrals will also shed light on mesh convergence. Integrals were introduced as a method for determining the area under a curve. Initially, estimates were made for the area under a curve by using a series of rectangles or polygons, as shown in Fig. 10.1.

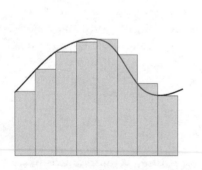

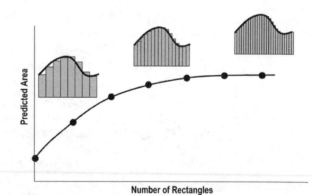

Fig. 10.1. Rectangle method for determining the area under a curve.

Fig. 10.2. Area estimate improves with the resolution of the rectangles.

Knowledge of the height and width of each rectangle enabled the calculation of its area, with the sum of the areas constituting an estimate for the total area under the curve. It is clear that the estimate for the area will improve as the number of rectangles increases. Fig. 10.2 shows this graphically.

Increasing the mesh density or the number of nodes in a given area produces a similar effect in a finite element model. If the curve in the previous figure represents the ideal deformed shape of a portion of a model, the images in Fig. 10.3 show a model's ability to capture that deformation with different mesh densities. As the number of nodes increases, the flexibility of the structure increases because it becomes less rigid. Consequently, for a given load, the model will undergo greater deformation or strain. In a linear analysis, stress is directly proportional to strain through Young's modulus (see Eq. 2.24). Hence, for a given load, stress will continue to rise as the mesh density increases, as the plot in Fig. 10.3 demonstrates, to a practical maximum where increasing the mesh density will have a negligible effect on the results. At this point, the model is considered converged. The degree of convergence can best be expressed by the percent change in the result of interest between the two most recent mesh refinements or convergence passes.

Fig. 10.3. As the number of nodes increases, the mesh can conform to the geometry more precisely.

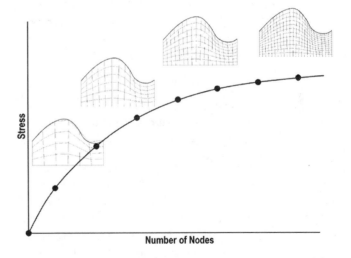

Historically, it was believed that the ideal model had just enough nodes or degrees of freedom to capture the convergence level desired without overmeshing, which would unnecessarily increase run times. The cost of overmeshing is becoming less of an issue as solvers and computers get faster. As mentioned previously, the time spent converging a frugal initial mesh may be more costly than the run time from a larger model. Today, the best definition of the ideal model is one that enables you to solve your problem in the time allotted. Remember, though, that even an excessive initial mesh may be underconverged. The only gain, in that case, will be in the number of convergence runs required. Regardless of the starting point, some understanding of the convergence level of the model is required to make intelligent design decisions.

How do you determine the correct convergence level? There are two schools of thought, but both approaches require quantification of the uncertainty in the model. Total uncertainty can be defined by Eq. 10.1.

$$Eq.\ 10.1 \qquad U_{Total} = U_{Geometry} + U_{Property} + U_{BC} + U_{Mesh}$$

One common position is that while the uncertainties driven by properties and boundary conditions are purely subjective and open to interpretation, the uncertainty contribution due to the mesh is totally under the analyst's control. Even the uncertainty due to geometry is not totally controllable by the analyst. In many cases, manufacturing variability, such as welding and certain sand casting processes, renders geometry capture a question of probabilities. With this in mind, some analysts believe that the mesh should be converged to the tightest levels that time and resources permit. This will reduce the controllable uncertainty to a minimum and theoretically narrow the spread of the solution extremes. No fault can be found with this approach, assuming the analyst has a realistic appreciation of the relative gains in solution quality versus cost in achieving this quality. Obviously, if the geometry, properties, and boundary conditions are well thought out and verified, much is to be gained from the additional convergence effort. This brings us to the second approach to convergence.

A case can be made to balance the uncertainties in the model against predetermined goals. Effort spent on resolving one uncertainty while the others are broadbased will not significantly enhance confidence in the results. This concept is grounded more in psychology than mathematics. While equations such as 10.1 are easy to toss out, quantifying the actual factors involved often ranges from impractical to impossible. Consequently, your confidence in the results and the design decisions based on them will be grounded more on engineering judgment as *any* uncertainty increases. Matching the convergence in one model to another with which it is being compared is probably more important than blindly converging every project to a tight value. This approach frees the analyst to spend time on convergence when it is warranted and relax when ballpark, gross behavior is sufficient.

It is also very important to remember that convergence is directly related to the load and constraints applied to a specific run. Consider the extreme case of a model with no load. Any mesh will be sufficient to capture the behavior or lack thereof. One of the dangers of using multiple loads and constraints on the same model is that convergence might be obtained for the first boundary condition scenario and then the same mesh will be used for all others. The analyst could be misled into thinking the other solutions are converged when the actual convergence levels are unknown. If you wish to use the same model for multiple load cases, converge it for all applied loads to ensure that the results are meaningful.

Uncertainty versus Error

More important than the actual convergence level is knowledge of the convergence level. When the uncertainty can be quantified, design decisions can be made to minimize the effects of uncertainty. The term "uncertainty" is preferable to the term "error" because error implies a mistake. Modeling a part at its nominal dimensions or choosing a material property based on a simple average does not constitute an error. The fact of the matter is that you cannot know the combination of parameters that go into every production part. While error suggests incorrectness, uncertainty properly positions the data as a product of probability and variability. Any FEA results that improve your understanding of the part or system under study are valuable and worthwhile. Correct known sources of error and then make decisions on inevitable levels of uncertainty.

Error Estimates

It is common and extremely valuable, nonetheless, to calculate error estimates on a single result case. Error estimates in FEA refer to the relative difference between results across an element edge or at a node. In an idealized continuum, the difference in any result quantity across an infinitely small volume would be infinitely small. As the sample volume or area increases, the relative change in the result quantity across its boundary will increase. As this change approaches a significant percentage of the local result magnitude, it can be said that the sample volume or element is too large. Error estimates attempt to quantify the convergence of a model by calculating the relative change in output between adjacent nodes or elements. Fig. 10.4 shows a simplistic nodal error estimate.

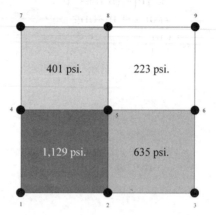

Fig. 10.4. Simple nodal error estimate.

The maximum stress on an element using Node 5 is 1,129 psi, and the minimum is 223 psi. Therefore, the reported error for Node 5 might be (1,129-223)/1,129, or 80.25%. FEA codes employ much more sophisticated error calculations that adjust the error based on the magnitude of the output and average over larger portions of the model. These adjustments allow you to use error estimates as a better gauge of convergence. A good example of the need for these adjustments can be illustrated with the simple error estimate method described previously. Fig. 10.5 shows possible stress scenarios.

Fig. 10.5. Two stress conditions with seemingly conflicting error estimates.

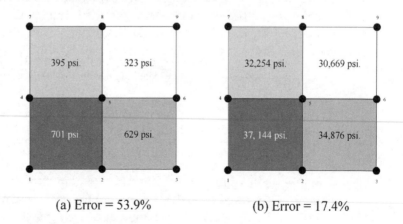

(a) Error = 53.9%   (b) Error = 17.4%

While the error estimate as defined earlier in case (b) is significantly lower than case (a), the local variation is a significant amount of stress. Convergence in regions of negligible stress is inconsequential and should be reported as such. The next two equations show more sophisticated error estimating algorithms. Eq. 10.2 is the normalized present maximum difference method, and Eq. 10.3, the probable error method. The *Max*, *Min*, and *Avg* values refer to the stress in elements attached to the node in question.

$$\textit{Eq. 10.2} \quad Error = \left| \frac{Stress_{Max} - Stress_{Min}}{Stress_{ModelMax}} \right| x100 \ \%$$

$$\textit{Eq. 10.3} \quad Error = \left(\frac{1}{\sqrt{N}}\right)\sqrt{\frac{\displaystyle\sum_{i=1}^{N}(Stress_{Element} - Stress_{Avg})^2}{N}}$$

Relating Error Estimates to Convergence

The following examples in Figs. 10.6 and 10.7 pertain to two of the fundamental stress states described in Chapter 2. Knowing how error estimates relate to convergence for your typical geometries and stress patterns will allow you to minimize the size and frequency of convergence runs.

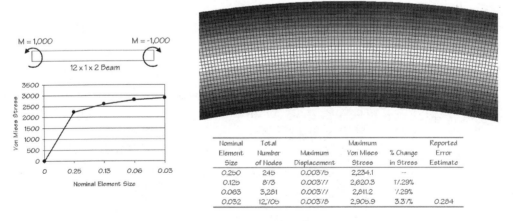

Nominal Element Size	Total Number of Nodes	Maximum Displacement	Maximum Von Mises Stress	% Change in Stress	Reported Error Estimate
0.250	245	0.00375	2,234.1	--	
0.125	873	0.00377	2,820.3	17.29%	
0.063	3,281	0.00377	2,811.2	7.29%	
0.032	12,705	0.00378	2,905.9	3.37%	0.284

Fig. 10.6. Convergence of beam in pure bending.

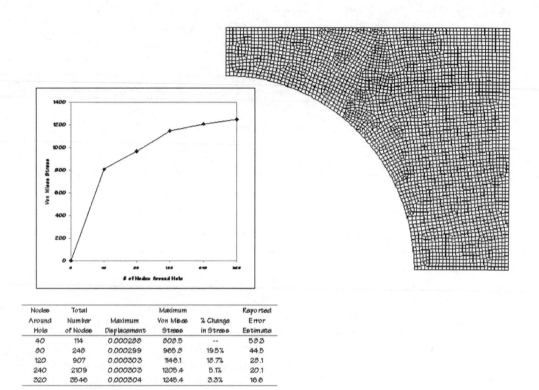

Nodes Around Hole	Total Number of Nodes	Maximum Displacement	Maximum Von Mises Stress	% Change in Stress	Reported Error Estimate
40	114	0.000288	808.5	--	58.3
80	248	0.000299	965.8	19.5%	44.5
120	907	0.000303	1146.1	18.7%	28.1
240	2109	0.000303	1205.4	5.1%	20.1
320	3548	0.000304	1245.4	3.3%	16.8

Fig. 10.7. Convergence of plate with a hole.

More Examples

Examples of automeshed parts in Figs. 10.8 and 10.9 show the mesh required to achieve a convergence of 10%. Corresponding plots show the local mesh size versus local stress and error estimates. The starting point on each plot reflects the default automesh size specified by the preprocessor.

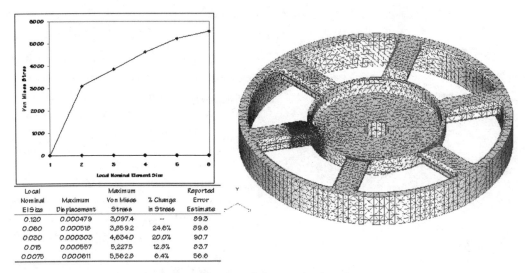

Local Nominal El Size	Maximum Displacement	Maximum Von Mises Stress	% Change in Stress	Reported Error Estimate
0.120	0.000479	3,097.4	--	89.3
0.060	0.000518	3,859.2	24.6%	89.8
0.030	0.000303	4,634.0	20.0%	90.7
0.015	0.000557	5,227.5	12.8%	83.7
0.0075	0.000611	5,562.8	6.4%	58.6

Fig. 10.8. Convergence plot and resultant mesh for a valve handle.

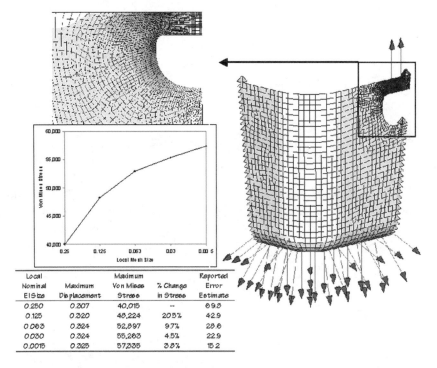

Local Nominal El Size	Maximum Displacement	Maximum Von Mises Stress	% Change in Stress	Reported Error Estimate
0.250	0.307	40,015	--	69.8
0.125	0.320	48,224	20.5%	42.9
0.063	0.324	52,897	9.7%	28.6
0.030	0.324	55,263	4.5%	22.9
0.0015	0.325	57,335	3.8%	15.2

Fig. 10.9. Convergence plot and resultant mesh for a parts bin.

P-elements versus H-elements

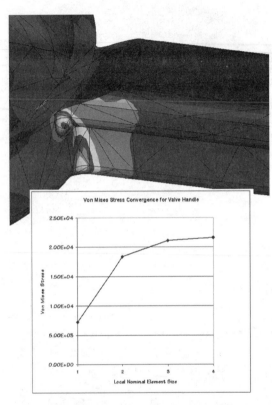

Fig. 10.10. P-element mesh and convergence plot of valve handle.

H-element codes and projects using them account for most of the FEA market. The convergence capabilities of p-elements, however, have been the key factor in the growth of their popularity. As described earlier, p-elements differ from h-elements in their ability to utilize more complex element edge definitions with higher polynomial functions. While convergence in an h-element mesh involves adding more nodes to capture high strain gradients, p-elements can represent similar behavior with a smaller number of elements using a more powerful mathematical definition. P-elements are also much more conducive to automated convergence routines. While the stress calculations on a high order edge are much more complicated than those on a linear or quadratic h-element edge definition, the user interface issues are comparably less complex. Few h-element codes can robustly and repeatedly calculate new mesh sizes based on error estimates and effect a local mesh refinement, whereas p-element technology routinely updates edge polynomial order based on various convergence parameters. The examples shown in Figs. 10.8 and 10.9 have been repeated in Pro/MECHANICA, a popular p-element based tool. The resulting meshes and related convergence curves are shown in Figs. 10.10 and 10.11.

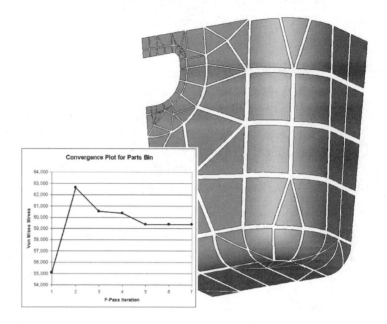

Fig. 10.11. P-element mesh and convergence plot of parts bin.

Despite their obvious superiority in capturing high strain gradients, p-elements are not immune to convergence issues. High aspect ratio elements, distorted elements, and overly large elements may require high polynomial orders to converge, if they indeed converge. Moreover, a visual review of the results is of equal importance to a convergence plot review because no automatic routine is infallible. A textbook convergence may exist on a mesh which has obvious inconsistencies in the stress contours. Both measures of solution quality should be reviewed. Use engineering judgment and your understanding of finite element behavior to make convergence decisions.

Element Quality versus Convergence

The shape of individual elements has a direct impact on the accuracy of local data and the resulting convergence. In addition to mesh size, element quality in areas of interest should be controlled. Dirty geometry or small features with respect to mesh size can cause poorly shaped elements to occur in automeshed parts. Follow the guidelines for meshing detailed in Chapter 7 to ensure that convergence is related only to mesh size.

Summary

Understanding model accuracy is extremely important in FEA. It is important to get a handle on the relative quality of the data resulting from a finite element study. The most carefully researched properties and boundary conditions with meticulously modeled geometry can be easily betrayed by an improper mesh. Convergence is load dependent. A mesh converged for one load case may not be the best mesh for a different loading scenario. Using relative convergence data and error estimates with visual examination and engineering judgment can ensure that the data provided by the analysis are the best indication of system performance. However, despite the most rigorous convergence processes, it is only through correlation to test data and reconciliation of all detected discrepancies that confidence in analysis results can be achieved.

11

Displaying and Interpreting Results

After working hard and diligently up to the point of solving a model, do not rush through the final step of examining results. You should spend as much time "debugging" the analysis results as in debugging the model setup and its mesh. Considering your results to be wrong until you have proved them right to yourself and your peers is not a bad idea. Be aware that there are many different ways of displaying and interpreting the output of a run. Hence, verify that what is displayed on your screen is really what interests you, is sufficiently accurate, and fulfills the original goals of the analysis.

Method for Viewing Results

Fig. 11.1. Flowchart of recommended method for viewing results.

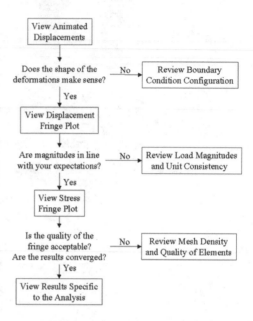

Although there are many different methods of proceeding through the review of solution data, a few guidelines are presented in this section that should make this process more robust and straightforward. Refer to the flowchart in Fig. 11.1 as you read through this section, and try to make use of it in your FEA process.

Displacement Results

It is recommended that you become accustomed to viewing displacement results before anything else. The basic quantity that FEA solves for is displacements, which are then used to extract all other quantities. Hence, if displacements are incorrect, you cannot trust the output data for anything else.

Animation

Animation of the model displacements serves as the best means of letting you visualize how your model has responded to its boundary conditions. As shown in Fig. 11.2 (b), most codes will give you the option of displaying this animated deformation overlaid on the undeformed geometry. In addition, you may also be allowed to define the number of frames to be used in constructing the animation. More frames will require more computational resources but will smooth out this type of display; hence, a compromise number should be selected. For a static analysis, you may also be able to choose between a repeating and an alternating animation. The former repeatedly shows the animated deformation of your model from its undeformed state to its fully deformed condition under the magnitude and direction of your specified loading. In a linear static study, the alternating display takes advantage of this linearity to show you how your model would deform in both the specified and opposite directions of the applied load. Hence, for the section of a suspension shaft end, loaded as in Fig. 11.2 (a), an alternating display would animate this section from its undeformed shape to its fully extended condition, back through its undeformed shape to a compressed condition, back to its undeformed shape and then start again. This type of display serves to accentuate and further exaggerate the resulting response. However, when making use of an alternating display, you must remind your public of its nature, lest you allow them to think that you have made a mistake and loaded the model in the opposite direction.

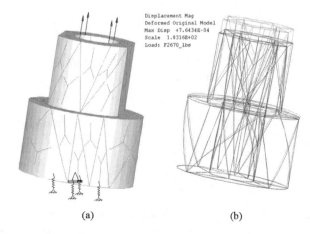

Fig. 11.2. Pro/MECHANICA FEM of an axially loaded section of a suspension component (a), and resulting animation display (b).

(a)   (b)

Regardless of the animation display method you choose, the first question you should ask yourself is whether the component is deforming as you expected. If the animation does not make sense to you, chances are your boundary conditions should be modified and the analysis rerun before you proceed. Let the animation help you sort through different ways of constraining and loading the model. Because displacements are usually small relative to the size of the model, the animation must be scaled for you to witness any motion. Most codes automate the value of this scale factor at 10% of the maximum model envelope dimension, although you can change it to either a different percentage of the envelope dimension or a numerical entry. Generally, you will have only a rough estimate for this envelope dimension. Hence, your code should communicate to you the equivalent numerical value of its percent magnitude–otherwise, you should get used to modifying the scale factor so that it becomes a known value. Get in the habit of ensuring that everyone viewing your results is aware of the scale factor. It is not uncommon for untrained eyes to become excitedly alarmed at the seemingly large displacement animation displayed on your screen. These displacement scaling comments apply anytime you display a deformed model, whether animation is the quantity of display or another quantity is used.

Note that for modal and linear buckling analyses, the magnitude of the resulting displacements, and thus the scale factor, is only a relative quantity. Generally, animation results of these analyses only serve to give you information regarding mode shapes and respective frequencies or buckling load factors at which they occur. You will usually not be interested in the magnitude of the eigenvectors. In fact, the animation of displacement results is probably as far as you will take these types of analysis. Also note that some codes will let you scale results according to the magnitude of a quantity other than displacements. Be wary of this type of scaling: the corresponding displays will not be very intuitive.

Magnitude of Deformed Shape

Even if the animation of displacement results does make sense, you must also check that their magnitude is not surprising. The value of the animation scale factor should help you notice gross behavior, but a filled contour or *fringe* plot of these magnitudes, such as in Fig. 11.3, will provide a more detailed picture. Unexpectedly high or low displacements could be caused by an improper definition of the loading and/or elemental properties. It is not unusual to see models that, in the rush to get them set up and solved, have mistakenly acquired a mixed set of units. If displacement magnitudes are in the least surprising, check the model's load and property values. Always look for order of magnitude discrepancies vis-a-vis your expectations first. You can then follow up in more detail by comparing display values to those obtained empirically or through a hand calculation or other analysis method. Are the resulting displacements so large that the linear assumption should be questioned? This is always a good point to check the linear assumption and consider following up with a nonlinear analysis.

Fig. 11.3. Pro/MECHANICA displacement fringe plot of suspension component section.

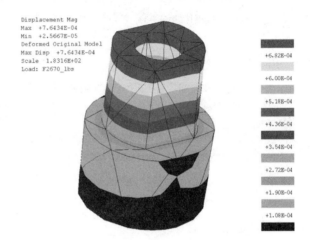

Of course, if the actual goal of the analysis is to solve for displacements, as would be the case in a stiffness study, it is very important to be thorough in this step of the results interpretation process. Even if the goal of the FEA is solely to report stresses, these quantities are derived from strains, which are in turn derived from displacements. Hence, the quality of reported displacements is crucial.

Stress Results

Stresses are by far the most common quantity required from FEA. Generally, interest in FEA arises from a desire to predict the possible failure modes of a part and use this information not only to prevent failures, but to be able to fully optimize the part's structure. Being competitive in today's engineering environment no longer involves simply ensuring that a designed component lives through its prototype testing, but also guaranteeing that it is the lightest, most economical sample of its kind. Optimization through stress analysis is probably the most powerful feature of FEA. Hence, a complete understanding of stress results display and interpretation capabilities is warranted.

Fringe Quality

The various stress quantities that may be displayed by your code are discussed later in this chapter. Yet, regardless of the type of stress, the first thing to check is the quality of its fringe plot. Examine unaveraged, noncontinuous tone results (more on this later) and verify that the stress contour is relatively smooth. Beware of hard discontinuities in stress levels across element edges and/or jagged distributions that follow these edges, both of which might point to a problem with one or more of the elements used in their calculation. "Bad" elements can be a sign of too coarse a mesh, poor element definition, or problems with the underlying geometry, the last being more of an issue for p-elements. Fig. 11.4 shows the difference in contour quality between a hex element hand mesh and a default tet element automesh obtained with Pro/MECHANICA. Note that a 10% percent convergence criterion in the default stress error estimate quantities was satisfied by both model solutions. Which one would you include in your report?

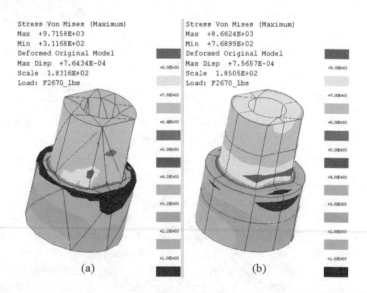

Fig. 11.4. Pro/MECHANICA von Mises stress results of a tet element automesh (a) and a hex element hand mesh (b) of a suspension component section.

A special mention of p-element plotting grids is warranted here. Whereas in h-elements the solution to having an overly coarse mesh is resolved only through its refinement, p-element codes may allow you to refine the plotting grid for each element instead. An element's plotting grid refers to the number of points within an element for which results are calculated and reported. Hence, if the fringe quality is poor within the boundaries of a p-element, a larger plotting grid may smooth it out. Remember that larger grids require more disk space and increase the solution time.

Convergence

Poor fringe quality is usually a sign of poor solution convergence. You should refer to the convergence concepts presented in Chapter 10. Suffice it to say here that if the mesh is too coarse, refining the mesh in problem areas should help the model converge. In the same manner, redefining poorly shaped elements resulting from high aspect ratios or edge or face angles that are either too small, will also improve results. If the underlying geometry itself is causing headaches, deletion and recreation of areas of poor convergence will typically take less time than modifying these areas in an attempt to correct them. Always check error estimate quantities that may be supplied by your code to ensure that they achieve acceptable levels. In p-element codes, investigate convergence curves if available to ensure that they leveled off properly to the reported solution values.

Stress Magnitude

Just as you should not be surprised by the magnitude of the deformations, the magnitude of the stresses should not be entirely unexpected. Globally high or low stress magnitudes could be caused by incorrect values of the model's loads and/or elemental properties. If overall stress magnitudes are in the least surprising, be sure to check them.

If the problem was set up correctly, locally high stress areas may still be found in your results. Sometimes they will be real and will have to be resolved through an improved design: this is the reason for performing FEA. On other occasions, these high stress areas will be caused by "bad" elements as described above. Most codes will auto-scale the initial display scale, and then let the user modify it as needed. If the auto-scale appears to show the entire model in the color your code has been set up to designate low stress (commonly blue or violet), it could be an indication of very small, probably fictitious high stress regions. The magnitude of stress in these regions has not only weighted the scaling to the high end, but most likely has driven up your run time and coarsened your convergence by making reported error estimates misleading. This last point results from the fact that stress error estimates are usually reported in terms of a percentage of the maximum stress. Hence, if the maximum stress on a few elements is indeed fictitiously high, a low reported error will actually be high compared to the stress magnitudes on the rest of the model.

For example, consider a model that exhibits stresses under 30 ksi for 99% of its elements, and that these stresses are in line with your expectations. If a few bad elements in this model display magnitudes in the 100 ksi range, a reported error estimate of 10% of maximum stress will actually mean a 33% error estimate in the stresses of interest. If your code reports this type of error estimate, make it a habit to compute what it means in terms of the actual stress levels of concern.

Centroidal versus Corner Stress

For the vast majority of FEAs motivated by the desire to obtain stress values, you will be most interested in the surface of the material, because this is usually where the highest magnitudes occur. For solid models, this becomes a bit problematic because stress errors tend to be greatest at these surface nodes, which are at extreme points of the element. Directly calculated at such nodes, these stress values are known as *corner stresses*. Alternatively, surface stresses may be extrapolated from more accurate values calculated in the interior of the element, known as *centroidal stresses*. However, because of error associated with the extrapolation process, the resulting accuracy of reported surface stresses is similar for both methods, although the value of the stresses themselves is different.

For h-element solutions, it is often useful to be able to display the non-extrapolated centroidal stresses on the surface of the elements. Ideally, the contour of this plot should be relatively smooth. Adjacent elements with colors two or three ranges away from each other usually suggest convergence issues that must be addressed.

If your code allows you to specify the manner by which you want it to extract stress results, remember that corner stress extraction is more resource intensive than centroidal extrapolation. In fact, for nonlinear analyses, the former is prohibitively expensive.

Entire Model versus Individual Components or Groups

When analyzing a component or assembly of components, you may be primarily interested in the results for only a portion of the model. Some codes will allow you to cut away components or regions of interest for separate inspection. Other codes will require you to preset groups in the modeling stage prior to analysis in order to turn groups on or off as desired when displaying results.

"Pregrouping" should not pose an inconvenience for investigating a single component in an assembly, given that it is always good modeling practice to create separate groups for each component in assembly modeling. However, if your postprocessor does not allow arbitrary cuts in a component for investigating results along a cross section through the component, you will have to place the elements on either side of the desired cut in separate groups prior to the analysis. This procedure will typically require modifying model geometry to include the cutting surface. If a review of stresses within a part is something you foresee being interested in, be sure to take it into consideration in the modeling stage. Fig. 11.5 shows a model that exemplifies these concepts. Note that if group display is supported, your postprocessor will either automatically adjust the max/min legends for the selected group or provide you with the option to do so.

Fig. 11.5. Stress results of an assembly (a), one of its members (b), and a cut section of the member (c). This was done using groups in Pro/MECHANICA.

Types of Output Data

Now that you are familiar with the recommended method for viewing FEA results, a more detailed discussion of each type of output data available is warranted. Note that the types discussed here are the most common in FEA codes. However, your code may not support some of these and/or may present you with others. Refer to your software documentation for a more specific description of available results. Note that for dynamic analyses, results are available at each step or master interval for which you requested full results. Also note that to report results in terms of a user-defined coordinate system, some codes require that the system be defined in the modeling stage prior to the analysis. Again, refer to your software documentation.

Displacements

In addition to the display of displacement magnitudes, you will generally be allowed to request the display of components with respect to either the global coordinate system or a user-defined coordinate system. This type of display is extremely useful when you are interested in the rigidity of the structure along a particular direction. Of course, if the direction of interest is not aligned with the global coordinate system and/or is not of the Cartesian type, you should create an appropriate coordinate system in either the modeling or results display stage, depending on the capabilities afforded by your code. Presenting radial displacement results of a pressure vessel analysis is a good example of this.

Rotations

You should be able to request the display of rotation results for non-solid elements only, unless your code supports rotational DOFs on solids. These results will generally be available as magnitudes of the rotation vector or as components about each coordinate axis of either the global coordinate system or a user-defined local coordinate system. When torsional rigidity is important, such as in vehicle frame structures, these are useful quantities.

Velocities and Accelerations

When reviewing the results of a dynamic analysis, velocities and accelerations at each requested master interval may be displayed. These results can be either linear or rotational, and are generally available as magnitudes or components in or about each direction relative to either the global coordinate system or a user-defined local coordinate system.

Strain Quantities

Elemental strains are the quantities extracted directly from displacement results that give rise to stress results. These quantities are not necessary nearly as frequently as displacements or stresses. However, certain phenomena, such as the assessment of rupture, require the determination of strain values as a goal of the analysis. Correlation to strain gauge test data may also be performed by investigating these results. Note that positive strain values typically indicate tensile strain while negative values indicate compression.

Max, Mid, Min Principal Strain

At every solution point in the model, the calculated strains may be resolved into their principal components. The *max principal* strain is the most positive of these, the *min principal* is the least positive, and the *mid principal* has a numerical value between the former two.

Normal and Shear Strain

Normal and shear strains should also be available for display at every point in your model. Normal strains are usually reported in XX, YY, and ZZ components, while shear strains are displayed in terms of XY, XZ, and YZ planes. These quantities will generally be displayed relative to global coordinate system directions, although you may be given the option to select a user-defined coordinate system instead.

Shell Membrane and Transverse Shear Strain

Other possible display options may be available to you specifically for shell elements. *Shell membrane* strains refer to a biaxial strain state along the mid-plane of the shell element. These strains may be reported as max/min principal, XX/YY normal, or XY shear quantities. This definition utilizes the local coordinate system of each element, where the Z axis is normal to the shell. By comparison, *shell transverse shear* strains occur along the mathematical thickness of the shell element, reported with respect to the elemental X and Y directions.

Strain Energy

Rather than the actual strains, you will often be more interested in the *strain energy* calculated throughout the model. This quantity is useful for predicting high stress areas in ductile materials, and is the basis for the derivation of the von Mises stress. Strain energy is generally available in many different display forms. For all elements, a *total* strain energy may be reported. For shell elements, a *membrane* value per unit area due to the stretching of such elements' mid-planes may be displayed. Shell and line elements may report both *bending* and *shear* strain energies per unit length. *Tensile* and *torsion* strain energies per unit length may also be available for line elements only.

Stress Quantities

Because stress results are generally used to predict and guard against failure, the choice of display output is directly related to the failure theory used for each particular analysis. Refer to Chapter 2 for a detailed discussion of failure theories.

Max, Mid, Min Principal Stress

By selecting the *max principal stress* for display, you are requesting the postprocessor to report the most positive principal stress for all elements. Conversely, the *min principal stress* reports the least positive principal stress for all elements. Both stress states are utilized as failure criteria by the *maximum normal stress* theory. The *mid principal stress* is the principal stress with a numerical value between the max and min principal. Note that positive values typically indicate tensile stresses, whereas negative values indicate compression.

Normal and Shear Stresses

For other stress state calculations, you may be interested in the *normal stresses* along each axis of either the global coordinate system or a coordinate system defined by you. Beware that some codes require that such user-coordinate systems be defined prior to the analysis. Regardless of the coordinate system, most postprocessors provide normal stress quantities for display as XX, YY, and ZZ stresses. Similarly, if you are interested in *shear stresses*, XY, XZ, and YZ values should also be available. Again, positive is typically related to tension and negative to compression.

Von Mises Stress

The most popular quantity for display when evaluating components made of ductile materials, the *von Mises stress* is defined by the *distortion energy theory* as a combination of all principal stresses, which are themselves a combination of all normal and shear stresses. The von Mises stress state is compared to its max and min counterparts in Fig. 11.6. When used to guard against yield of ductile materials, this quantity will generally best match experimental data. Note that von Mises stresses are by definition always positive and do not provide an indication of a tensile or compressive state.

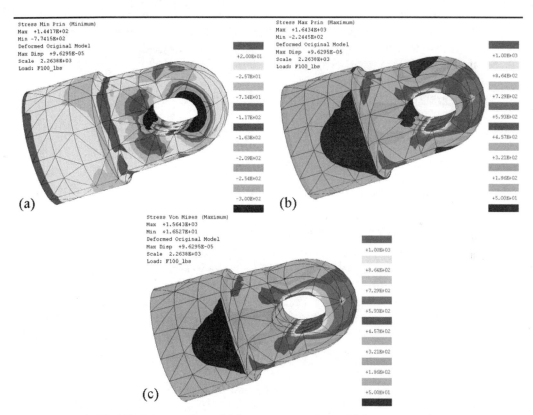

Fig. 11.6. Pro/MECHANICA min principal (a) versus max principal (b) versus von Mises (c) stress results of a lug analysis.

Reaction and Resultant Forces and Moments

Analysis goals will often include determining the *reaction forces and moments* occurring at each model constraint. These reactions may then be used as input loads to separate models of the supporting attachments that your constraints have attempted to simulate. On other occasions you will be more interested in resultant forces that occur internally in the model. These in turn can be used to evaluate the strength of internal structural joints in assembly modeling, or as load parameters in a simplified model of a section of the original. Both reaction and resultant forces and moments should be available in your results database, relative to either a global or user-defined coordinate system. However, some codes will require you to specifically request these results prior to the analysis. Be sure to check your software documentation.

Line Element Results

Because of their mathematical nature (result of the absence of full geometrical representation), some of the line element results are often categorized separately.

Strain and Stress

The one-dimensional geometry of a line element makes it impossible to display a complete state of strain or stress all at once. Hence, line element strain and stress results are usually divided into axial, bending, torsional, and total by the postprocessor for display purposes. For any of these quantities, you will generally be allowed to choose to display either the maximum or the minimum value across the element's mathematical cross section. Alternatively, some codes will allow you to specify the exact recovery point for which you are interested in these results. Note that recovery points are defined when the line element property is created. You will not be given the option of changing the recovery points after the results have been calculated. Always remember to note their defined location to avoid possible interpretation errors.

Axial display quantities in Fig. 11.7(a) will report only tensile/compressive strain or stress values along the axis of the line element. Bending and torsional values will usually report strains and stresses due to bending and torsion, respectively. A beam bending stress display is shown in Fig. 11.7(b). Some codes will also report a total strain or stress state of the line element, which is a linear combination of axial and bending strains or stresses, shown in Fig. 11.7(c). The latter quantity is generally reported for line elements in a max principal strain or stress display of a mixed element model.

Fig. 11.7. Axial (a) versus bending (b) versus total stress (c) for the same beam model solution.

Resultant Force and Moments

Similar to stress and strain results, line element resultants may be displayed in several ways. Resultant forces can be displayed in either axial or shear components. Resultant moments may also be separated into bending and torsional. These force and moment displays make use of the local beam coordinates. Alternatively, you may request the postprocessor to report resultant forces or moments with respect to either the global or a user-defined coordinate system. For any of these quantities you should be able to choose the display of maximum, minimum, or specific recovery point values.

Some codes will also allow you to graph shear and moment diagrams along a line element. These graphs are very useful for correlation purposes against user-derived or published diagrams for different boundary condition states.

Rigid Element Forces

By definition, rigid elements couple one or more degrees of freedom of two or more nodes in a model. If requested, codes that support these elements can report the force required to maintain this coupling. Because these forces are reported at the nodes that define the rigid element, results at each node of a two-node rigid link should be equal and opposite.

Shell Element Results

Similar to line elements, shells lack a full geometric representation and make use of mathematics to interpret their thickness property. Hence, results of shell elements must be further specified.

Location

For a given strain or stress state, the postprocessor must be directed as to the location along the thickness of the shell for which to report the results. The choices generally available are top, bottom, maximum, and minimum. Certain codes will also allow the specification of an intermediate recovery surface. For displaying results along either the top or bottom of the shells, you must ensure that the shell normals are properly aligned and that you understand the system's interpretation of the top and bottom geometry of each shell. This comment also applies for other intermediate recovery surfaces, with the exception of the midplane. However, due to their nature, maximum and minimum results are not sensitive to geometric misinterpretation.

Measures

In addition to the system-defined set of result quantities, some codes allow users to define their own measures that are more model specific. Possible measures include a maximum stress quantity over a given portion of a model or a minimum deflection at the extremity of a component in an assembly model. If available, these measures are very useful quantities with which you should become familiar. They will often save you time by obviating the need to hunt for specific information within the raw database. Remember that these measures will generally consist of a single numerical value result.

Types of Results Displays

Now that you are familiar with the various types of output data you can extract from the results database, you should become familiar with the types of displays you can use to present them. Hence, this section will focus on display types most commonly found in FEA codes. Keep in mind that your particular postprocessor may not offer some of the features discussed here and/or offer additional ones. Be sure to check your software documentation.

Animation

Scaling parameters, overlaying the undeformed model, number of frames, and being able to choose between repeating and alternating animation displays have been presented thus far in the context of animating displacement magnitudes. However, other animation types may be available to you. Most current codes will allow you to animate strains and stresses along with deformations, thereby giving you the ability to present how these quantities vary as the loading is applied from zero to its full magnitude. You may also be able to animate directional components along either the global or user-defined coordinate system axes, instead of overall magnitudes for any of these quantities.

Nonlinear results can usually be animated as a sequence of the load step results, thereby showing the true development of displacement and stress in a component. Dynamic results may be animated across time steps or input frequency phase increments to provide a better understanding of the time or frequency dependent data.

Fringe

Filled contour or *fringe plots* are the display type most commonly associated with FEA. To report a fringe plot, the postprocessor divides the selected data type into a user-specified number of value ranges. Each range is then assigned a particular color. Hence, the model is displayed, in either its undeformed or deformed shape, with color fringes indicating the value range associated with each location. Your code should allow you to edit the scale of the display within the results window to reflect ranges of interest. Note that although all intermediate ranges should generally be of equal size, the lowest and highest ranges may be larger in order to distinctively display all values below and above the overall range of interest. You may also be able to redefine the colors used for each range. Blue is generally associated with values close to zero, and red with values at the highest positive range of the scale. It is good practice for the scale to progress through color shades from less to more intense as the values get farther from zero.

Averaged versus Unaveraged versus Continuous Tone

In reality, strain, stress, and resultant force values are continuous through the material of the structure. By comparison, due to the discretization inherent in the technology, these FEA calculated quantities are virtually never continuous across two elements. In a well-behaved model, this discontinuity is often minimal and is hidden by a fringe plot's discrete ranges. However, especially in high gradient areas, discontinuities can be quite large and show up as a jagged fringe distribution. This display choppiness may be useful for debugging bad elements in the model as described previously in this chapter. However, because these quantities should be continuous, most codes allow for *averaging* in fringe plots. For each element connection, the values across the separating element edge are averaged to report a value that will make the transition continuous between the elements. Assuming that the mesh has been cleaned up and has converged to an acceptable level, a reported average value will almost always be closer to the real value because continuity is satisfied. You should nonetheless always look first at unaveraged results so that errors in the mesh are not hidden by the display. Note that an averaged display will still show discrete changes in fringe colors at range transitions. Appearing below are a few important notes on averaging (assuming your code even allows the output format).

- Never average between elements of different type
- Never average between elements of different material property
- Never average between shells of different thickness
- Never average between line elements of different cross sections

Any of the above will cause smearing of the modeled discontinuity across the transition elements.

Whereas averaging mathematically changes the transition values between elements to force a continuous stress field, a *continuous tone* display option simply alters the fringe display quality so that the color tones transition smoothly between fringes. This option has the visual effect of a continuous stress field, although the transition values remain discontinuous. Once you have successfully debugged your results, continuous tone displays provide very nice presentation pictures. Fig. 11.8 shows the difference in the selected mode of display for the results of an analysis of a drilled plate under tension.

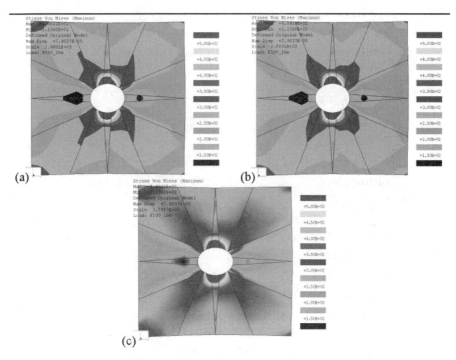

Fig. 11.8. Pro/MECHANICA von Mises stress results of an intentionally underconverged analysis of a drilled plate under tension: standard (a), averaged (b), and continuous tone (c).

Isolines and Isosurfaces

When dividing the results of a given quantity into ranges, there is another option for display besides a fringe plot: a contour plot. Whereas fringe plots will color fill the areas between the limiting boundary values of each range, a contour plot will color outline the boundaries themselves. In other words, contour plots show *isolines*, differently colored curves on your model that follow a constant quantity value. An FEA contour plot, shown in Fig. 11.9(a), is the equivalent of a topology map that shows constant altitude isolines for a given geography. When these curves are far apart from one another, they indicate a relatively low gradient region; when they are close together, a relatively high gradient region is implied.

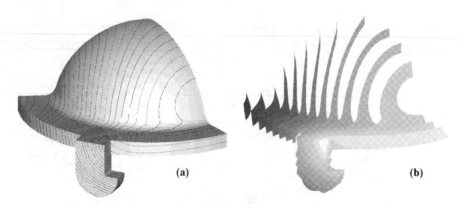

Fig. 11.9. Isolines (a) and isosurfaces (b) of the same model solution.

When examining results on three-dimensional solid elements, you may choose to display *isosurfaces* of a constant quantity value which will allow you to investigate how results vary inside the elements [see Fig. 11.9(b)]. Regardless of your choice of isolines or isosurfaces, you should be able to select the number of display levels and colors, just as in a fringe plot. You should also be given the choice of averaging the display quantity, as described above for fringe plots.

Query

Query may or may not be a separate display option in your code. Your postprocessor should nonetheless allow querying of every solution point in your model for its exact value, and you should be able to query specific locations of a fringe or contour plot, as shown in Fig. 11.10. In addition, some codes provide a dynamic query that allows you to scan the display by dragging your mouse's cursor over each solution point while its value is displayed in a side data form. This type of query lets you avoid cluttering the screen with many individual display values. Other possible, useful query options are "model max, min" and "view max, min," which display the maximum and minimum numerical values of the display quantity over the entire model or in a specific screen view, respectively.

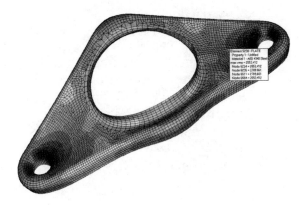

Fig. 11.10. Querying within a fringe plot.

Vector Plots

For solutions that are vectorial in nature–that is, they have a resultant direction in addition to a magnitude–a *vector plot* may prove very useful. Quantities that may be displayed with vectors include displacement, velocity, and acceleration magnitudes and their rotational equivalents, and max and min principal stresses.

Orientation of Principal Stress

The most useful vector plot is arguably that of max and min principal stresses, which displays the directions of both principal quantities (see Fig. 11.11). This is extremely helpful in determining strain gauge placement and orientation when conducting correlation testing.

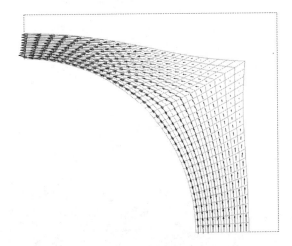

Fig. 11.11. Maximum principal stress vector plot.

Graphs

The least fancy of all presentation displays is the oldest and sometimes most numerically useful: the *graph*. Your postprocessor should allow you to graph the values of a chosen quantity of interest as a function of location along a selected number of nodes, edges, or individual line elements. Fig. 11.12(b) shows how the von Mises stresses vary along shell edges that follow a surface line. Graphs are very useful in determining the continuity of results, and therefore, results quality. As your solution reaches better convergence, these graphs should become smoother.

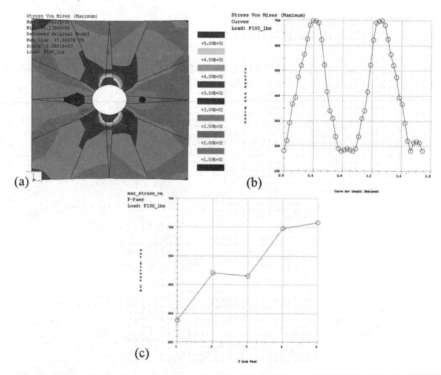

Fig. 11.12. Von Mises stress results for an analysis of a drilled plate: fringe plot (a), graph along shell edges following the hole definition curve (b), and maximum of the model as a function of polynomial order (c).

For purposes of investigating convergence curves, p-element codes should also allow you to graph the value of a given measure as a function of polynomial order. These graphs should show all quantities of interest leveling off nicely to a solution value as the p-order increases, thereby indicating properly converged results. Would you consider the convergence curve of Fig. 11.12(c) to be satisfactory? Do you see a relation between this curve and the quality of the corresponding fringe plot of Fig. 11.12(a)? Many users view these convergence plots as the real strength of p-element codes over their h-element counterparts.

Graphs showing the change in a results quantity over multiple load sets of a nonlinear or dynamic analysis are critical to interpreting total system behavior.

Results Interpretation: Quality Inspection and Verification Techniques

Once you have sorted through the many ways you can display FEA results, you must be able to correctly interpret them. Results interpretation is the last step in an FEA, where the goals set up at the beginning of the study may finally be met. As mentioned at the beginning of this chapter, you should never rush through this step. You have come a long way to get to this point, and you should dedicate the proper time to it in order to conclude your analysis as correctly as possible with the data made available to you.

Convergence Issues

The first thing you must determine in FEA results interpretation is whether the solution to the analysis has converged. The quality of your results is an indication of, and depends on, this convergence criterion. Be on the alert for very high gradients in selected displays. All quantities of interest should flow relatively smoothly and continuously throughout the model. Look for "bad" elements and/or coarse mesh regions that may require modification. Graphs of results along entities in areas of interest should be relatively smooth. In p-element codes, convergence curves should level off nicely to the final reported value. Refer to Chapter 10 and the review process section at the beginning of this chapter for additional information on convergence.

Correlation to Expectations (Common Sense)

Once you have determined that your solution has reached an adequate level of convergence, verify that the results correlate with your expectations. Of course, you will not always be right–this is the reason you are performing a high level analysis–yet the results should not generally be entirely unexpected. Take the time to reconcile any differences between the results and your assumed system response. Play the devil's advocate. Consult with your peers. Take a closer look at your boundary conditions, element properties, and all other assumptions. You should never continue past this point until you are convinced that what you are seeing on the screen is real. The farther you progress in FEA, the more projects you take to completion, the better tuned your common sense or engineering judgment will become. As you gain experience with this technology, it will become easier to quickly spot regions of concern in the results. Take many notes and share your insights with your analysis peers and the design group in general.

Qualification and Review of Assumptions

Each one of your loads, constraints, properties, and modeling and solution choices carried an assumption. You should always keep a record of assumptions as you progress through the setup process. When interpreting your results, take the time to review your assumptions and qualify (and possibly quantify) each of their effects on the solution. Again, if your assumptions have not provided what you expected, you must reconcile the differences.

Correlation to Closed Form Equations

Think about the fact that until recently, engineering analysis took place without the aid of FEA, and that it was performed exceptionally well in many cases. FEA should not shut your eyes to other analysis techniques. There are many systems that can be simplified and approximated with closed form equations. Many classical equations were developed empirically and carry quite a bit of correlation testing behind them. Refer to Chapter 2 to review some of the fundamental equations and concepts as they relate to your analysis.

You should attempt to use hand calculations to predict and verify your model. Unless the model is very simple, you should not expect to see a strict correlation to the hand calculations; otherwise, why would you need FEA? What you are looking for, however, is to be in the ballpark. If your rough hand calculation is an order of magnitude off, this could be a sign of a problem in your analysis. In such a case, you should investigate and resolve the discrepancy.

Use of Test Models

In the world of FEA, test models are your best friends. If you are attempting to use a new modeling technique, evaluating it in a small and simple test model is much more efficient than in a full, and probably complex and resource intensive, analysis. Generally, with test models you can quickly perform many iterations, and thus be able to not only validate your new technique, but to fine tune it. A good rule of thumb to use with test models is "one question, one answer." Develop test models to isolate the technique or quantity of interest.

In results interpretation, test models are often just as helpful as they are in modeling. If you do not understand what you are seeing on the screen, you can quickly set up a small test model that will present the results more clearly through simplification. Test models are also good for correlating resulting deformation magnitudes. An extremely simplified model that just captures the general structure of a part should not deform much differently than the model of the part when subjected to the same boundary conditions. Hence, a quick analysis will serve to verify that deformation magnitudes of the full model are in line with those allowed by the general shape of the structure.

Correlation to Testing

Although not always possible, it is recommended that you correlate your results to reality through a testing program. This should be brought up at the beginning of the project so that no one is surprised by your request upon presentation of your results. Because of its perceived power, FEA has of late attained a reputation among project managers as being the sole panacea of the design process. Indeed, this technology will allow the group to get much further along this process up front than was previously possible. However, due to all of the uncertainties involved in model setup, meshing, and model solution, you must attempt to clarify for all involved that results should still be correlated to ensure a successful design solution. Once you know that your model accurately represents reality, its power to achieve an optimized design becomes virtually limitless. Moreover, you may then be able to use this knowledge to validate other similar models.

Of course, testing itself may differ from reality. A guideline for correlation with FEA is that you always analyze what you test, and test what you analyze. The boundary conditions of a test fixture will often be different from those of the actual application, and this difference must be accounted for in your model. As a good design practice, you should always try to be involved in development of the test fixtures to be used for the component you are analyzing. If this development occurs after you have had the chance to run a few analyses, you may typically have obtained insights that will be key in the attempt to mimic reality with testing. You may also learn that some of the assumptions you made require closer examination.

Relation to Problem Goals

In addition to verifying the quality of your data, you must verify at this point that the solution you have obtained indeed satisfies the goals outlined at the onset of the project. There is no real use in presenting beautiful, correlated stress results for a part, when the goals of the study were to provide component deflections at a certain location on the part. Hopefully, you have not lost sight of the original goals. If you have lost sight of the goals, this is the final place to catch yourself before you file an incorrect report.

At the beginning of the study, you should have outlined the required results data. Again, verify that you have solved for such data. There is nothing wrong with providing additional data that you think might be useful or important to the quality and efficiency of the design. However, it is recommended that you first present what was requested and is expected by the design group and then complement your report with additional information.

Bulk Calculations on Results

While a single analysis can provide reams of data that can help you make better product decisions, there are many occasions when these data must be adjusted or combined with other data to provide the performance information in the format best suited for project requirements.

Combined Load Sets

The most basic calculation on a set of FEA results data is the scaling of linear static results to allow the presentation of the applied load at a greater or lesser magnitude. The next most common calculation is the combination of output sets from multiple solutions along with linear scaling. This capability actually makes it convenient in some projects to apply each of the model's independent loads as its own unit load set. The unit load might be 1, 10, or 100 force units, depending on the expected scale of the actual applied load. Using unit loads allows you to scale and combine the loads by entering a straightforward scale factor. An example of this might be scaling the results of a 10 lb load by 3.5 to represent a 35 lb load, 5.7 to represent a 57 lb load and so on. If unit loads are not used, an alternative is to scale the results of a 35 lb load by 57/35, or 1.629, to represent a 57 lb load.

Beyond output set scaling and combining, the same tools can provide endless capabilities for manipulating results data to better reflect the goals of the analysis or project needs. This capability is not available in all codes; check your software documentation or consult your support organization. As you become more proficient with the technology and tackle more complex problems, having access to these tools will greatly expand your capability to provide engineering data on a simulation project.

Fatigue Estimates

If your analysis is to relate primarily to a fatigue condition due to a single load set of sinusoidal nature, Goodman or modified Goodman calculations can indicate areas of fatigue concerns relatively easily. The modified Goodman relation is provided by the following equation:

Eq. 11.1 $$\frac{\sigma_m}{S_{ut}} + \frac{\sigma_a}{S_e} = \frac{1}{n}$$

where σ_a = peak stress amplitude, σ_m = mean stress, S_e = endurance limit or fatigue strength for specified life, S_{ut} = ultimate tensile strength of the material, and n = factor of safety in design.

The modified Goodman relation suggests that if the result of the left side of the equation is greater than or equal to the reciprocal of the design safety factor, the component being analyzed should last as long, if not longer, than the number of cycles specified by S_e. Otherwise, the part can be expected to last a fraction of that number of cycles.

To calculate an output set for the modified Goodman relation, you will need two output sets representing the most positive and most negative loading conditions on the model, or the two extremes of the sinusoidal loading. You may then create a new output set called *fatigue data*, in which three new output quantities are defined: *mean stress, peak stress,* and *fatigue criteria*. Calculate mean and peak stresses using Eqs. 11.2 and 11.3.

$$Eq.\ 11.2 \qquad \sigma_m = (\sigma_{max} + \sigma_{min})/2$$

$$Eq.\ 11.3 \qquad \sigma_a = (\sigma_{max} - \sigma_{min})/2$$

Once these output quantities are obtained, the modified Goodman relation can be calculated using Eq. 11.1 and stored. The best way to evaluate the new output quantity is to set the color legend to two colors where everything above $1/n$ is white and everything below $1/n$ is red. The areas of fatigue concern will be highlighted. It is recommended that you refer to Chapter 2 and other sources to review the fundamentals of fatigue analysis.

Mohr's Criterion

Another excellent use for bulk calculations on FEA output is to generate a fringe plot of the Coulomb-Mohr criterion for brittle material failure. As described in Chapter 2, this criterion provides a better indication of brittle fracture than the maximum principal stress alone. The Coulomb-Mohr theory states that fracture will occur when indicated in Eq. 11.4.

$$Eq.\ 11.4 \qquad \frac{\sigma_1}{S_{ut}} - \frac{\sigma_3}{S_{uc}} \geq 1$$

In this theory, σ_1 and σ_3 refer to the maximum and minimum principal stresses and S_{ut} and S_{uc} refer the ultimate tensile and compressive material strengths, respectively. Your code may let you reference the material database for the ultimate strengths in the calculation. If not, you will simply have to type in the appropriate values. Again, you can edit the color legend as described previously to quickly flag any area of fracture concern.

Summary

This chapter is focused on the final step of the FEA process: results display and interpretation. Although this process may be carried out in many different manners, an efficient and proven method has been presented here. If you are new to FEA, following this method and adjusting it to fit the specific needs of your organization are recommended. If you have been conducting FEA for a while, review this method and compare it to yours–you might find something useful that you had not previously considered. However, regardless of the method you use, you must always ensure that your results are displayed in terms of the goals outlined at the beginning of the project. These results should be qualified in terms of accuracy and correctness. Accuracy is related to the convergence level and quality of the solution, which should meet your project needs. Correctness refers to results verification through the validation of all your modeling assumptions. Validation may be conducted with the aid of additional FEA in the form of test models or other analytical means. Validation should always include correlation testing to ensure the success of the design.

Once results have been qualified and validated, you should be ready to file your report. The nature of each presentation should be clear to everyone in the design group. With the exception of modal and linear buckling analysis results, do not leave out range legends in fringe or contour plots. Verify that the range scale used reflects the values of interest. When displaying the model in its deformed shape, again with the exception of modal and linear buckling analysis results, you should indicate the display scale factor so that your audience is not startled by the seemingly large deformations. Use titles to indicate different design iterations and other relevant information. In short, make sure that people reviewing your presentation on their own have enough information at their fingertips, so that they do not have to infer anything. The last thing you want is for your interpretation efforts to be misinterpreted.

With the appropriate and diligent use of all concepts presented in this chapter, you should be successful in completing your FEA process and be ready to start all over again with a new FEA challenge.

12

Optimization: Tying It All Together

Performance simulation must be considered a process versus merely a task in a product design schedule. When treated as elements of an interactive process, the simulation models become refined and honed as the design materializes. While this book has focused primarily on the subprocess of gathering high quality data regarding a design using finite element methods, working this data into the big picture must now be addressed.

Engineering versus Analysis

In Chapter 1, a distinction was made between the design analyst and the analysis specialist. While most of the information in this book is applicable to all levels of FEA users, it is the design analyst who can, in the end, reap the greatest benefit from improved integration of simulation into the design process. This is primarily because the basic definitions of roles are different. An analysis specialist is tasked primarily with calculating the behavior of a system with one or more specified parameters. He/she rarely has the freedom to stray beyond these parameters. A design analyst, on the other hand, typically defines the parameters on the fly while pursuing more tangible product goals. In developing a part or product, the design analyst can start with simple, conceptual geometry to establish baselines and data points. In the hands of a designer, FEA truly becomes a powerful tool.

With solid modeling and rapid prototyping, rapid product development can be achieved by cutting through many questions before much time or significant resources are wasted. A key facet of design engineering is that the design analyst can make use of coarse data with relatively high uncertainty in making design decisions. Of course, engineering judgment must be applied. That term, *engineering judgment,* is tossed about rather casually in the analysis industry. The fact of the matter is that making an engineering judgment on finite element results is difficult, if not impossible, without a firm grounding in the fundamentals of structural mechanics and the behavior of finite element models under similar conditions. Review Chapters 2 through 11 periodically as you grow into FEA to ensure that your understanding is solid so that your engineering judgment is not misdirected.

Avoiding "Emotional Commitment"

The concept of *emotional commitment* was introduced and explored in Chapter 5. It was shown that emotional commitment, and its partner, *concurrent commitment,* can significantly reduce the benefits to be gained from up-front simulation. While concurrent commitment may not always be in your control, emotional commitment is. Avoiding the latter can be difficult because most good design engineers develop one or two "gut feel" concepts very quickly after being presented with a challenge. While these should not be dismissed, it is best to come to the proverbial drawing board with no preconceived notions about the final geometry. If possible, make no premature decisions about materials or manufacturing processes. Do not start the design on the tube. As mentioned previously, shaded CAD models become real once they appear. It will be difficult to backtrack once the first few features are modeled. However, it is easy to erase or scribble out entire parts on a sketch. Use sketches to brainstorm. When the initial ideas are flushed out, begin the simulation with simple geometry that is easy and painless to change. With this start, optimization will come naturally.

Optimization

There are two primary rules that apply to any optimization project. First, do not expect an optimization routine to design your parts for you. Second, if you do not ask the question, you will not get the answer. As powerful as some of today's shape optimization routines are, forgetting either of these rules will most likely result in a suboptimal design and much wasted time. While it is possible to get lucky and hit on the optimal configuration without following the proper methodology, remember that lightning rarely strikes in the same place twice. Following these guidelines will provide you with the optimal chance to optimize.

Ask Questions

One common pitfall most designers fall into when approaching a new part or system is to get caught up in one of the commitments described earlier. Frequently, one or maybe two concepts or configurations are explored and optimization efforts, if employed, are spent making evolutionary changes to the initial concept. Good old fashioned brainstorming, which takes place in *phase I optimization*, is the best way to reduce the probability of getting locked into a design prematurely. Taking the best and most viable concepts to completion, with the best possible cost and performance, occurs in phase II optimization.

In phase I optimization, concepts are being explored. Various tests are applied to these rough ideas to determine if they warrant further investigation. Some of these tests might include marketability, manufacturability, performance, and cost. From a performance standpoint, FEA can quickly provide a wealth of information. Phase I models should be coarse, and phase I assumptions should be broad. If the assumptions and model detail are consistent across the various concepts, comparisons are valid. Ideally, many phase I models will be evaluated, comprising various materials, manufacturing methods, and geometric configurations. After a review, two or three concepts should appear to hold more promise and warrant further exploration.

The following problem was posed to a group of design analysts in a workshop on optimization. Some of the phase I concepts and assumptions are included to illustrate the process.

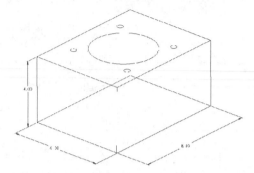

Fig. 12.1. Initial design of winch support machined from a block of steel.

The part shown in Fig. 12.1 is mounted on I-beams and supports a lifting device or winch. The winch has an internal motor and is bolted to the top of the mount so that its lifting cable runs down through the mount's central bore. The manufacturer wishes to redesign this part in response to increased demand. The original design, a machined block of steel, is not cost effective in large quantities. Assume that the following restrictions and known environmental conditions were specified by marketing.

1. Bolt hole patterns and sizes must remain the same for retrofit and compatibility.

2. The main bore diameter must remain the same.

3. Mount must support 1,500 lbs cable tension; some swinging may occur.

4. Mount/winch system must not resonate in the operating motor speed range of 0 to 6,000 RPM.

5. Quantities warrant an investment in tooling.

6. Weight of the winch is 100 lbs. Its center of gravity is three inches directly above the center of the main bore.

These specifications allow for optimization on part height, material, manufacturing process, and general geometry. Four manufacturing processes were suggested: (1) casting, (2) extrusion, (3) injection molding, and (4) sheet metal fabrication.

Phase I Optimization: Brainstorming

To illustrate the concept of phase I optimization, the casting process will be explored. In a real design program, all four manufacturing processes should be considered because the loose marketing specifications left room for each of these processes to provide its own benefits. Three unique geometry configurations were suggested for a cast mount. Material has yet to be discussed. Figs. 12.2 to 12.4 show the different options as you might envision them in their final form. The phase I assumption set listed below was outlined for evaluation of these castings.

1. Material properties will remain linear, isotropic, and homogeneous.

2. A 300-lb side component will be added to the load, oriented toward the long side of the mount, to account for swinging.

3. Mount is rigidly attached to the I-beams by the bolts.

4. Motor/winch mass is rigidly attached to its mounting holes.

5. Temperature variation will have a negligible effect on the part's behavior.

6. For comparison of the various concepts, a shell model will be sufficient.

7. Results due to gravity are negligible.

8. No secondary machining will be considered.

Fig. 12.2. First design option for casting the winch mount.

Fig. 12.3. Second design option for casting the winch mount.

Fig. 12.4. Third design option for casting the winch mount.

Figs. 12.5, 12.6, and 12.7 show the finite element models of these concepts. Note that a decision was made as to the initial guess for a part height, which is common across these FEMs.

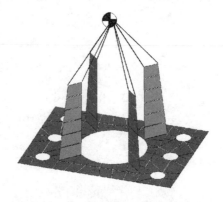

Fig. 12.5. A finite element model of the first design casting option.

Fig. 12.6. A finite element model of the second design casting option.

Fig. 12.7. A finite element model of the third design casting option.

All corresponding FEAs converged to less than 10% change in von Mises stress and first natural frequency. From convergence studies for the same material, nominal wall thickness, and part height, results were noted as they appear in Table 12.1.

Table 12.1. Summary of casting design option preliminary analysis results

Option	Weight	Maximum von Mises stress	Maximum displacement	First natural frequency
1	7.50 lbs	2,030 PSI	0.00019 in	71 Hz
2	12.8 lbs	1,700 PSI	0.00018 in	86 Hz
3	7.10 lbs	3,000 PSI	0.00050 in	41 Hz

At this point, it is usually possible to discard certain concepts as being significantly less qualified than others. In this case, option 3 shows much higher stress and displacement than the other two options and its first natural frequency is much lower. The design specification stated that the part must not resonate below 6,000 RPM, which equates to 100 Hz. It will be much easier to push the first frequency of options 1 and 2 above this operating range than to do the same on option 3. This is important to note because the first mode shape, or the shape in which the part wants to deform upon resonance, is sympathetic to the static loading deformation. What this means is that the part wants to vibrate in the same direction it is pulled by the loading. This makes the likelihood of noticeable vibration much greater. Performance of options 1 and 2 is comparable. While option 2 has slightly better results, its weight is over 70% greater than option 1. The stress results for all options are well within the allowable range for steel, so that stress does not turn out to be the driving factor in design approval. From these data it appears that option 1 shows the most promise for an optimal design.

The next step in phase I is to perform a *parameter response study* on the key variables defined previously: material, part height, and nominal wall. A parameter response study essentially tells you how a quantity of interest changes as a parameter changes. The simplest form of a parameter response study is the *design point sensitivity*, or simply *sensitivity* study. This study is also called a *local sensitivity* or a *perturbation sensitivity* study. Sensitivity studies adjust the variable or variables being evaluated to slightly lower and greater values. The amplitude of this perturbation should be no greater than 1.0% of the nominal to avoid any chance for redistribution of stress "hot spots" or coupling with other features. The change in the response quantity is usually reported as the slope of a line through the different response magnitudes. Figs. 12.8 through 12.10 show sensitivity study results for the first natural mode of option 1. These studies were conducted for a ± 0.5% variation in nominal parameters of 3.75 inches for the part height, 0.25 inches for its wall thickness, and standard steel properties for material elasticity and density values.

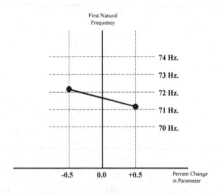

Fig. 12.8 Sensitivity plot for the part height of casting option 1.

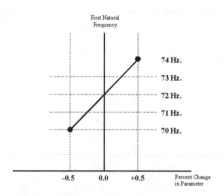

Fig. 12.9 Sensitivity plot for the nominal wall thickness of casting option 1.

Fig. 12.10. Sensitivity plot for Young's modulus and density combined.

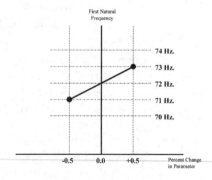

Because these results indicate that an increase in height reduces the first natural frequency, a reduction in height may be warranted. The natural frequency rises with an increase in thickness and material stiffness when the density is held constant. The relative slope of these plots indicates a particular design point's sensitivity to a small change in a parameter. Consequently, it can also be concluded that for small changes, an increase in nominal thickness will have the greatest effect, followed by a modulus change, and finally a reduction in part height. These data must be placed in the proper context. An increase in thickness is less desirable than a decrease in part height as the former increases material cost while the latter reduces the same. Furthermore, increasing the material stiffness beyond that of steel is probably not practical. Even though the similar decrease in the density should have caused a slight increase in the first mode, the change in weight was small compared to the total weight of the system. Therefore, the stiffness change dominated that parameter. Because changing material from steel appears to provide no benefit at this point, this option will not be explored.

It is important to remember that a sensitivity study only indicates the sensitivity to change at a specific combination of parameters. While this information is valuable for tweaking a design to optimize it, as will be done in phase II optimization, it can be misleading if used to predict gross trends. The primary reason for this is the fact that many parameter responses are not linear over the entire range of the parameter. A good example is a plate or beam's bending stiffness response to thickness. This response is a third order function of thickness. Higher order responses with consistent slope sign (i.e., constantly increasing or decreasing) can be predicted with some degree of confidence by performing sensitivity studies. It is best to perform these sensitivity studies at a couple configurations of the allowable parameter range to improve your understanding of the overall parameter response, and to minimize the chance that the data will be misunderstood. Although these studies can quickly predict trends to allow for the elimination of certain variables, they can be misleading; be careful with their use.

One hazard in relying on these local sensitivity studies is that sometimes the response trend to a parameter change inverts one or more times. If Fig. 12.11 represents the continuous stress response to a changing parameter, it can be seen that local sensitivity studies at points A and B might mislead you to conclude an erroneous trend. Although the slope at both these points may be approximated as being zero, extrapolating outside this range will get you in trouble right away.

The study shown in Fig. 12.11 is called a *full parameter response* study or a *global sensitivity* study. It provides a complete, unambiguous picture of a system's response to a changing parameter.

Fig. 12.11. Global sensitivity plot of von Mises stress versus the angular position of the second hole in the plate appearing in Fig. 12.12.

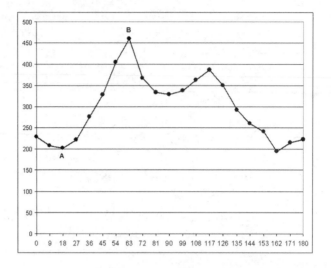

Fig. 12.12. Plate with two holes used for the global sensitivity study.

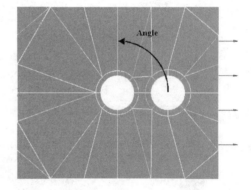

The study of Fig. 12.11 is an actual global sensitivity study of the plate model shown in Fig. 12.12. In this study, the von Mises stress at a location on the edge of the center hole is investigated versus the relative angular position of the side hole. It is clear from this single plot that stress is at a maximum when the hole is approximately in line with the diagonals of the plate, and that the best positions are at approximately 0° and 180°. Global sensitivity studies are completed by building a matrix of the quantities of interest at increments between the minimum and maximum parameter values. The size of the increment controls the resolution of the study. It is possible to miss trend changes with too coarse a resolution. However, because each increment requires a separate analysis, too many increments can be cost prohibitive. While some FEA tools, such as Pro/MECHANICA and ANSYS, can automate the process of developing a global sensitivity study, you will most likely need to set up and run the incremental analyses manually. Figs. 12.13 through 12.15 show a full response study of the first natural frequency for the parameters shown in the local sensitivity studies of Figs. 12.8, 12.9, and 12.10. The ranges in Table 12.2 were used for each parameter.

Table 12.2. Range of parameter values used in the full parameter response study

Parameter	Minimum value	Maximum value
Part height	2.75 in	3.75 in
Nominal wall	0.125 in	0.500 in

Fig. 12.13. Global sensitivity study of the first natural frequency versus part height of casting option 1.

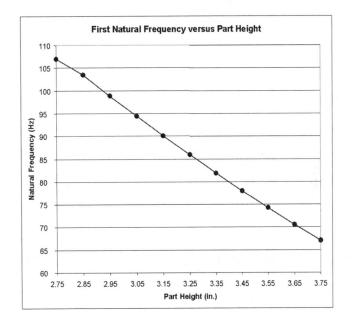

Fig. 12.14. Global sensitivity study of the first natural frequency versus nominal wall thickness of casting option 1.

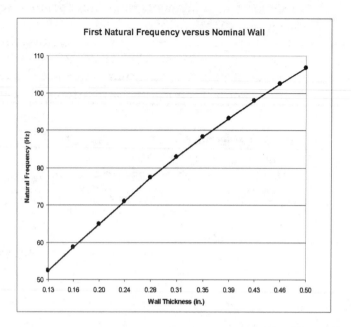

The full parameter response studies indicate that a one-inch drop in the height alone will not get the natural frequency above 100 Hz. Increasing the nominal wall to 0.50 inches will achieve this but may be excessive from a weight/cost standpoint. It is clear that the ideal configuration for this particular concept is a combination of the two parameters. Similar conclusions should be drawn for all the other concepts available to you in phase I. Some concepts will drop out, some parameters within an otherwise good concept will be discarded, and a handful will make the first cut. All these decisions can be made without spending too much time and resources detailing any particular design.

Phase I assumptions for the other manufacturing processes are listed below for reference.

Extrusion

1. Material properties are homogenous, isotropic, and remain linear.

2. To account for swinging, a 300-lb side component will be added to the load and oriented toward the long side of the mount.

3. Attachment to I-beams is modeled by a constraint on 0.25 inches of each bolt hole.

4. The motor/winch mass is rigidly attached to its mounting holes.

5. Temperature variation will have a negligible effect on the part's behavior.

6. Results due to gravity are negligible.

7. No secondary machining will be evaluated.

8. Load from the winch is applied to a surface patch on top of the extrusion.

9. Stress risers due to threads or thread forming can be neglected.

10. Profile radii are not required for feature definition will be neglected.

11. The extruded wall thickness remains constant.

Injection Molding

1. Material properties will remain linear, isotropic, and homogeneous.

2. All nonlinear stiffening effects will be neglected.

3. To account for swinging, a 300-lb side component will be added to the load and oriented toward the long side of the mount.

4. Attachment to I-beams will be modeled by a vertices constraint on the entire interface surface and planar constraints on each bolt hole.

5. The motor/winch mass is rigidly attached to its mounting holes.

6. Temperature variation will have a negligible effect on the part's behavior.

7. The part is modeled as a uniform wall thickness shell model.

8. No secondary machining will be evaluated.

9. Stress risers due to threads or thread forming can be neglected.

10. Residual stress and warpage form molding will be neglected.

11. Embrittlement due to solvent or oil interaction is neglected.

12. Creep effects are neglected.

Sheet Metal Weldment

1. Material properties will remain linear, isotropic, and homogeneous.

2. Weld and heat affected zones have the same properties as the base metal.

3. To account for swinging, a 300-lb side component will be added to the load and oriented toward the long side of the mount.

4. Attachment to I-beams is modeled by a vertical constraint in surface patches at bolt heads.

5. The motor/winch mass is rigidly attached to its mounting holes.

6. Temperature variation will have a negligible effect on the part's behavior.

7. All sheet metal parts are the same wall thickness.

8. Wall thickness bend radii can be neglected.

9. Weld bead geometry will be neglected.

10. Results due to gravity are negligible.

11. No residual stress due to forming will be included.

12. Load is applied to the top of the weldment on surface patches where the motor is mounted.

In short, phase I optimization is a process for quickly sorting through a variety of options and configurations. The assumptions made in this process are broader than a verification analysis, and the techniques for compiling data are more qualitative than quantitative. The results of each study are valid only as they compare to similar studies. At the conclusion of this phase, two or three concepts should rise above the rest for cost, quality, manufacturability, and function.

In addition to the coarse models and parameter response studies described earlier, phase I optimization may require testing of simplified prototypes for data that cannot be easily obtained with FEA. Simple test models can also be used to ensure that the assumptions employed are not excessively broad. Test models may also be required to develop the initial assumption set as well. In the previous example, a 300-lb static side load was assumed to account for swinging. This could be confirmed and/or adjusted with a simple test. If the assumptions at this stage are grossly incorrect, a valid concept may be rejected and less desirable options may appear qualified. The actual usability of a design cannot be simulated with FEA in many cases. Samples may be required for market perception studies and a review of available options by management.

Another FEA tool that is becoming more reliable for phase I optimization involves automatic topology optimization. Altair Computing and The MacNeal-Schwendler Corporation have released products of this nature. Topology optimization seeks to calculate a shape of constant stress or strain energy based on inputs and constraints. A simple analogy to this process is the sculptor's response when asked how she makes her work so life-like. The sculptor stated that she merely chips away anything that does not look like her subject. Topology optimizers seek to remove material that does not appear to contribute to the load path.

The initial design of the winch mount was processed through the topology optimizing algorithm of Hyperstruct from Altair Computing. Hyperstruct computed the shape shown in Figs. 12.15 and 12.16.

Fig. 12.15. Winch mount geometry suggested by Hyperstruct.

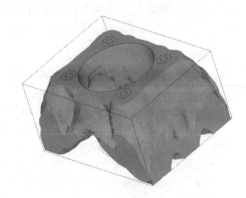

Fig. 12.16. Cross section of Hyperstruct solution of the winch mount.

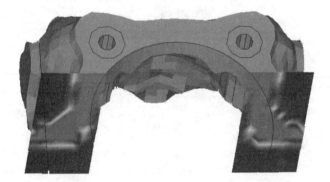

This technology is relatively new compared to general FEA. It currently supports a linear static and model solution in both 2D and 3D geometries. Its intent is to suggest structures, not to design parts for you. Based on the computed topology and the relative size of various portions of the structure, the design engineer can develop a more manufacturable component.

The topology optimization suggested a more rounded version of the casting option 1. The wall thickness in the topology optimized design must be thinned to be manufacturable, and the part must be reconfigured to suggest a more recognizable material flow. These suggestions must be taken into account in the final design concept.

Phase II Optimization: Fine Tuning

The project now enters the second phase of optimization where a specific concept is refined to produce the most performance for the lowest cost. Whereease phase I optimization focused on qualitative information, in phase II, the geometry, boundary conditions, and other assumptions must be reconsidered. At the completion of phase II, a final design should be ready for documentation and prototyping.

Phase II optimization is ideally suited for automatic optimization routines. These tools are excellent for tweaking a design to generate the lowest weight within specified stress or other performance limits. Most optimization programs require three sets of specifications in addition to the initial geometry: goal, limits, and part parameters. The goal specification is simply the measure by which the success of an optimization is judged. In most cases, the goal is to reduce weight, volume, or cost. While these three are usually directly related to each other, multiple material models or assemblies might have more expensive or heavier parts or materials. If the goal is to reduce cost, the optimizer should try to minimize the volume of the more expensive components first, and then work on optimizing the less expensive ones.

After the goal, the performance limits must be specified. This can be one of the most difficult parts of setting up an optimization because an unrealistic set of limits can cause the optimization to fail. If you are lucky, it will fail quickly. If Murphy has anything to say about it, the optimizer will crunch for hours or days trying to reconcile the limits before giving up. Typically, limits are specified in terms of maximum allowable stress, minimum allowable displacement, or minimum allowable first natural frequency. Weight limits, moment of inertia limits for rotating components, or reaction force limits are also common. The best way to ensure that the limits are reasonable is to have performed a series of sensitivity and parameter response studies prior to starting the optimization. These should suggest at least one solution within the specified limits. In the case of the winch mount, if natural frequency was the only limit, a wall thickness of 0.50 inches will get the first mode above 100 Hz. This is probably not the most desirable option but at least the optimizer can tell you if there is a better one. If the optimizer cannot satisfy the limits, it may not report any useful data.

Finally, the part parameters must be specified. These are the same parameters used in the previous phase I studies. You should learn from prior work if certain parameters are not worth pursuing, such as material stiffness. Reducing the number of parameters that the optimizer has to consider will greatly reduce optimization run time. In addition, the phase I studies should suggest a best-guess starting point for the optimizer. It is more efficient to start the optimizer close to your estimate of the optimal configuration and let the software refine the model. While it may seem like you are doing the software's job, remember that an optimization algorithm does not have the intelligence or foresight to understand the intricacies of parameter space. It is quite possible that the system may inadvertently identify a false minimum or local valley as the optimal location when the true minimum is quite different.

Fig. 12.17. 3D parameter space for two parameters.

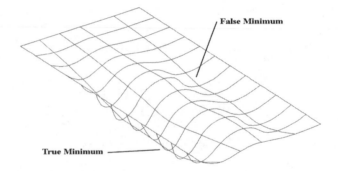

Fig. 12.17 illustrates this scenario in a depiction of parameter space. In this illustration, the two horizontal axes represent two different parameters that can be varied and the vertical axis represents the goal. As the two parameters combine differently, an infinite number of responses are possible. It is the optimizer's job to navigate this terrain. You can help the optimizer by using the data from the phase I work to select a starting point close to the true minimum. A good way to look at the optimization process is that phase I should get you 80 to 85% of the way there and phase II should take you the rest of the way. It may be that the estimate you can make after phase I is sufficient for the level of uncertainty in the product and the assumption set. If this is the case, the lengthy process of phase II may not be necessary. Use your judgment to make this call.

A shell model of the winch mount design from the completed phase I studies was optimized using Pro/MECHANICA.

Fig. 12.18. Pro/MECHANICA setup form for the winch mount optimization.

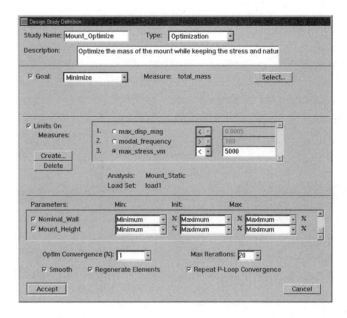

Fig. 12.18 shows the Pro/MECHANICA optimization setup form with the goal, limits, and part parameters specified. The specified goal was to minimize mass to keep costs down. The limits used in this optimization were to maintain the first natural frequency above 100 Hz, the controlling quantity in the design in previous work. Limits were also set on von Mises stress and maximum displacement of 5,000 PSI and 0.0005 in, respectively, to prevent the occurrence of an inadvertent combination of parameters finding a higher natural frequency at the expense of the other requirements in the design. The bottom of the form shows the part parameters used. Both part height and nominal wall were allowed to vary the full range of the parameter as specified in the sensitivity studies. However, the optimizer was instructed to use the shortest part height (minimum) and the thickest nominal wall (maximum) first. Based on the parameter response studies, this initial configuration will satisfy all limits. The optimizer can now begin to reduce weight while maintaining these limits.

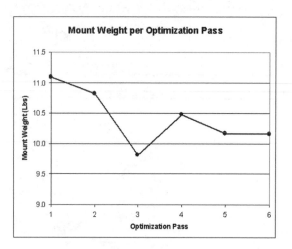

Fig. 12.19. Weight versus optimization pass showing the weight improvement as features are adjusted. At pass 3, too much material was removed and a correction can be seen in the plot.

Fig. 12.19 shows a plot of the part mass as various combinations of parameters were tried. Note that a significant amount of weight was removed from the initial design. The mass of the original design, the block with the holes, was approximately 39 lbs. The original phase I concept was only 7.5 lbs, but an additional 2.6 lbs were required to allow the part to meet specifications. It is not uncommon to actually add weight when a product is designed using up-front optimization. Instead of overdesigning and cutting back, you will learn to let the system add the features and the mass required for the problem at hand. On the other hand, optimization programs on existing components will generally yield weight savings.

Summary

The chapter describes the phases of an optimization project that has made the best use of the FEA tools at hand. Phase I models may be built and discarded relatively quickly before you have a chance to invest much time or emotion in their design. The best ones are evaluated more thoroughly using slightly more complex sensitivity and parameter response studies, and parameters once thought to be significant in the decision making process may be found to be minor players. Finally, the designs that make it through the screening process were taken into phase II optimization to identify the best configuration of the concept. While not every new product or part design will warrant such a rigorous evaluation, these concepts should be employed more often than not. You have the capability to make a significant impact on the cost and quality of the products you design using the advanced technologies at hand. Try to make the biggest splash you can.

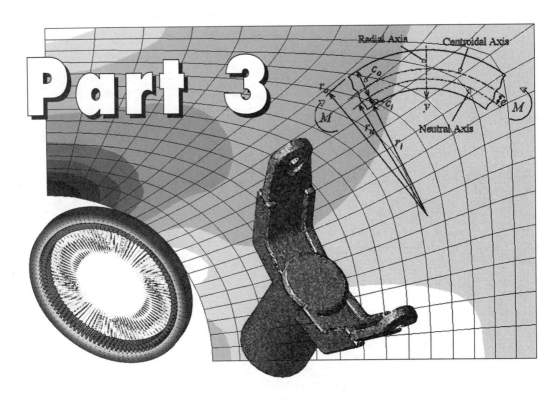

Part 3

Advanced Modeling Techniques and Applications

13

Modeling Assemblies and Weldments

Analyzing the structural response of assemblies is a natural extension of the part modeling techniques described previously in this text. In reality, few parts exist and operate in isolation. While the interaction between assembly components can often be approximated by boundary conditions, in many cases the entire assembly or subassemblies must be modeled to fully understand the system. This is readily apparent when multiple parts are joined together such that they behave as a single, continuous load-bearing structure, as in most weldments and heavily bolted castings. Another reason to model assemblies is when part deformation is not definable in a fixed coordinate system. The mating components may deform such that the final position, deformation, and stress state of the part under investigation are functions of displacement. Modeling and evaluating assembly interactions via additional modeling is as challenging as selecting boundary conditions to represent these interactions. The same amount of thought and care must be used in modeling assemblies as in modeling boundary conditions. In most cases, several equally viable options are available. It is your responsibility to qualify the accuracy of a chosen technique, and it is in your best interest to document that technique for future reference. This chapter begins by outlining means for identifying various types of assembly interactions and some general modeling methodologies. More specific techniques for modeling assembly interactions will then be presented. Many assembly models will need to mix and match these techniques depending on the goals of the analysis.

Design Model versus Analysis Model

While the compatibility of the design/CAD model and the analysis model should always be considered, the two will most likely diverge irretrievably in assembly modeling. Some continuous weldments might lend themselves to the use of a single geometry database for both design and analysis. However, as soon as fasteners are introduced, special handling and adjustment are required. You should probably assume that your CAD models will require substantial modification if they are to be used as the template for an assembly model. This assumption may spare you considerable disappointment at a later time. In reality, most assembly models will require that the geometry be developed specifically for the FEM. With this approach, you may be happily surprised when you find an opportunity to use existing CAD data for the analysis.

Assembly Type Controls Usability of CAD Model

Most assembly models can be categorized as either *continuous load bearing structures* or *jointed assemblies*. Continuous assemblies provide the greatest opportunity for utilizing CAD data because load effects at joints or fasteners are typically ignored. A jointed assembly, on the other hand, has uncoupled degrees of freedom at fasteners such as pins, bearings, or slots. Therefore, special handling is required to model these joints and, additionally, a large degree of simplification is usually required to make these joints efficient. If possible, make the decision regarding the type of assembly before starting the geometry. If it is not possible to make such decision prior to beginning the geometry, decide prior to finite element modeling. This decision will greatly impact how the geometry is laid out. The characteristics of each type will be discussed below with tips for improving modeling efficiency.

Continuous Load Bearing Structures

As the name implies, these assemblies do not have, or can be approximated as not having, any free, internal degrees of freedom. Instead, they behave as a single part. Plate weldments are good examples of such structures. Assemblies with many fasteners can often be approximated as being continuous unless the integrity of the fastener is called into question. Due to frictional effects, a pipe fitting flange with several bolts tightened to spec will most likely behave as if the part is welded to its mount. If your initial calculations indicate that the bolt preload can carry any axial load, this assumption is usually true. An exception would be when the load path and/or lack of stiffness combine to allow lift-off of one assembly surface from the other. The two bolted flanges in Fig. 13.1 illustrate this phenomenon.

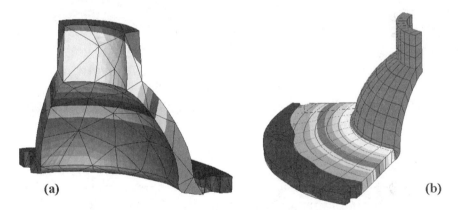

(a)   (b)

Fig. 13.1. Two bolted housings under an axial tensile load. The narrow flanged part (a) can be coupled at the entire interface surface without any loss in model integrity. The wide flanged part (b) undergoes enough deflection in the flange to warrant some thought as to where to join the interface surface with its mating part.

Under an axial pull, the flange in case 13.1(a) will probably remain in contact over the entire mating surface as long as the bolts do not fail. However, this assumption cannot be made so easily for case 13.1(b). The center of the flange may, in reality, lift off its mating surface. Therefore, if the assumption was made that the interface was continuous, the model may be overstiffened and the results may suggest that the part is much stronger than it actually is. In this case, the mesh should be assumed continuous only outside of the bolt circle.

In many cases, it is not convenient to build the active model as a continuous mesh of shells or solids. In others, flexibility to evaluate the load on fasteners at a later time must be built into the model. In cases where the individual components of a continuous assembly are separate continuous meshes, rigid links or multipoint constraints (MPCs) can be used to attach nodes across the interface.

Fig. 13.2. Shell model of housing assembly constructed with different nominal mesh sizes at the interface surface. If the interface can be assumed continuous, rigid links can bridge the two parts for proper load transfer.

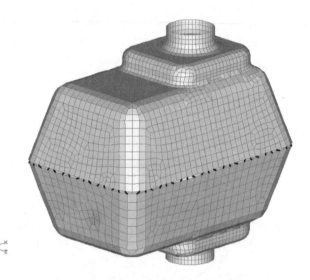

Fig. 13.2 illustrates this on a portion of a housing model consisting of two separate halves. The dark lines are the rigid links that ensure load continuity between the two halves. In this model, if the mesh had been planned to provide a matching node pattern on each half of the interface, the meshes could have simply been merged. The use of rigid links provides some flexibility in modeling in that the meshes at interfaces do not need to match in all cases.

In fact, if there is no concern about detailed fastener information later in the study, the mesh across assumed continuous interfaces should be constructed as continuous using merged nodes to transfer the loads. This keeps the model size and complexity to a minimum while ensuring that the model behaves per the assumption. When there is concern that fasteners may be of interest, plan the mesh so that fastener elements, usually beam elements, can be properly located.

Jointed Assemblies with Free Degrees of Freedom

When some relative motion at an interface or fastener is expected in the assembly, that interface is considered *jointed*. A degree of freedom is left uncoupled at some point in the assembly. Joints may be the result of translational or rotational guide pins, hinges, bearings, slides, or slots. These joints are included in the assembly because load transfer across the one or more coupled degrees of freedom is important to system behavior and the load transfer may be affected if the free component is coupled. A good example is a shaft in a journal bearing as shown in Fig. 13.3.

Fig. 13.3. Solid model of shaft in journal.

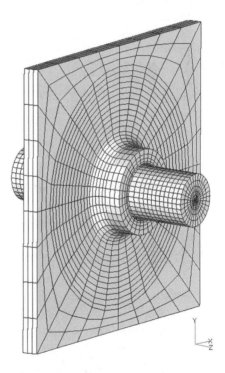

The shaft can transmit forces in the radial direction and moments in the plane of the frame, the XY plane as shown in the figure, if the clearance is tight. However, axial moments and forces cannot be transmitted through this joint. If this interaction cannot be modeled with boundary conditions, the assembly joint must be modeled. One of the many options for modeling this joint is shown in Fig. 13.4.

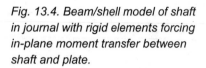

Fig. 13.4. Beam/shell model of shaft in journal with rigid elements forcing in-plane moment transfer between shaft and plate.

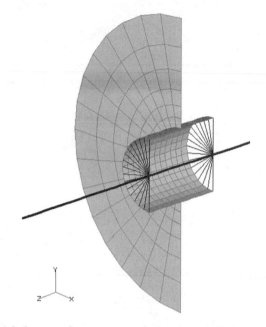

The shaft is modeled with beam elements, and the frame/boss/journal assembly is modeled with shells. A spider web of rigid elements ties the contact points at the front and back of the journal to the shaft in X and Y only. Beam releases are used at the nodes where the rigid elements attach to the shaft to ensure that no axial rotation or translation is coupled.

The modeling techniques for joints differ according to type, expected load path, and analysis goals. Some modeling methods for common joints and fasteners are detailed toward the end of this chapter. The same care taken for defining boundary conditions must be employed when modeling joints. In effect, the two are very similar. Joint modeling must remove spatial rigid body motion from the system. As results local to the joint become more important, more care must be taken with the joint representation. Contact with friction, the extreme modeling technique for a joint detail, requires the fewest assumptions but it may be difficult to solve these connections and the run times may be excessive. However, as discussed in Chapter 8, this may sometimes prove to be the most efficient technique.

Make use of the diverse elemental degrees of freedom available to you when considering joints. If local rotation about a bolt or rivet is desired without requiring detailed joint data, a beam element can simply connect two properly positioned nodes in a shell or solid model. Rigid links or multipoint constraints can also provide a great deal of flexibility in joint modeling. In most cases you have the ability to define the coupling between the degrees of freedom of any two nodes in a model with an MPC. Two plates can be modeled with a hinged interface if the nodes directly across from each other are only coupled in the translational degrees of freedom. Fig. 13.5 shows the difference in resulting deformation when the edges of the two plates are fully coupled (a) and coupled with a hinge (b) as described previously.

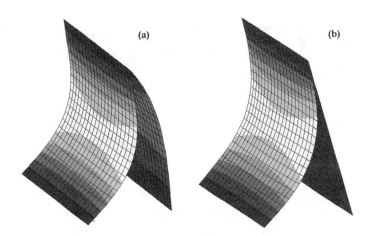

(a)   (b)

Fig. 13.5. Two plate models pinned at bases and loaded with a uniform pressure on the left face. The assembly in (a) is modeled as if the coincident edges are welded and (b) is modeled as if they are hinged. The resulting deformation is noticeably different.

Mixed Assembly Modeling

Many assemblies will contain some combination of continuous and jointed interfaces. Furthermore, many assembly models will have combinations of different element types. Mixing element types in an assembly rather than in a part is much more common and often more efficient because assemblies can often contain shafts or beam-like parts, thin-walled components, and chunky components. Since assembly models can grow quite large, it is absolutely in your best interest to choose the most efficient element type for the part represented. When mixing assembly or element types in a single assembly, take care to couple all appropriate degrees of freedom where required. Next, the load transfer between the idealized geometry of one element must be accounted for in another type. The bolted plate model shown in Fig. 13.6 is a good example. For one reason or another, it was determined that the bolt should be modeled with beams of an equivalent cross section. However, a beam can only transfer load through an end point or node, while the actual bolt will load the plate through its head. To correct for this, a spider web of rigid links is included to transfer the load from the bolt end to the diameter of the actual bolt head or washer. Note that the limitation on this technique as presented is that it does not account for the bolt preload. If the local effects caused by the preload are a desired output, a modification of this technique or a different technique must be employed. Refer to the "Bolts, Rivets, and Pins" section later in the chapter.

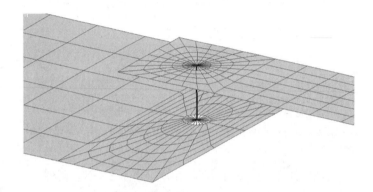

Fig. 13.6. Two plates attached with a beam element representing a bolt. The load at the end of the bolt is transferred to the plates with a spider web of rigid elements to increase the effective surface area.

No special consideration is required when mixing continuous and jointed assembly types beyond the specific needs of each interface. The more interfaces included in a model, the greater the chance for error. Build assemblies slowly and carefully to reduce the probability of modeling errors.

Realistic Interpretation of Local Joint Results

As mentioned previously, an assembly interface must be treated like a boundary condition. With that in mind, the results local to an assembly interface must be interpreted in the proper context. It is best to assume that an interface in a large assembly model is only good for transferring loads. Local results should at least be suspect and at best disregarded. If local results are important, consider making the interface more detailed, possibly with contact, in a submodel focused on that interface alone. Assembly models can become extremely large. Consequently, using assembly models for macro behavior, and smaller, dedicated models for micro results is recommended.

Component Contribution Analysis

The concept of component contribution analysis (CCA) for assembly modeling is introduced in Chapter 4. That discussion will be reiterated here with more detail as it pertains to each step. CCA is most applicable to jointed assemblies although more complex, continuous structures may benefit from a step-by-step study. CCA allows you to evaluate each part or continuous subassembly on its own without letting the complexity of a full assembly skew the results or confuse the situation. The process will require more patience as the instant gratification of simply throwing all the parts together and seeing what happens is set aside for a more systematic approach aimed at improving accuracy and understanding. In the short term, it will take somewhat longer than the quick-and-dirty approach, but the payoff will be reduced iterations, fewer incorrect prototypes, and a shorter product development cycle–all the reasons your company invested in simulation in the first place.

Treat Continuous Subassemblies as Single Component

The first step in CCA is to differentiate continuous subassembly interfaces from jointed ones. Due to the load path in a continuous interface, parts sharing these interfaces should be considered and modeled as a single component. At the initial stage, symmetry idealizations should be identified as well. Model each part or continuous subassembly individually with the most appropriate idealizations for the particular geometry and load path.

Use Free Body Diagrams or Kinematics to Develop Boundary Conditions

Develop boundary conditions representing the jointed interfaces between components. This process should entail the use of all techniques described throughout this book for determining and qualifying the effectiveness of a boundary condition scheme. Document the assumptions underlying each boundary condition and bracket the solution as required.

A free body diagram (FBD) or kinematic simulation can prove valuable in determining the reactions between all parts in the assembly. By performing one or both of these studies correctly you can be confident of a consistent load distribution throughout each assembly component. It should be clear by this point that if your boundary condition assumptions are incorrect, especially in load magnitude, all downstream results will suffer. Do not be afraid to ask a peer to review your FBD and calculations.

Optimize Each Component on Its Own Merits

With each continuous component modeled separately, the limitations and benefits of particular part designs can be evaluated. Approach each part as a separate optimization project. In effect, when only one part is analyzed or improved in an FEA study, a CCA analysis is carried out but does not combine the parts with modeled joints.

You may learn enough at this stage to dispense with further assembly modeling. This is where the time savings begin to emerge. Things to look for at this stage are parts that are so rigid that their individual performance is no longer of interest. If part cost is an issue, these components might be optimized further. Next, seek to identify the weak link in your assembly. Which components will contribute disproportionately to overall system displacement? Which parts are most likely to break first? By the time you proceed with assembly modeling, you should have a good feeling about how each part will contribute to the overall displacement proportional to its size and/or designed stiffness. You should also have optimized each component in order to avoid an obvious weak link, and individual components should be evenly designed.

Combine into Assemblies One Component at a Time

Once each part has been taken as far as possible on its own and you have confirmed that further assembly modeling is required, it is time to begin evaluating modeling of the joints. Combine components one joint at a time. It is best to build off the part that will be the assembly's link to ground. This part should have the most predictable performance. As you build up the assembly, keep the following points in mind.

Use Test Models of Joints

Build test models of all joints. There are several ways to model any joint. Whether the joint is rotational or translational in nature, some methods will provide better results than others for your particular problem. The best way to evaluate options is by using test models. The example of the shaft in a journal bearing described earlier is one example of a test model. The description of each particular joint later in the chapter will include more test model ideas. Use these models to ensure that the load transfer is as expected, and that it is the most efficient means of making the assembly transition.

Identify Fasteners Requiring Detailed Results

As you evaluate joint modeling techniques, consider whether you will need information at a fastener such as shear load or axial force. If these data are required, use a joint technique that will assist in extracting the information. If axial forces of a bolt are of interest, model the bolt using a beam or solid mesh of the appropriate cross section and material. In this fashion, the axial response will be improved, and the stress, load, and displacement should correlate using basic beam and spring calculations. On the other hand, if detailed fastener data are not required, search for the most accurate and efficient joint model possible.

Consider Submodeling for More Detailed Local Results

When detailed results near a joint are important, it is best to plan to model the joint with a submodel, thereby allowing greater attention to modeling and results detail. Plan for the submodels when the mesh near a joint is constructed. If your code supports automatic submodeling, familiarize yourself with the available techniques. If submodeling is not supported automatically, you can accomplish the same thing by using a node pattern and numbering scheme at the boundary of the area being detailed, and transferring the displacements and rotations from the assembly results to the submodel boundaries. Again, certain codes handle this better than others.

Examine Results Local to Each Joint

While the results local to each joint should not be considered accurate in magnitude, the displacement and stress distribution can provide valuable insight on the quality of the joint assumption. As with boundary conditions, the joint should not allow or prevent any deformation that the actual interface would not allow or prevent. The local results can then indicate the quality of the load transfer developed by the joint. If you can identify a highly localized discrepancy between the modeled joint and the actual joint, you may be able to ignore it. However, if the local behavior suggests that the force transferred between the parts is incorrect, an adjustment should be made.

Review Performance of Each Part at Each Step

After each component is added, compare the behavior of the part in the assembly model to its performance in the component analysis. If they do not match, resolve the difference before proceeding. The difference may be the sole result of improved load transfer in the assembly model. However, you may learn that the assumptions on which you based your component analysis boundary conditions were incorrect. Similarly, the difference might suggest that the joint understiffened, overstiffened, or provided the wrong load transfer to the part. Again, resolve the discrepancy before proceeding. A difference between results and expectations is acceptable; failure to understand the difference is not.

Review Complete Assembly for Conformance to Global Expectations

When the complete assembly is finally modeled, you should have complete confidence in the results because you will have qualified all assumptions along the way. Now you are in a position to evaluate the assembly response as indicative of the design. The final displacement should not surprise you but may show a need for design improvement. As with the response of a single component, if the final response does not meet your expectations, resolve the discrepancy. If it does behave as expected, you can proceed with improvement of the components as required.

Submodel Portions of System Using Assembly Reactions

Upon review of the final system response, you may identify areas of local behavior that warrant a more detailed study. You will have to make the call, based on the size and complexity of the model and the time you have to complete your study, as to whether you refine the model in the assembly or build a submodel. Staying in the assembly model in which you invested so much time is attractive and may be the best bet. However, when detailed contact modeling or a nonlinear technique is required to provide the refinement you need, a submodel will nearly always make sense.

Modeling Continuous Load Bearing Assemblies

With the overall methodology in place, a discussion of the modeling techniques for the two types of assembly interfaces will be discussed. A continuous interface is the most straightforward because it is the most like modeling a single part. However, an assembly model carries issues of interface strength and local results. These cannot and should not always be ignored. Although techniques for solid assemblies and thin-walled assemblies are similar, they are also different enough to warrant separate discussion.

Solid Assemblies

Solid parts are typically joined continuously via welding or fastening. A fastened solid could use bolts, staking, press fits, dowels, or rivets. These fasteners are all handled similarly because it is assumed that they completely and continuously couple the interface between the two parts involved.

Weldments

A key concern in modeling welded solid interfaces is whether and how to treat the weld specifically. If the weld is far from the area of interest, no special treatment is required. In practice, most solid weldments are stronger at the weld and failure usually occurs just beyond the bead. Assuming that the material properties at the weld and heat affected zone (HAZ) are the same as those of the base material is typical. Deviating from this assumption should only follow detailed testing of actual parts.

Modeling of the weld geometry is warranted only when detailed results near the weld are required. It is common practice to manually mesh brick/wedge elements in these areas to improve accuracy. The danger here is that it is difficult to manually converge a hand meshed model. Although the results from a similarly dense hand meshed model are likely to be better than an automeshed one, the time it takes to remesh for convergence usually discourages new users. The catch 22 here is that if the local results were important enough to warrant hand meshing in the first place, convergence is doubly important. These situations are ideally suited to adaptive p-elements, which allow convergence without remeshing.

Another reason to use adaptive p-elements is that when full weld penetration is not achieved or not possible, as in the cross section shown in Fig. 13.7, the stresses at the root of the weld are nearly singular. High stresses in this area are a common cause of solid weld failure, or fracture from the inside out. Adaptive p-elements are better suited to capture these high gradients.

Fig. 13.7. Plane stress model of fillet weld in which vertical plate was too thick for full penetration.

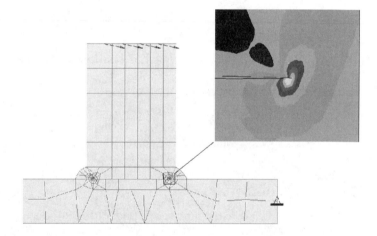

Another concern in modeling welded solids lacking full penetration is the possibility of relative motion and contact within the interface itself. As the noncontinuous area enclosed by the weld increases, the likelihood that contact interaction may affect local weld results increases. Using the cross sectional model previously shown, Fig. 13.8 demonstrates the stress distribution in the rightmost weld with contact between the mating parts (a) and without contact (b) for comparison. Adaptive p-elements in Pro/MECHANICA were used to capture the detailed results.

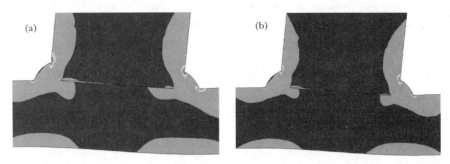

Fig. 13.8. Stress distribution in welded section can be affected by contact conditions between unwelded surface areas. Case (a) shows a contact interface, and (b) shows the joint with no contact.

The example shows that contact conditions prevent the surfaces on the right side of the joint from passing through each other, thereby resulting in relatively low stress compared to the left side of the joint. On the left side, the contact and no contact examples have very similar stress distributions since both allow lift-off. This example suggests that failure to use contact conditions has resulted in an overprediction of root stress in the compressive condition, but had little impact on the tension. Look for trends like this in test models of your systems before committing to contact. The contact run in this simple model took 7.75 times longer than the no contact run.

Detailed weld modeling is not recommended in a large assembly model. A small assembly involving just the welded parts or a submodel of the area in question should be used.

Bolted Castings or Machined Parts

The primary modeling concern in a continuous fastened interface is the amount of surface area that should be coupled. Certainly a region local to each fastener should be able to be merged without a problem. If this is not true, the continuous interface assumption may not be valid. How far to expand the coupled area depends on the stiffness, loading, and expected deformation of the interface surfaces. Any surface area expected to separate from its opposing surface should be left free, while any surface expected to be in compression should be joined. Making these determinations may require a few iterations. Any detail specific only to the fasteners, which are now assumed to be part of the continuous interface, could probably be removed to improve model efficiency. If detailed data in that area are needed, a continuous interface may not be valid.

Sheet Metal Assemblies

While practically any shape of weld or fastener could, theoretically, be modeled as a solid in a solid model, it is difficult to model a bolt with a shell element. Most sheet metal assemblies are best modeled with shell elements. Shell models will usually require mixed element modeling if fastener or weld detail is required. If solids are used to model the sheet metal parts, the model could follow the abovementioned guidelines. The choice of modeling techniques is again dependent on study goals and requirements.

Weldments

Sheet metal weldments can be extracted from certain CAD models when the software provides tools for mid-plane extraction. Even then, the benefits of using a CAD model are questionable because the CAD model must be built using just the right techniques. These techniques are not always compatible with the best techniques for part design or modeling for downstream modification. Consequently, building a CAD model specifically for FEA is common even when mid-plane extraction is available. When this is the case, it may be just as convenient to build the 3D wireframe and surfaces directly instead of building a solid first. Even if the CAD model can be used efficiently to develop the mid-plane surfaces, it is important to remember that the resultant weldment will have only continuous welds at the seams. Some weldments have other types of welds and you may have to deal with them. In all sheet metal weldment analyses, it is important to remember that the actual geometry of a weld is very difficult to predict in most manufacturing environments. Consider all weld detail a gross approximation unless a detailed study of a specific case has been undertaken. It is good practice to use the analysis model to find a design that depends on the weld for fastening only, and not for structural integrity.

Continuous Welds

Continuous welds are the easiest and most common way to assemble a sheet metal weldment in FEA. Essentially, the system is modeled as a continuous plate with no special treatment for the seams. If more detail is needed at the weld, it is possible to build solid weld elements at the seam where the weld geometry is required. This will help distribute the load to the actual area of the weld and adds the additional stiffness provided by the added material. If only the additional bending stiffness is required, you can add beam elements at the seam with the proper cross section.

Skip or Partial Welds

A common question posed to a design engineer that can be answered with an FEA model is how much weld the structure really needs before integrity is compromised. Consequently, techniques for skip or partial welds are used to tie different plates together in varying amounts. The most common method is to simply merge only the nodes in the areas of interest for load transfer. If the plates are offset and parallel or the edges are otherwise not coincident, rigid links or plate elements can be used to bridge the gap. If plate elements are used, they are usually given the material properties of the base metal and a thickness that is average for the thicknesses in the area. One danger of modeling skip welds with plates is that the results local to the ends of the partial welds are near singular and guaranteed to be incorrect. This is not usually a problem if taken into consideration when interpreting the final results, unless the hot spots created by these weld elements link up with other, more believable high stress regions. If this occurs, more detail in the weld area should be modeled, possibly by going to solids with a submodel.

Plug or Spot Welds

Spot welds will be differentiated from plug welds in this discussion as being closer to point attachments, whereas plug welds will have a significant area. Spot welds can be modeled several ways. A stiff beam can attach two nodes across the plates being welded. Similarly, a rigid element can be used or local nodes can simply be merged if the plates are coincident. Plug welds, on the other hand, may be modeled with a line element as well, but will probably solve more realistically if patches on the opposing surfaces are aligned and the plug is modeled with solid elements between them. If the connections are far from the area of interest, you then have more flexibility in the choice of techniques. Remember that there are always several ways to model any connection and you should choose the one which will provide the required degree of accuracy.

Riveted or Bolted Assemblies

Riveted or bolted plates that provide continuous load transfer can generally be treated as welded assemblies. If the fasteners are numerous enough and the joint is far from any area of interest, the model can simply be built as a continuous mesh. It is equally common to use the same techniques described for spot welds to model these fasteners. If the fasteners are not numerous enough to assume the components are perfectly continuous, it may be valid to attach the parts using stiff beams or rigid links. The result will then behave more like a jointed interface. These techniques are described in greater detail later in this chapter.

More on Continuous Assemblies

There are many other types of continuous structures in addition to sheet metal weldments. While the different types are numerous, a couple of the most common warrant discussion. Use these techniques as examples to help you develop your own solutions to the many problems that will arise in product development challenges. As mentioned previously, the only limitation to the modeling techniques available to you is your creativity.

Plastic Assemblies

Due to the relative flexibility of plastic assemblies, it is not fair to assume that much of the interface surface beyond the actual connection is continuous. Plastic components can be adhesive bonded, sonic or thermal welded, heat staked or fastened. In each of these cases, couple the mesh local to the fasteners only and run a linear study. If a significant surface area goes into contact, you can probably merge those areas as well. However, if you expect significant sliding after the contact, use multipoint constraints or constraint equations to couple only the normal translation. Use contact as a last resort unless you plan to use a nonlinear analysis anyway. In that case, some contact may not add much run time to the already long run time required for the nonlinear study. Reviewing the section in Chapter 15 on nonlinear analysis methodology is recommended.

When fasteners such as screws and rivets are used, remember that these steel or aluminum components are much stiffer than the surrounding plastic. Fastener stiffness may need to be accounted for with a beam or solid elements of the appropriate cross section. Verify that the beam or solid elements are tied into the plastic mesh properly in order to avoid overstiffening or understiffening the model.

Printed Circuit Boards

Printed circuit board (PCB) analysis can be made very difficult or relatively easy. Again, the goals of the study must be considered. Many analyses of PCBs are concerned with gross behavior such as overall deflection under dead weight or in assembly. For these studies, most of the components on the board add no stiffness and can be omitted. Model the components with significant mass and those that will directly affect the stiffness. Transformers, large integrated circuit (IC) chips, and heat sinks will most likely affect the bending properties of the board. The connectivity of the components should also be modeled. If the part is soldered to the board at only the leads, this attachment should be modeled. Many large components are adhesive bonded prior to soldering or fixed with screws, rivets, or other fasteners. These connections will provide much better coupling of the loads and stiffnesses. Discuss the existing or planned assembly with the electrical engineers in charge of designing the PCB before rushing into the analysis. Build a simple test model of an existing sample of the intended PCB base material. The material, thickness, and number of layers will vary from application to application and the stiffness, or equivalent Young's modulus, should be verified with test data.

The natural frequency of a PCB is often of interest. The frequency will provide an indication of the board's ability to withstand a vibrating environment. For these problems, the board should include the components required for stiffness as described above, but it must also include any additional components with a high mass moment of inertia in any direction. The example in Chapter 6 showing the change in the first modal frequency with the actual mass distribution included, versus the point mass value alone, should provide enough incentive to consider inputting the mass moments numerically or modeling a rough shape of the component.

Beyond these gross behavioral studies, you may need to consider the stress levels in connections or solder joints. These studies are very difficult and usually require a nonlinear approach with detailed material data. In addition, consideration of solder joint quality will suggest that cold joints or voids in the solder will have a significant effect on connection integrity. Initial studies for an overall feel of the stress levels in various parts of the PCB are often warranted but do not approach a detailed joint evaluation unless you have taken the time to fully prepare for the challenge.

Construction Assemblies

While the focus of this book has been on product design, many design engineers must address structures that fall into the realm of civil or structural engineering. These structures are usually beam models, but could be plates or even solids depending on the problem. Some of these systems can be thought of in terms of the product modeling techniques discussed throughout this text. The need for thoughtful boundary conditions and careful results interpretation is still important.

A few techniques for large beam structures are not really applicable in other systems. When the size of the structure is significantly greater than the cross section of the beams involved, the model becomes less sensitive to the precision of neutral axes. Consequently, offsets are probably not warranted. Remember, however, that proper calculation of the rotational constant and shear center can still affect gross behavior. Another area of consideration is the local stiffness at joints. While the joint data in a beam model are not detailed enough to make judgments about local stress without a submodel, some means of connecting beams in the actual structure add bending stiffness to the system. Two pipes welded cleanly to each other in a T can be approximated by a simple beam to beam connection. However, when mount plates, splices, and large connectors are involved, the local bending stiffness will increase as illustrated in Fig. 13.9. The local stiffness can be adjusted by increasing the cross section of the local beams or by adding a plate or beam as a gusset to brace the corner. Correlation testing is recommended before choosing a stiffening technique.

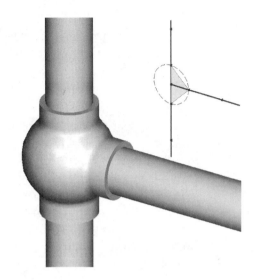

Fig. 13.9. Fittings or mounting hardware can stiffen the interface of two beams. The additional stiffness of this coupling was modeled with two plate elements.

Modeling Jointed Interfaces

Jointed interfaces provide one of the more interesting challenges in FEA. They provide somewhat of a bridge between rigid kinetic analysis and elastic structural analysis. Both types of analyses require that the proper degrees of freedom are coupled or left free. Redundant constraints in either type of model will cause the resulting data to be incorrect. In FEA, the user may also be concerned about local stresses near a joint. This section will describe techniques for building in the various degrees of freedom at joints, as well as techniques for improving the joint model to allow a reasonable estimate of local behavior.

Identify Critical Information Up Front

As with any concept in analysis, if you do not ask a question, you will not get an answer–at least not the right one. Therefore, it is important that you decide up front if stress or deformation is required in the fastener or other component directly included in the joint. The method for modeling the joint is directly affected by this decision. Review the goals and assumptions that are valid for the problem at hand before plunging in. Much time at the end of the process will be saved if you do not need to redo the models because an error in a joint was found. It is not always easy to spot an error in a complex assembly. Use the CCA methods described previously and always build test models of the joints you decide to use. Test models are especially important if the technique or the particular application of the technique is new to you. Document these test models for later reference.

One good test to determine if the joint you modeled truly provides the freedom you expect is by running a modal analysis. If you build a test model so that the degree of freedom needed is the only free component, the first mode should have zero magnitude and show the displacements of that DOF. In the case of rotational joints, the animation of the rotation may not be clear depending on the frames your postprocessor chooses to display. Therefore, it might be advantageous to add a weak spring stiffness to that degree of freedom so that the motion is numerically constrained but still the weakest one in the model. This will limit the calculated displacement to some finite magnitude and it should show better in postprocessing.

Rotational Joints

Modeling rotational joints requires an understanding of element degrees of freedom as well as coordinate systems and multipoint constraints or constraint equations. While merging two nodes on elements of different types will couple the translational degrees of freedom, the different element types–beam, shell, and solid–will not couple rotational degrees of freedom when joined to each other. You can use this to your advantage when developing rotational joints. Similarly, if you do not have a good grasp of the fundamental concepts of degrees of freedom and you do not build test models to verify your assumptions, you can build in joints which do not free up the rotations desired. Once again, if the modeling is incorrect, the deflections in the full assembly may not make it obvious.

A classic example of an incorrect joint derives from the tendency for new users to simulate the connection of a cylindrical or circular hole in a solid or plate to a grounded pin with a constraint that fixes all degrees of freedom except the rotational one in the direction parallel to the hole axis. For a solid model this will not work simply because rotational DOFs are not supported by solid elements. In a shell model, because it will only free the local and not the spatial rotations of the hole's edge, the joint will not work as desired. An appropriate technique is to position a cylindrical coordinate system coaxial with the hole and fix all degrees of freedom on the hole except for the tangential translations. This technique was presented in Chapter 8 in greater detail.

Bolts, Rivets, and Pins

The rotational degrees of freedom of a fastener should be considered only if the spacing between the fasteners is large enough to warrant expectation of this type of deformation. If the bolt or rivet pattern is tight, assuming a continuous interface may be best. If you have doubts, build a simple test model of an equivalent structure. Fig. 13.10 shows a narrow plate bolted to a frame with a load in the center. Case (a) has one fastener on each end and case (b) has four.

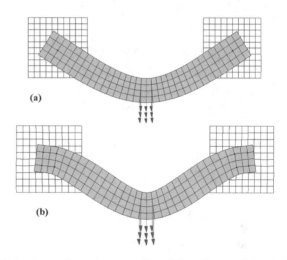

Fig. 13.10. The thin plate in these models is attached to the stationary frame with a single fastener on each end (a) and four fasteners (b). As the number of fasteners increases, the interface behaves more as if it were continuous than jointed.

With the two fasteners, it is clear that the rotational freedom of the joint affects the system stiffness. However, with a few more bolts, the interface becomes more rigid, and detailed modeling of the rotations may not be required. In this model, beam elements were centered on the two opposing holes and tied to the perimeter of holes with a rigid element spider web. The axial (X in the beam C.S.) rotational degrees of freedom of each beam end were released to provide the rotation. Friction between the parts is not accounted for with this technique.

The method described above is the most common technique for modeling a fastener that allows rotations when the actual hole is modeled. An alternate approach, when the hole is not modeled but filled with plates or solids, is to simply use an MPC that couples all degrees of freedom except for the appropriate rotational component. Merging nodes on coplanar shell surfaces as described in the section on spot welds will not work because the shell-shell interface will remove the rotational components as well as the translational ones at the merged nodes. Remember that if bolt preload must be considered, preloading a beam with a temperature load that shortens its length or a stiff spring element that allows preload may be required. If the latter is still insufficient, the following techniques using contact may be utilized.

Detailed Fastener Interfaces

In some systems, the clamping force on a part due to a torqued fastener is very sizeable and may sometimes be the limiting factor in a design, especially for connections that involve cylindrical clamping via pinch bolts or pipe clamps. The contact pressure at these interfaces is very complex and is dependent upon the geometry of the clamp, the distance between the bolt and part centerline, and the system clearance. For these systems, solid contact models can provide excellent data with very few, if any, assumptions. One difficulty in setting up these models is determining how to apply the appropriate bolt preload and distribute the contact pressure under the bolt head, as it would be developed by the resulting deformations. One successful technique is to model the clamp and fastener with initial penetration between the bolt head and counterbore, as shown in Fig. 13.11.

Fig. 13.11. Symmetry model of pipe clamp with initial penetration of fastener. All contact regions noted in the figure are important to total system response.

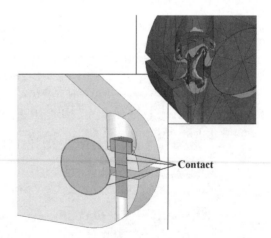

The contact region, defined between the bottom of the bolt head and its corresponding load surface, is used to pull the model into compression.

This technique is only valid if your solver supports initial penetration of a contact region. In this model, it is assumed that the interlocking threads behave as a continuous interface and the mesh at the thread engagement is merged. If the bolt is threaded into a part that can be assumed rigid or grounded, a cross section of the bolt at the root of the interlocking threads can be constrained instead of having to deal with merged volumes. In fact, if system geometry permits, in some cases using a constraint with enforced axial displacement in place of the initial penetration is effective. Thus, usage of this technique is possible in several codes.

The key to the accuracy of these models is correct input of the initial penetration distance. Because both the bolt and the fastened component will deflect to reach an equilibrium configuration that is generally unknown in advance, it is extremely difficult, if not impossible, to make an educated guess as to how large this penetration should be. However, because most of the model, except for the contact interface itself, will behave linearly (assuming small deflections and linear material properties), it is possible to determine this penetration in a single iteration. First, you must know the expected axial force on the bolt. The following equations can be used to estimate the penetration:

> *Eq. 13.1   Torquing of the bolt head $F = T/(K*D*1.2)$*

> *Eq. 13.2   Torquing of the nut $F = T/(K*D)$*

where F = the axial force in the fastener, T = applied torque, K = friction factor, and D = major diameter of the fastener.

Note that the factor of 1.2 in Eq. 13.1 is used to account for the flexibility of the bolt shank, which springs back a small amount after the torque is applied. This torsional spring effect is avoided when the nut is torqued instead–hence, the missing factor in Eq. 13.2. Some typical values for the friction coefficient, K, are provided in Table 13.1.

Table 13.1. Typical friction factors for calculating bolt loading

Condition	Friction factor
Nonplated, nonlubricated connection	0.30
Zinc plated	0.20
Cadmium plated	0.16
Anti-seize lubricated	0.12

When using any type of locking nut, the K factor will generally be greater than 0.3; the exact value should be readily available from the nut manufacturer. Note from the equations that as K decreases, the axial force increases because less torsional energy is lost through friction. Therefore, if there is any doubt as to its exact value, using a lower K to assume a higher bolt force for the same applied torque is conservative.

Once you know the expected force on the fastener, you can create a resultant force measure on a transverse cut surface through the bolt shank; otherwise, request these element or node forces so that the resulting axial force on the bolt may be obtained from the analysis. You can then begin with a small bolt head penetration. Once the model has solved, compare the resulting force to the expected value and adjust this initial penetration as a linear extrapolation of these numbers. Rerun the analysis and compare the new resultant force. This single iteration should provide what you need. However, if deemed necessary, adjust one more time, rerun, and check again.

Although more than one iteration might be required to complete this exercise, if these forces are important to part performance, you must be thorough in this process. The reward will be a very complex, real stress state solution that could not easily be obtained through any other means.

Rotational Bearings

Modeling rotational bearings requires consideration of the radial and axial stiffness of the component. In most cases, it is assumed that the bearing is rigid in comparison to the rest of the system. This is not always true and the bearing elasticity may have a significant impact on the deflection of shafts using those bearings. The most basic bearing model with a beam mesh representing the shaft simply ties a point on the shaft at the center of the bearing surface to the mounting features of the bearing with a rigid link. The axial rotation of the beam should be released at this point to provide rotation.

The above technique transfers in plane moments, or moments that might be generated by deflection of the shaft, to the supporting plate. While this might be applicable to journal bearings, tapered bearings, or other bearings with rotational stiffness, ball bearings should be able to take small rotations or small displacements without appreciable moment transfer. To free this up, the rigid element should be replaced with stiff springs. The technique will also now let you dial in the radial stiffness of the bearing. Figs. 13.12 and 13.13 show the difference between these two methods. The test model in 13.12 has rigid links tying the bearing to the plate. You can clearly see that the moment created by the centrally located load is carried through the plate. In the adjusted model of Fig. 13.13, the shaft is free to deform without imparting any bending moment on the plate.

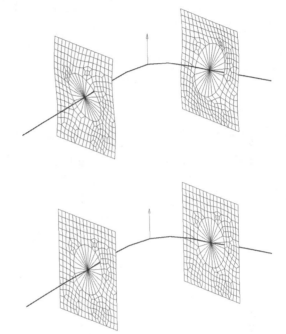

Fig. 13.12. Beam model of shaft tied to plate with rigid elements to induce moment on plate.

Fig. 13.13. Beam model of shaft tied to plate with stiff springs so that no moment is transferred to plate.

When the shaft is comprised of solids, the bearings can be handled a couple of different ways. One option is to tie a node at the center of the solid shaft to the bearing mounts as described for a beam element shaft. Because solids have no rotational degrees of freedom, this will not transfer any moments to the mounting area. To achieve a moment transfer as indicated in Fig. 13.12, it is best to attach the shaft to both the front and back of the bearing bore using rigid elements or stiff springs. You can still attach a centrally located node to the bearing bore to achieve the required moment transfer. However, if you use the nodes on the perimeter of the shaft, some local compression of the shaft diameter will come into play and it will be much easier to switch to a contact solution later. This technique requires that you plan in advance and build the mesh with aligned nodes on the shaft and bearing bore. Figures 13.14 through 13.16 show a symmetry model of a shaft in a journal.

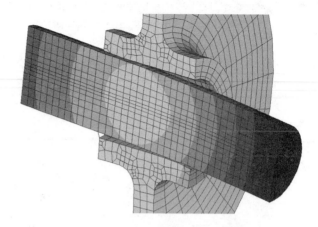

Fig. 13.14. Solid model of shaft in journal with contact conditions between shaft and ends of bore.

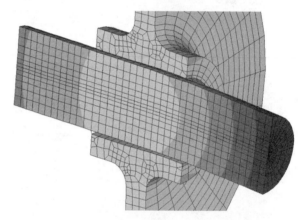

Fig. 13.15. Solid model of shaft in journal with stiff springs coupling shaft and ends of bore around entire perimeter of shaft.

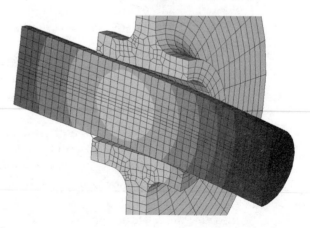

Fig. 13.16. Solid model of shaft in journal with stiff springs only coupling portions of interface expected to be in compression.

In Fig. 13.14, the lines connecting the front and back of the journal to the shaft are actually gap elements and a contact solution was used. Fig. 13.15 shows the same model with the exception that the gap elements were converted to stiff springs (136 lbs/in) and solved as a linear problem. The total displacement is somewhat less than the gap solution. This is due to the fact that the springs in the portion of the model that should show the shaft separating from the journal are actually pulling the journal with the shaft, overstiffening the system. Fig. 13.16 shows the spring model revised so that only the springs in the contact area are stiff. The others have been converted to very weak springs; they were not removed altogether just in case the load direction is changed. Leaving them in but rendering them ineffective will save time in later revisions. This model matches the contact model very closely and ran in a fraction of the time. For this simple test model, the contact solution took 26 minutes and the linear solution took only two minutes.

For small models, the contact option may be efficient enough. Consequently, developing a creative solution to avoid contact would not be warranted. However, for larger models, a nonlinear solution might be prohibitive.

Translational Joints

A translational joint allows two nodes or meshes to slide past each other or translate relative to one another. A sliding joint might represent a linear bearing, a shaft in a journal, or a pin in a slot. The joint might simply represent two surfaces that are in sliding contact. As mentioned previously, if the model size is small enough or you intend to introduce nonlinearity in some other way, using contact may be the best solution to the translational joint model. However, many assembly models are too large to cast into the nonlinear arena without a compelling reason. Therefore, it is important to understand the options available for two parts that react with a translational connection.

Sliding Contact

Modeling sliding contact between two nodes in one or two directions is relatively straightforward in most codes. A multipoint constraint or a constraint equation can enforce the relative position of the nodes to keep them colinear or coplanar. The difficulty comes from constraining groups of nodes, possibly on nonplanar surfaces, to slide without using nonlinear contact. The solution in some cases is to use alternate coordinate systems. A cylindrical coordinate system in conjunction with MPCs can force two cylindrical surfaces to stay concentric while allowing axial sliding just as the action of a ball joint can be simulated with a spherical coordinate system. It is important to remember that the local results may not be valid, depending on the system, and that these joints are used primarily to ensure load transfer between two parts. If your solver supports spring elements that can provide stiffness in one or more degrees of freedom (DOF springs) versus simply point-to-point translational or rotational springs, consider using these instead of MPCs.

An alternate technique for modeling sliding contact on planar surfaces that undergo very small amounts of sliding is to use orthotropic material properties. A thin layer of solid elements can be inserted between the two interfacing surfaces to provide a continuous solid mesh. The interface solids can be assigned a material property with high normal stiffness but extremely low in-plane stiffness. Because this material will only transfer normal loads but will not be able to carry in-plane loads, sliding-like behavior results. While this technique has been used with limited success, using MPCs or DOF springs is more straightforward. In addition, if you decide that contact is warranted later, line elements such as two-point rigid links or springs can easily be converted to gap elements without any additional modeling.

Linear Bearings

Most systems attached to linear bearings can be modeled with the assumption that the linear bearing is grounded. If this is the case, and the elasticity of the bearing components is not a factor in the stiffness of the model, boundary conditions should be used instead of trying to develop an assembly joint. However, if the linear bearing is mounted to an elastic member with a stiffness similar to the component being guided by the bearing, a sliding joint may be required. These joints can be modeled with MPCs effectively, but there will be no bearing stiffness. Degree-of-freedom springs can achieve both goals. The stiffness of the springs can essentially be dialed in to match the elastic properties of the bearing.

Guide Pins and Rods

Guide pins usually have both rotational and translational freedom. Because they are typically cylindrical, the interface can be modeled with MPCs or DOF springs and a properly positioned cylindrical coordinate system. It is also important to have meshed the mating components with aligned nodes on the opposing mating surfaces. These elements can also be converted to gap elements later if such is warranted. Remember as well that most of these techniques are valid only for small deformations. First, if deformations are large, nonlinear effects might stiffen the model and alter the solution. Second, coordinate systems remain fixed in their initial spatial positions. While their orientation might be correct for the undeformed model, the accuracy of the solution will begin to degrade as the guide pins deform off their original axis. This error will be small if deformations are minimal, but could be large if the parts move dramatically.

Slots

Slots can be modeled much like guide pins and linear bearings. Slots that are not lines or arcs can be modeled in a limited fashion by using constraint equations which relate the position of one node to another. If the path of the slot can be described with an equation, a single node can be forced to follow it. This technique should be used with care because if more than a small part of a slot which could easily be approximated as a line or an arc comes into play, the displacements and strains in the model may be too large for a linear solution.

Summary

Assembly modeling can be one of the more challenging and enjoyable tasks in FEA. The more complex the model, the better it feels when it is complete. However, with complexity comes additional uncertainty. The possibility increases of introducing error that is unaccounted for. Make the decision to enter into assembly modeling carefully. If the part in question can be easily defined in a fixed reference frame, consider using boundary conditions to model the interface with the rest of the system. If an assembly is required, build slowly using component contribution analysis.

Verify that the results from the assembly correlate with your expectations and that the joints genuinely represent the physical system. Detailed behavior in assembly models will be more difficult to correlate to test data due to the degree of variability from one assembly to the next and from the actual model. This is another reason why test models of joints and subassemblies are so important. If you can develop a high degree of confidence in the parts, the evaluation of the whole will be more positive. Do not overmodel an assembly. Know when to stop. Get the answers you need and get on with the design. If you were careful and patient in your modeling, you can be confident that the design will be based on the best possible data.

14

Thermal Expansion Analysis

Thermal expansion analysis of parts or assemblies provides stress and displacement results for a variety of scenarios. The most obvious one is the evaluation of effects on geometry resulting from cooling or heating. Some common uses include parts meant for an engine or freezer, or parts that carry extremely hot or cold fluids (see Fig. 14.1). In addition, thermal expansion analysis can be used to verify parts that must be designed to take advantage of varying expansion rates, such as the bi-metal thermostat blade shown later in the chapter. Finally, several not-so-obvious uses for thermal expansion analysis are covered at the end of this chapter.

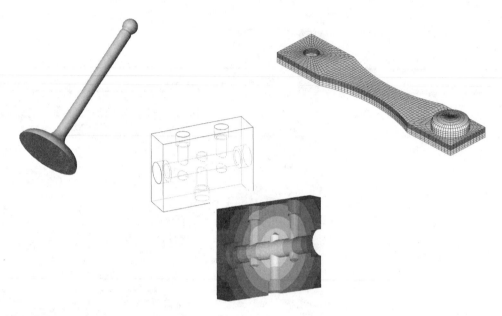

Fig. 14.1. Sample parts intended to work in extremely hot or cold conditions.

Thermal Expansion Basics

Thermal expansion analysis is a structural rather than a thermal solution. This can cause some confusion when evaluating FEA software. A linear static system will be able to model stress and deflection due to a change in temperature. The best way to differentiate thermal stress analysis from heat transfer analysis is by output quantity. If temperature is the desired output, use a heat transfer solver. If stress and displacement are required, use a structural solution.

Material Properties

Analyzing for thermal expansion requires the inclusion of an additional material property, the coefficient of thermal expansion, or α. The table below lists selected coefficients of thermal expansion for commonly used materials.

Table 14.1. Representative thermal expansion coefficients for commonly used materials

Material	Coefficient of thermal expansion (in/in/°F x10⁻⁶)
Aluminum 319F sand cast	11.70
Aluminum 380 die cast	11.50
Brass, cartridge	11.10
Brass, cast	10.10
Brass, yellow	11.30
Bronze, commercial	10.20
Bronze, naval	11.80
Bronze, phosphor	9.90
Carbon steel	6.50
Copper	9.80
Gold	7.78
Grey iron	5.60
Lead	16.30
Magnesium, die cast	14.80
Nickel	7.00
Plastic, ABS	54.00
Plastic, acetal	45.00
Plastic, nylon 6	47.00
Plastic, polycarbonate	36.00
Plastic, polypropylene	45.00
Silver	10.30
Stainless steel 305	9.60
Stainless steel 410	5.60
Stainless steel 430	5.80
Steel 1010	8.40
Tin	15.00
Titanium	4.90
Zinc	15.20

In addition to α, a reference temperature, T_{REF}, must be input for the material. The strain induced on a part with a given property due to increasing temperature, ε_{TE}, is calculated by the following equation:

$$Eq.\ 14.\ 1 \quad \varepsilon_{TE} = \alpha (T_{APPLIED} - T_{REF})$$

where $T_{APPLIED}$ is the temperature input as a load by the analyst. If T_{REF} is set to zero for all materials, the temperature change affecting the part will simply be the applied temperature. To increase the ε_{TE}, you can increase α, $T_{APPLIED}$, or decrease T_{REF}.

Boundary Conditions

The boundary conditions used in a thermal expansion analysis are the same as in a linear static analysis, with the inclusion of a temperature load. A temperature load can be applied as an ambient temperature or as a local temperature. An ambient temperature will act uniformly on every element in the model. A local temperature, on the other hand, is applied to one or more nodes or element faces. These might represent the temperature load of a power transistor on a heat sink or the effect of a hot steady state fluid on the inside of a valve. A local temperature load should be considered more like a constraint than a load.

Local Temperature Loads

Developing a nonuniform temperature profile for a part can be difficult. Since conduction is not accounted for in a structural analysis, a nodal temperature only affects the elements using that node. Multiple local temperatures apply a step loading to a part. Local surface regioning is typically utilized to better control the areas where temperature will be applied. A more continuous distribution can be obtained by utilizing a thermal (heat transfer) analysis to preprocess the structural analysis. In a heat transfer analysis, the known local temperatures can be applied and the system can then be solved for the total steady state volumetric temperature distribution. Most preprocessors and postprocessors provide the option to convert nodal temperatures in a heat transfer output set to nodal loads as input to a structural analysis. This will provide a more gradual, continuous transition between known or applied temperatures.

Because a thermal expansion analysis is still a structural analysis, structural loads can be combined with temperature loads. It may be worthwhile to place the thermal and structural loads in separate load cases so that the effects of each may be independently reviewed.

Constraints

The FEM must also be fully constrained to prevent rigid body motion. Constraint choices must be made with even more care than in a structural analysis without thermal loads because thermal strain will be induced between any two points in an isotropic material. If two nodes are constrained in the same direction at different locations in the body, a strain-induced reaction force will occur at those nodes. Consequently, high local stresses will be calculated at these points, even if applied structural loads do not generate reactions in the DOFs of the constraints. Refer to Chapter 8 on modeling unconstrained parts for tips on applying constraints that do not add stiffness to the part. These techniques will be important if thermal expansion modeling is used.

An important consideration in thermal expansion modeling is that if the temperature field is uniform, a totally unconstrained body will expand from its volumetric center without resultant stress. Use this baseline to evaluate the displacement and stress behavior of an expanding part.

Other Uses for Thermal Expansion

Thermal expansion has been used to simulate behavior beyond its obvious intent. A few of the more useful techniques are described below.

Modeling Weld Stress

As a weld bead cools, it contracts. This contraction generates a residual stress field in the part. These stresses may then combine with operational stresses to induce a failure earlier in product life than might be suggested by a study of operational loads alone.

Fig. 14.2 shows a plane stress model of two fillet welds and the stress fields calculated by the thermal contraction of the beads. For this approximation, the α was assumed to be the same for both the welds and the steel plates. The T_{REF} represents the only difference between the weld and plate material. T_{REF}, in this simulation, was estimated to be the temperature of each region during the welding process. T_{REF} for the steel plate was set to 120°F, and for the weld bead material, 2,000°F. The applied load was an ambient temperature of 70°F, the storage or operating temperature of the part. As would be expected, a stress field is generated by the different amount of contraction.

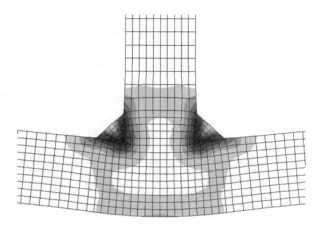

Fig. 14.2. Plane stress model of two fillet welds and stress fields calculated by thermal contraction of beads.

This approximation makes a couple of gross assumptions. First, most material properties are typically temperature dependent enough that you should expect values at weld temperature to be different than at room temperature. Second, the initial thermal distribution is much more continuous than modeled. Consequently, the strain would more gradually transition between the weld and the plate. This would most likely lower the stress calculated. Finally, the welding process is not instantaneous. Consequently, cooling may actually start in different portions of the part at different times. Multipass welds only exaggerate this. Chill blocks can also complicate the approximation. Utilization of thermal expansion for the approximation of residual weld stress should be considered only a rough indicator of the actual system. A more detailed study, combined with a transient thermal analysis, could provide excellent results using the same general concepts. Use test geometries with a rigorous correlation process before trusting results for a complex model.

Injection Molded Part Shrinkage

It should be noted immediately that assumptions and limitations described for weld contraction apply to this technique as well. However, when considered as a rough approximation, an engineer can draw some valuable conclusions from these data. The plastic note tray shown in Fig. 14.3 was modeled with isotropic properties, uniform wall thickness, and no draft.

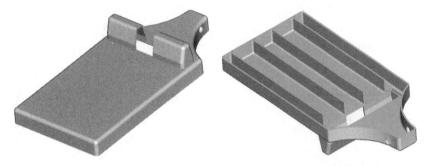

Fig. 14.3. Tray modeled with isotropic properties, uniform wall thickness, and no draft.

Fig. 14.4. Displacement results of tray modeled as shell.

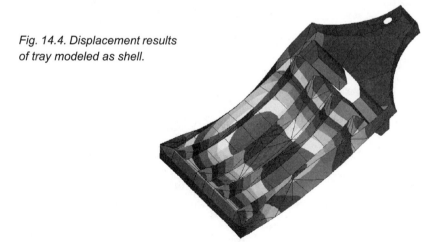

An exaggerated deformed plot of a shell model of the tray is shown in Fig. 14.4. The T_{REF} was set to zero for the material. The point of ejection from the mold was chosen to be the simulation starting point because, prior to that, it would be even more difficult to assume that the material properties, primarily stiffness, would be constant. Taking the start point back to the melt phase would require a highly nonlinear study and should only be tackled with the use of specialty injection molding software. The applied ambient load was the expected temperature drop between the part's temperature at mold ejection and room temperature. Again, contraction, not expansion, is the desired response. This is achieved by a negative ΔT or $(T_{APPLIED} - T_{REF})$. Note that warpage has indicated a potential problem.

Another interesting study along the same lines is the difference in residual stress that occurs when a part is cooled in the mold versus after its ejection. In the simple study of a plate with a hole in it, the plane stress model in Fig. 14.5 was cooled in the same manner as in the previous example with no restriction on the outer edges or hole. The stress plot of Fig. 14.5 shows that an unrestricted part deforms with no stress.

Fig. 14.5. Stress plot for a plane stress model of plate containing an unrestricted hole.

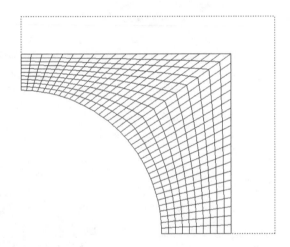

The model in Fig. 14.6 was cooled with the hole radius fixed using a radial constraint with a cylindrical coordinate system. This approximates cooling around a steel core pin that would be much more rigid than the plastic. Note how the residual stress field is significant in this model.

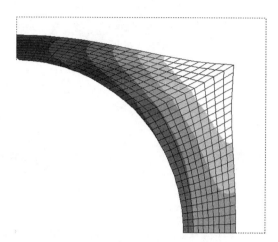

Fig. 14.6. Stress results for a model cooled with hole radius fixed using a radial constraint on a cylindrical coordinate system.

Press-fit Analysis

When press-fits are analyzed on simple systems, an enforced displacement can be used quickly to impart an estimated deflection. However, the analyst is then required to calculate or estimate the contribution to the resultant displacement from both the inner and outer parts. This calculation is well documented for a circular press-fit but can be nearly impossible for more complex shapes. A more reliable method is to use a finite element model of the actual, unassembled shapes and a contact interface between the two parts. This method will allow both the inner and outer parts to reach equilibrium naturally.

Contact with initial penetration can be quite tricky in some codes. If the initial position of a gap element is in penetration, the solver assumes that the closure vector, or contact side, is opposite the expected one. One workaround is to use an initial clearance between parts and apply a thermal expansion coefficient to the inner part. Combining this α with an ambient temperature calculated to expand the part to its preassembled size would allow the press-fit to engage more naturally. Two options are available for calculating the input parameters. In Fig. 14.7, a 0.001 radial press-fit is desired.

Fig. 14.7. Desired radial press-fit of 0.001.

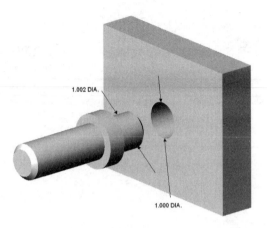

The actual parts are to be modeled with the dimensions shown in Fig. 14.7. For a desired final part diameter of 1.002", the part could be modeled with a 0.998" diameter to provide clearance inside the 1.000" diameter bore. A diametral increase of 0.004" is now needed from the thermal expansion. Because the thermal expansion is not actually modeling a physical phenomenon, the choice of parameters is somewhat arbitrary. Using an applied ΔT of 100 degrees Fahrenheit, the α can be backed out using equation 14.1.

$$\varepsilon_{TE} = \Delta diameter/diameter = (0.004\ in)/(0.998\ in) = 0.00401\ in/in$$
$$\alpha = \varepsilon_{TE}\ /\ \Delta T = (0.00401\ in/in)\ /\ 100\,°F = 4.01e\text{-}5\ in/in/°F$$

Another option is to use the α of the material, which for steel is 8.40e-6 in/in/°F and then backing out the applied ΔT from equation 14.1.

$$\varepsilon_{TE} = \Delta diameter/diameter = (0.004\ in)/(0.998\ in) = 0.00401\ in/in$$

$$\Delta T = \varepsilon_{TE}\ /\alpha = (0.00401\ in/in)\ /\ (8.40e\text{-}6\ in/in/°F) = 477\ °F$$

Either method will produce the same results. If contact is not used, orthotropic properties are recommended. In this fashion, the inner part can be given radial stiffness without a tangential or axial stiffness. The fictitious stress caused by the expected sliding interface between the two parts will be minimized. Even if orthotropic properties are used, be prepared to ignore the behavior right at the interface.

Spring Preload

Techniques similar to those mentioned in the press-fit example could be used to preload a spring. To achieve this, however, a rod element must be used instead of a scalar spring constant in order for a material property to be assigned. The rod can be given an initial tension or compression by playing with applied temperatures, T_{REF} and α.

Example of Incorrectly Used Thermal Expansion

The motor end cap shown in Fig. 14.8 was clamped in a four-fingered vise at its mounting holes to turn the bearing bore in production. A high vise pressure caused the bore geometry to collapse elastically so that after the precise boring operation was complete, the sprung-free part exhibited eccentricity at the bore. The design team wished to use FEA to evaluate the sensitivity of this bore deformation to different loading and ribbing configurations with a goal of improving roundness after the boring process.

Fig. 14.8. Motor end cap.

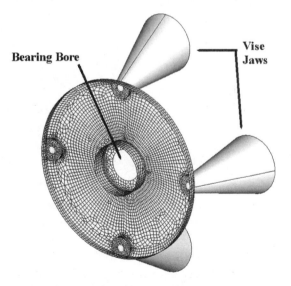

Bearing Bore

Vise Jaws

Two approaches were suggested. The first was to fix the mounting holes and adjust thermal expansion parameters to produce a reaction force at the constraints equal to the known vise loading.

Fig. 14.9. First approach.

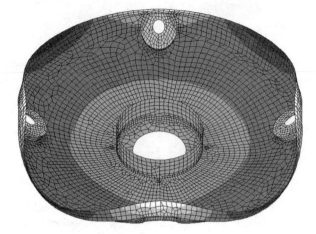

The second condition used a combination of constraints and radially oriented loads to actually compress the end cap.

Fig. 14.10. Second approach.

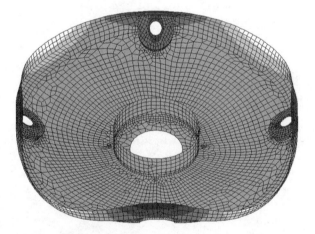

Both methods produced realistic looking results but the second method was the correct one. In reality, the bearing bore was forced to collapse somewhat by the vise loads. The thermal boundary conditions placed it in an expansive state. The differences between the two results are subtle but the run-out at the bore of interest was only 0.0005" or 0.0192% of the desired 2.60" diameter. Small differences could mean the distinction between useful information and pretty pictures.

Summary

Thermal expansion modeling should be understood by all analysts. Benefits extend beyond simple temperature load studies, as shown in the last few examples. Remember that when any technique is used creatively to simulate nonstandard behavior, great care must be taken to ensure that you can distinguish correct from false results. Test models are critical when using any technique that is new to you. Even variations of familiar concepts should be tried out on simple geometries to gain confidence in their usage.

15

Nonlinear Analysis

A linear approximation is a fast and efficient solution for many engineering problems. However, most of the world is nonlinear. The degree of nonlinearity, or the variance between a nonlinear solution and a linear approximation, will determine how valid the linear results are. In many cases, simply understanding the effects of the nonlinearity can enable a design engineer to make sound design decisions on linear results. To understand these effects, however, the design engineer must have a working knowledge of nonlinear concepts and terminology. While some nonlinear techniques are only slightly more complicated than their linear counterparts, others require a deep understanding of both the physical and numerical phenomena involved in the nonlinear calculations.

In this chapter, the basic concepts of nonlinear analysis are reviewed. Guidelines for identifying nonlinear situations will be presented and workarounds for evaluating nonlinear systems with linear techniques will be offered. This chapter is meant to provide an overview of the nonlinear world rather than an in-depth treatise. The needs and terminology of each software package, as well as every problem, are different. If you are ready to pursue the more advanced nonlinear capabilities available within FEA, such as nonlinear material behavior, plasticity, or nonlinear buckling, to name a few, be prepared to move slowly, research your tools' capabilities and have an internal or external coach ready and willing to assist. A complex nonlinear problem can be very rewarding when completed but, as with any new technique in FEA, should be approached gradually.

Basic Concepts in Nonlinear Analysis

In the physical world, there is a fine line between linear and nonlinear behavior. Parts you design, tools you use, and events that surround you transition between linear and nonlinear behavior smoothly and imperceptibly. It is possible, albeit impractical, that all problems be solved with a nonlinear algorithm. So why is it that transitioning a model into the nonlinear world is such a big deal in FEA? The primary reason is the solution methodology used by the software. Just as continuous volumes are represented with discrete pieces called elements in FEA, a continuous or smooth physical process must be modeled in discrete steps. The abrupt transitions between stiffnesses in a point-to-point, stress-strain curve representation, or the abrupt change in load path when two parts contact each other, force the solution to try to numerically reconcile an instantaneous change. The fact of the matter is that even a nonlinear solution is still only solving many linear steps iteratively and successively. The smoother the nonlinear behavior being modeled, such as stress stiffening, the easier it will be to numerically adjust to the changes. With the current state of computing and FEA solving technology, even smooth nonlinear behavior makes most simulation tasks time- and/or cost-prohibitive in a hectic design schedule. Therefore, as a design analyst, you must make the decision as to when to switch to a nonlinear solution if the linear approximation is not sufficient.

Why Use Nonlinear Analysis?

If nonlinear analysis were as fast, easy to set up, and inexpensive as linear analysis, the "to use nonlinear or not to use nonlinear" question would not arise. A nonlinear solution will be as accurate as a linear solution, if not more so, for most problems that fit well in the linear assumption category. However, nonlinear runs do take significantly longer to set up and solve than similar linear studies. The data required to fully justify the nonlinear analysis can be time consuming to collect and require special training and added cost over and above the cost of the solution itself. Consequently, it is recommended that the goals and needs of each particular problem be evaluated for its nonlinear components. The decision to use nonlinear analysis should derive from a need that cannot be satisfied by a linear approximation. Some common reasons to use nonlinear analysis are described below.

Exact Performance Data Required

While performing comparative or trend analyses, it may have been perfectly valid to assume that the plastic parts being analyzed had linear material properties or that large displacement effects were similar between concepts. However, in the latter stages of a design, there may come a point where you must question many of the simplifications and approximations made during the development studies. For product or user safety reasons, a check run solving for one or more nonlinear behaviors may be the last simulation necessary before committing to a prototype.

Another reason for obtaining exact performance data is a postmortem study. A part or system is failing and you need to know why before you can fix it. In this case, the possibility of nonlinear behavior must be considered because it can have a dramatic effect on stresses or displacements.

Contact Is Inevitable

While many part-to-part interfaces can be modeled with linear conditions and tricks to approximate contact, there are many instances that simply require the ability for the parts' contacting surfaces to impact, slide, and/or lift off one another. This is most common when the loading on the system causes a portion of the contact interface to be in a compressive state while the rest wants to disengage. As contact in most FEA codes become easier to use and faster to solve, the arguments for avoiding it are beginning to fall away.

Large Displacement in Flexible Parts

In many thin-walled plastic parts, geometric or stress stiffening plays a critical role in the final response to realistic loading. Failure to account for these large displacement effects can yield catastrophic-looking displacements or stresses. It is not uncommon to see the calculated displacement in a thin-walled structure come down to one-fifth the corresponding linear result when large displacement effects are included. It is nearly impossible to estimate this reduction in displacement by simply examining the linear results.

Detailed Stress Input to Fatigue Analyses

Fatigue analysis codes are becoming increasingly popular in FEA. These codes typically take a sequence of maximum principal stresses structured by the loading history of the system, and calculate damage or life estimates. However, to provide meaningful data, these tools require precise stresses. There is enough uncertainty in the other facets of the analysis that, if fatigue results are to be believable, diligence in properties and reported stresses is required. The exception is a trend analysis using fatigue criteria. In this case, the simplifications described earlier for linear trend analyses are valid, as long as no one forgets the quality, or lack thereof, of the individual design data.

Manufacturing and Forming Simulation

Another growing segment of the finite element analysis industry is the use of simulation to predict manufacturing processes. The most common are metal forming or forging and plastic or casting filling simulations. In these tools, the event of interest is nonlinear by definition. Attempting to use a linear solution to approximate metal forming would be pointless. Most of these problems are best handled by specialized codes that incorporate the nonlinear requirements specific to these problems directly into the basic interface.

Identifying Nonlinear Behavior

Before attempting to identify nonlinear behavior, it is important to understand the underlying assumptions of linear behavior. These are the conditions a system must exhibit to make an attempt at modeling with a linear solution even possible. Once they are understood, examining most real world problems to determine if the basic assumptions are violated becomes possible.

Fundamental Conditions of Linearity

Stress-strain

The relationship of stress to strain over the strains being studied must be linear and elastic. For all practical purposes, stress is proportional to strain via Young's modulus. If the strain state is known, the stress state is easily determined. In addition, the part or system must return to the starting state elastically when all external loads are removed. Most materials will exhibit a change in stiffness or modulus before inelastic or plastic behavior is encountered.

Strain Displacement

The displacements and rotations must be small, such that the relationship of strain to displacement can be approximated as linear. This is a difficult concept to grasp for new users of nonlinear tools. In an introduction to mechanics class in college, the following standard assumptions were common, albeit their importance was not fully appreciated: "*Displacements must be small*" or $sin(\theta) \approx \theta$ or $tan(\theta) \approx 0$. When local displacements or rotations are small, higher order terms in the constitutive equations are negligible and only the linear terms are required to calculate a precise solution to a static problem. However, when displacements become large, these nonlinear terms begin to become significant and must be accounted for.

In addition, as displacements become large, the tensile uniaxial or biaxial stress fields begin to induce resistance to further deformation. The parts stiffen over and above material properties and, consequently, alter the final structural results.

Load Continuity

The magnitude, orientation, and distribution of loads must not change between the unloaded (undeformed) and deformed conditions. This essentially means that no loads, such as resulting contact reaction forces between two surfaces that are initially separated, can be introduced after the initial load is applied, and that the loads applied do not change direction as the feature they are applied to begins to deform. This assumption is usually valid if the assumption of small displacement is valid.

Common Symptoms of Nonlinear Behavior

Stress Levels Approach the Yield Point

You are asking for problems if you assume that your structure will not yield because the yield point for the material in your college mechanics book is 15,000 PSI and your FEA results show a maximum stress of 14,870 PSI. Ignoring the obvious fact that a safety factor of one is being used in this design (*although your customers will not be so quick to ignore this!*), there is the likelihood that nonlinear material behavior is beginning to affect your results. Most materials will exhibit a significant range of nonlinear elastic behavior long before the yield stress is reached.

Another important consideration as stresses approach plasticity is that the reported yield stress is typically an estimated quantity, derived graphically by offsetting the linear slope of the stress-strain curve 0.2% and noting the stress where the offset line crosses the test data. This yield stress value may be higher or lower than the elastic limit, which is the theoretical transition between elastic and plastic behavior. Therefore, the actual point where permanent set occurs may be appreciably lower than the reported yield stress. Use the reported data as a guideline on the stress state, not as gospel.

When evaluating the stress state of a part, remember that extremely localized behavior will most likely not cause gross yielding or plastic collapse if the local area is supported by regions of lower stress on all sides. An extreme example of this is a pen tip pressed against a table. While the stress just below the tip probably exceeds the elastic limit of the table material, local flow will result in a redistribution of the stress, and plastic behavior will be contained and transitioned back to an elastic state. Therefore, this localized yielding will probably not affect the gross behavior of the system.

Do not simply look at the reported maximum stress in your linear analysis results and assume that a nonlinear solution is required. Look at the areas of high stress, the distribution of the stress approaching and exceeding yield, and determine the cause of these stresses. If the cause is a point load/constraint or a Poisson effect, a nonlinear run is probably not warranted, although remodeling may be. Next, if the high stress in your linear run is real but highly localized, consider the possibility that it may redistribute and dissipate due to the support of the less stressed geometry around it. If you are at all uncertain in this situation, consult an expert.

Coupled Displacements Are Restrained

When a long, flat plate is supported at its ends and loaded with a uniform pressure, it wants to assume some curvature. For the length of the neutral surface to remain constant, longitudinal translation of one or both ends must occur. In this case, longitudinal displacement is coupled to the normal displacement by the shearing forces in the plate. When the ends of the plate are restrained, the plate can bend only if the neutral surface stretches. This stretch will generate significant in-plane tensile stresses, which will serve to stiffen the plate due to the geometric stiffening mentioned earlier. This is an excellent example because it highlights the contribution of axial or in-plane tensile stress to system nonlinearity.

While large displacements should prompt you to examine your system more closely, the degree of nonlinearity due to these displacements will be small in a lightly constrained case and larger as the constraints restrict the natural flow of the material. Consider the entire system before opting for a nonlinear solution.

Large Displacements Are Expected

In many cases, the desired performance of a part involves large displacement. Parts such as wire ties and certain springs must deflect substantially just to achieve their purpose. In these cases, it is best to assume that a nonlinear solution will be required. The guidelines specified later in this chapter will suggest that one or more linear studies on the part be made to debug the model. However, the results of these analyses should not be considered indicative of anything besides a valid or invalid mesh and/or constraint scheme.

Unreasonably High Deflections Are Observed

In contrast to the condition described above, most structures are not expected to deform significantly. In fact, excessive displacement is usually considered a failure condition, regardless of the stress levels. Consequently, if a part is designed using sound engineering judgment and based on some historical experience of similar parts, grossly unacceptable displacements should not be expected. If these deformations are observed in the linear analysis—an especially common development with plastic parts—a nonlinear run should be considered before panicking. While the deformation calculated by the nonlinear solution may still be unacceptable, it should be much more manageable and indicative of the real problem.

Two Surfaces or Curves Penetrate

Because contact is a nonlinear solution and therefore more time consuming, modeling a system with potential contact as a noncontact linear approximation and examining these results before proceeding is common. If penetration occurs in one or more of the suspect regions, a contact condition should be modeled and a nonlinear solution employed. By approaching a contact problem in this manner, you will save the time of modeling gap elements where they may not be needed. Moreover, because the solver must evaluate the state of every contact element at each load step, even if some contact regions are never engaged, their existence in the model may increase run times.

Of course, there are many situations where contact cannot be avoided. In fact, contact and its resultant reactions may be key items of interest in the study. In these cases, a linear run prior to the study is still recommended, as outlined in the nonlinear methodology below. If the model is large, estimates of the reaction forces could be applied as loads for the purpose of debugging and converging the mesh. Although convergence should be checked with contact engaged, some time may be saved by up-front convergence in the linear world.

Direct versus Iterative Solutions

A linear model is solved just as the name suggests, linearly or sequentially. It is analogous to solving a set of equations where the number of unknowns matches the number of equations. This is part of the reason why linear solutions are so fast. When there are more unknowns than equations, you must resort to some iterative approach, such as Newton-Raphson, to complete the problem. In this type of solution, a guess at one of the unknowns is made and the problem is solved. Based on the results of this solution, an error estimate is made on the initial guess. Using this error estimate, the guess is adjusted and the solution is repeated. This cycle continues until the error estimate is below some predetermined minimum. A nonlinear solution requires an iterative approach.

Overview of Nonlinear Solution Algorithm

It is easy to understand that a material model requiring a nonlinear stress-strain curve is nonlinear. However, another key feature of any nonlinear analysis is a nonsequential solution method. A guess at the solution for each load step is made and the error is calculated. This error is usually in the form of an energy balance error. When the error is within a certain tolerance, the solution considers this load step converged and moves on to the next. The efficiency of the iterative algorithm plays an important role in the speed of a nonlinear solution.

All nonlinear calculations involve a series of incremental solutions. The total load is applied in steps, called *load steps*. At each step, an intermediate solution is "predicted" using the incremental load magnitude. The final orientation and distribution of the loads, the deformation, and the stress levels from the previous intermediate solution are used as the initial state of the next step. An error estimate or residual is calculated based on force and energy balances, as well as displacement continuity. This is checked against user-defined tolerances. If the error is too high, the load step may be automatically bisected or the stiffness matrix may be rebuilt and the solution recalculated. This process continues until the total load is applied or the solver identifies a diverging condition.

Load Cases, Load Steps, and Convergence

The primary enabling mechanism underlying nonlinear analysis is the ability to apply the total load incrementally to reach successive, intermediate solutions. While most high end FEA codes provide some degree of automation to the load incrementing, they still require the analyst to get the ball rolling with good initial estimates of load step size. Loading can be applied incrementally to the problem using two similar yet uniquely important tools: the load step as described earlier, and the load case. Knowing when to use one versus the other is part of the "art" behind nonlinear FEA and can only come with experience because every problem is different.

Load Steps

Load steps are the building blocks of nonlinear load incrementing. They define the amount of change in the system that the iterative process must reconcile. If the load steps are too large, the solver may not be able to adjust the model to reduce the error and convergence will fail. If they are too small, the solution may take an excessive amount of time and try your patience and the project schedule. Ideally, the load steps should be just small enough to allow convergence. A linear analysis, with no change in any property or load, can be solved in a single load step. As the rate of change of behavior increases, the size of the load step should decrease. A few general guidelines for evaluating the load steps follow.

- The first load step should not cause plasticity.

  A load step size that satisfies this condition is usually easy to determine. The results of the linear preanalysis can be used as a guide. Because the linear analysis can be scaled linearly without loss of data integrity, the factor by which the load must be divided to bring all stress levels within the linear material range can be used as your first pass at the number of load steps required.

- Load steps at an abrupt transition should be small.

  A good analogy to this guideline is the convergence requirements of a finite element mesh. When deflection and stress gradients are gradual, larger elements will capture the behavior quite adequately. However, in local areas of high stress or strain, smaller elements are required to model the abrupt changes in state. Similarly, your nonlinear model will have a much greater chance to converge if the load stepping is designed so that a large step does not span a sharp change in the model state and the steps that do span it are reduced. Some examples of these conditions follow.

  - Contact with high forces or velocities.
  - Abrupt transitions in the stress-strain curve as might occur with a bilinear or trilinear material model.
  - Unloading of a plastically deforming part.
  - The onset of buckling.

- Gradual or moderate nonlinear problems can contain larger load steps.

  Parts initially in contact or in moderately large displacement scenarios can usually be solved with larger or fewer load steps. If your software has the capability of automatically adjusting load step size, it is usually in your best interest to take a chance and use a larger step. Which brings up a good point...

- A solution will typically be faster if software is not forced to automatically adjust the load step.

  While debugging a nonlinear run the first few times through, it is not uncommon to adjust the load step size. Adjustments resulting from failed convergence are common. However, if more than one run is required due to design improvement or boundary condition adjustment, it is in your best interest to examine the run summaries for instances where the solver had to take matters into its own "virtual" hands. This data can help guide you into adjusting the load step scenario so that automatic adjustments are minimized. Your subsequent runs will be more efficient. Again, you must balance the value of your time versus computer time. If the runs are going to solve overnight anyway, fine tuning your load stepping on a converging model may not be worth it. Some solvers will be more efficient for contact solutions if left to adapt load step size automatically. Check your software documentation.

Load Cases

You may have noticed that some of the suggestions above indicate that the load step size should change depending on behavior at a particular point in the loading history. Varying step size usually requires that the load be broken down into separate load cases. By using multiple load cases, the load can be applied in large steps up to a point of transition. The next load case can then have smaller steps to get through this difficult region, and be expanded again to larger steps in a third case as necessary. In addition, these separate load cases provide a hard stopping point for the solution to achieve convergence. Many analyses will converge better if they have stable points, almost like platforms to launch the next loading step.

Load cases in nonlinear analyses also provide a great deal of flexibility in the way loads are applied, removed, or changed. Multiple load and constraint cases allow you to break up the model into distinct operations, such as those that might be used in the assembly process. Nonlinear analyses use the results of one load case as the initial state of the next load case. Therefore, one load case can initiate a deformation with temporary loads, and then, when the behavior of interest is developed, the temporary loads can be removed. This technique can also allow a static analysis to be solved as if it were time dependent. Although all the loads are applied gradually, they can be viewed in a sequential animation that shows the correct "time line" of the events. Selected examples of using multiple load cases follow.

- Conditions where contact provides constraint in one or more directions, but the initial state is unconstrained.

- Developing normal forces prior to requiring friction to resist sliding.

- Initiating the imbalance to start a nonlinear buckling solution.

- Ensuring that thin pieces, such as switchblades, start deforming in the proper direction.

Methods for Updating Model Stiffnesses

There are two primary methods for iterating through the solution of changing material stiffnesses. At each load step, the iterative process can progress in one of two ways: a Newton-Raphson solution or a modified Newton-Raphson solution. Figs. 15.1 through 15.3 illustrate these methods on a load versus displacement curve. Each peak represents a separate stiffness update iteration within a load step. Your solver probably has a limit that you can adjust for the maximum number of iterations within a load step.

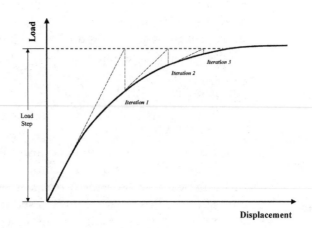

Fig. 15.1. Newton-Raphson method for stiffness updates.

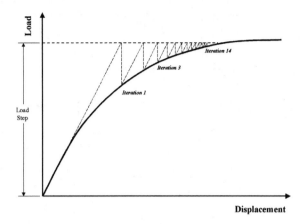

Fig. 15.2. Modified Newton-Raphson method can fail when hard.

The Newton-Raphson solution uses the tangent modulus corresponding to the previous iteration to calculate the next deformed position. The calculated displacement is projected back onto the load displacement curve parallel to the load axis. The difference in load between the load step force and the projected position is subtracted from the load step and the next displacement is calculated at the tangent modulus of the projected position. The stiffness continues to update until the difference between the load step magnitude and the projected position is within some tolerance.

An alternative to the straight Newton-Raphson method is called the modified Newton-Raphson method. In this method, the stiffness update uses the tangent modulus of the previously converged load step for each iteration. Hence, many more iterations are required to complete a load step but the solution is much more likely to converge under many more conditions.

The modified Newton-Raphson solution method is efficient if there is no strain hardening or an unloading path is not required. Fig. 15.3 shows how this solution algorithm can get confused and a diverging condition can result in a hardening situation.

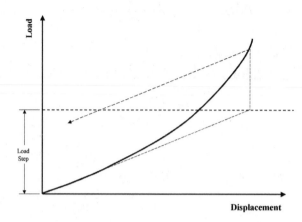

Fig. 15.3. Modified Newton-Raphson method can fail when hardening is involved.

There are other means for solving specialized nonlinear situations; availability will depend on your solver. If you expect to use nonlinear methods regularly in product design, it is highly recommended that you learn the capabilities of your tools so that you can use the most efficient means for completing a problem. Your solver may provide an automatic means for detecting the best solution algorithm for a given loading scenario. This will alleviate some of the need to know all the "ins and outs" of each algorithm, but it is still worthwhile to know why your solver has specific options.

Three Common Types of Nonlinear Behavior

Most nonlinear behavior in product design can be categorized into one of three common types: material, geometric, and boundary nonlinearity. Material nonlinearity is the type most commonly thought of when the topic of nonlinear is suggested. A stress-strain curve is typically known to be nonlinear; therefore, it requires a nonlinear analysis. However, the other two types are probably more common in design analysis for one simple reason: they are easier to set up. They require no special research on properties and converge more quickly. In many cases, if material nonlinearity is present, one or both of the other two types will be required as well.

Material Nonlinearity

A solution requiring a nonlinear material model is probably the most difficult type of nonlinear problem. The degree of difficulty is proportional to the degree of nonlinearity. A nonlinear elastic problem is easier to solve than a moderately plastic problem. Even the latter is easier to complete than a model with significant plasticity and relaxation. Approach these more complex problems only after you feel very comfortable with linear solutions and results interpretation. Never make assumptions about the quality of material data that was not specifically developed for the problem at hand. If a nonlinear material model is required to provide the data necessary to verify or improve a design, test data or careful research should be employed to qualify generic data found in data sheets or a vendor's reference database.

Material Definitions

The success and efficiency of a nonlinear material solution are dependent on the choice of material model. Some models require very little input while others require detailed stress-strain data. A simple model does not usually mean a simple solution. However, even the most complex material models are still significant idealizations of the real situation. This must be remembered at each step of the process. The following review of material models also includes guidelines and commonly encountered pitfalls.

Revisiting Young's Modulus

A linear analysis, as stated previously, requires that stress and strain be proportional through a constant called the *modulus of elasticity*. If this modulus is the initial slope of the stress-strain curve, also required by a nonlinear analysis, it can be called Young's modulus. In a nonlinear material model, two, three, or more modulii are required to define the stress-strain relationship. A key feature of a nonlinear solution is the need to continually reevaluate the strain state to determine which modulus, or stiffness, should be applied to an element at any given load step.

In the nonlinear world there are many types of modulii you should be familiar with. Some of the types overlap with different names. Some of these names are *tangent modulus*, *secant modulus*, *elastic modulus*, *plastic modulus*, *hardening modulus*, and even *first*, *second*, or *third modulus*. The name of the modulus used is dependent upon the material model chosen and the terminology used by your software or your peers. Each type will be identified with a material model for which it is appropriate.

Yield Criteria

When plastic behavior is desired or expected, you will need to tell the solver which criteria to seek to initiate yielding. *Yield criteria* are a function of the stresses in the model. Materials, or local elements with the assigned material model, are assumed linear if the criteria are less than 0 and plastic if the criteria are greater than or equal to 0. Some of the more common yield criteria follow.

- *Tresca,* which looks at the maximum shear stress in the model and provides a reasonable calculation of the brief plasticity in more brittle materials.

- *Von Mises* looks to the von Mises stress to determine yielding and is the best criterion for crystalline plastics and ductile metals.

- *Mohr-Coulomb,* which evaluates a combination of maximum and minimum principal stresses to determine yielding, is somewhat more accurate for plasticity in moderately brittle materials.

- *Drucker-Prager* combines data from the first and second stress invariants (see Eq. 2.21) and is better for problems involving materials such as soil and concrete. It provides the best model for yielding that is first invariant dependent such as at crack tips and in amorphous plastics.

The above criteria are primarily the same as the failure criteria for ductile material described in Chapter 2. Their validity for predicting yielding is also the same in that the von Mises criterion best represents most yielding conditions.

Hardening Rules

A *hardening rule* determines how the material model responds to repeated stress reversals, or switching between tension and compression. In ductile material that has never experienced plasticity, the yield point in tension can be expected to equal the opposite of the yield stress in compression. However, once tensile yielding has occurred, some materials experience a phenomenon known as the Bauschinger Effect, which causes the yield point in compression to be somewhat less than the compressive equivalent of the initial yield stress. Consequently, nonlinear solutions have implemented hardening rules to allow for adjustment of the yield point in stress reversals.

An *isotropic hardening* model does not take the Bauschinger effect into account and the compressive yield always equals the tensile yield; the absolute value of both equals the initially defined yield stress.

A *kinematic hardening* model will take into account the reduction in the compressive yield point after a stress reversal. A kinematic model is more computationally intensive and should not be used unless a portion of the model is expected to yield in tension and then in compression. Most materials do experience strain hardening and will see an improvement in accuracy if the kinematic hardening model is used when needed.

Commonly Used Material Models

An important decision to make when considering a nonlinear material analysis is the material model used. Various material approximations are available for use in your model. The choice of approximations is dependent upon the goals of your analysis and the material data available. The most common material models are shown in Figs. 15.4 through 15.7.

The elasto-plastic material models warrant some additional discussion. Regardless of your choice of models, three pieces of data must be specified: yield stress, yield criteria, and the hardening rule. All have been defined previously.

Fig. 15.4. Bilinear material model.

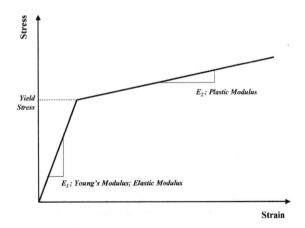

In a bilinear material model, there are two required modulii: the elastic modulus and the plastic modulus. The elastic modulus is typically the linear Young's modulus. The plastic modulus is activated when stress in an element exceeds the specified yield criterion. As in a linear analysis, the plastic modulus is interpolated for all strains in excess of yield. Many materials experience noticeable hardening after the onset of plasticity. In these materials, the response of the system to large strains will diverge in a bilinear model. Therefore, a trilinear model contains a third, hardening modulus to account for this. If your preprocessor supports trilinear material models, you will need to specify three modulii (E1, E2 and E3) and two transition stresses (yield stress and a hardening transition strain).

Fig. 15.5. Trilinear material model.

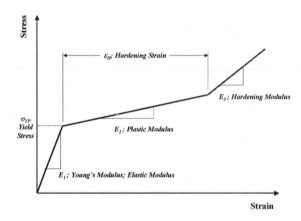

All material models with more than two modulus transitions are lumped into the multilinear classification. A multilinear model is input using data pairs of stress-strain (S-S) values. The first point should be zero stress and zero strain. Fig. 15.6 shows a typical S-S table and the resultant multilinear S-S curve.

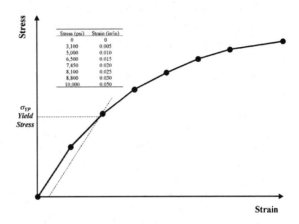

Fig. 15.6. Multilinear stress-strain curve and data points that generated the curve.

The multilinear S-S data are typically extracted from a standard tensile test. The source of this data should be scrutinized carefully. For ferrous materials, the S-S curve is relatively independent of temperature and strain rate. However, most other materials, especially plastics, will show a shift resulting from strain rate and temperature. As a general rule, stiffness increases as strain rate increases. This will result in an upward shift of the curve. Similarly, as the testing temperature increases, the material tends to soften and this is illustrated by a downward shift in the S-S curve. Fig. 15.7 illustrates this shift, once more in the form of a general rule.

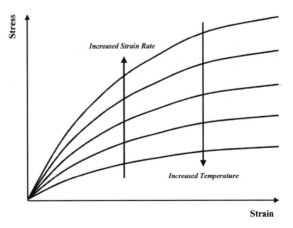

Fig. 15.7. Stress-strain curves offset by temperature and strain rate.

The importance of understanding this change in properties should be clear if correct results are the goal. While material testing under conditions similar to your actual problem is ideal, most S-S curves are obtained from the database of a material supplier. It is recommended that you identify the testing temperature and strain rate of your data. If these parameters do not match your problem somewhat, and material testing is not in the budget or timeline, an alternative is to obtain several curves from your supplier. If you can get curves at a couple of temperatures and strain rates, you may be able to estimate the appropriate shift. Again, the goals of the project must dictate the time and effort put into the material search.

There are a few guidelines for finding, tweaking, and inputting S-S data to improve efficiency and analysis convergence. The initial slope of the first segment in your S-S curve should match the elastic modulus assigned. It is common to find a slight difference between the initial slope or modulus of the S-S data and the published modulus of elasticity. It is your call as to which one to adjust if they are within 5%. A larger difference may indicate some inconsistency in the data set. Either the material or testing conditions might have been reported incorrectly or the initial specification of the linear elastic properties was wrong. Take the time to reconcile this difference.

Be wary of too many data points in the S-S curve because more is not always better. As the number of data points increases, the possibility that the slope may change sign increases. Ideally, the slope between successive pairs of points on an S-S should decrease gradually, but remain positive unless stress softening is intended to be modeled. In this case, a smooth change between positive and negative slopes should occur over a few points. If the slope of the S-S curve fluctuates, it may cause hardening and softening solutions that could drive the solver to many more iterations than necessary at best, but convergence failure is more likely. Limiting the number of data points to the minimum that still represents the data with a minimum amount of deviation error is worthwhile. This is often a judgment call. Plot these points in a spreadsheet or other graphing tool to check the consistency of the curvature. Undesired fluctuation, positive to negative to positive, can be identified by examining the sign of the slope between successive points.

Confirm with your supplier that the data constitute a true S-S curve versus an engineering S-S curve. Refer to the section on common material properties in Chapter 2 if you need to refresh your memory on these terms. The data entered into a nonlinear material model should reflect the true S-S data. Some preprocessors provide tools to convert between engineering and true S-S data, but the conversion assumes a standard test specimen cross section. Refer to your software documentation for details and obtain the test standard used in generating the data.

Geometric Nonlinearity

Geometric nonlinearity primarily refers to stiffness changes that are independent of material properties. These stiffness changes can be related to geometry constraints and/or the magnitude of strains. Geometric nonlinearity is as common in certain types of parts or systems as it is hard to identify. Unlike material nonlinearity, which has a fairly well-understood and visible point to estimate the transition from linear to nonlinear behavior, geometric stiffening has no hard transition marker. Unfortunately, even with all the guidelines presented in this book or offered by your peers and coaches, the only real way to know the actual degree of nonlinearity is to compare linear and nonlinear analysis results. In addition, large displacements may indicate that a nonlinear material model is warranted in order to accurately capture the stiffness in the system.

As mentioned previously, the common characteristics of most geometric nonlinear scenarios are high in-plane or axial tensile stresses induced by the deformation. The following example illustrates this point. The flat plate shown in Fig. 15.8 is subject to a uniform pressure with two distinct constraint cases. In the first, one end of the plate is pinned and the opposite is allowed to travel horizontally or guided. In the second, the opposite end is pinned as well.

Fig. 15.8. Flat plate under uniform pressure with two constraint cases: pinned-guided (a) and pinned-pinned (b).

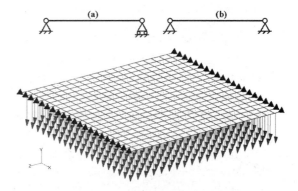

In Fig. 15.9, the plot on the right shows the nonlinearity of each with respect to displacement, and the plot on the left, with respect to stress. For this comparison, nonlinearity, *R*, is defined by the difference between the linear and nonlinear results.

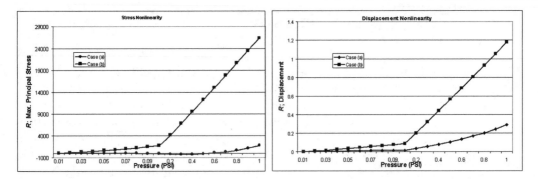

Fig. 15.9. Plots of nonlinearity for two constraint cases described above.

It is interesting to note the maximum displacement at maximum pressure of the two plates. The displacement of the pinned-guided is 25.5 times the 0.05 plate thickness and the stress nonlinearity is 1,700 psi. The pinned-pinned plate, on the other hand, only displaced twice the plate thickness at the maximum pressure and the maximum nonlinearity was 26,300 psi.

The moral of this story is that displacement is not the only characteristic of a geometric nonlinear situation. While various references indicate displacement of one-third a wall thickness as indicative of the transition to nonlinear, others put this number at one-half or one wall thickness. The bottom line is that the resultant stress field, both magnitude and orientation, is as important to note as the magnitude of the displacement.

Using Geometric Nonlinear FEA

Fortunately, geometric nonlinearity is relatively easy to analyze compared to material nonlinearity. In most cases, your solver will only require the specification of default nonlinear parameters and a small number of load steps. No special modeling or property information is required. The number of load steps may require adjustment to obtain convergence, but experience with similar systems will help guide you through this.

As mentioned previously, the onset of geometric nonlinearity is not marked by any clear measure. If you suspect that the displacement and stress levels combine to cause the linear and more correct nonlinear results to diverge, then a geometric nonlinear, large displacement run should be made. Once the difference is understood, optimization may be performed in the linear realm using trend analysis methods. However, a capping or verification run should be conducted in the nonlinear realm.

Boundary Nonlinearity

There are two common occurrences of boundary nonlinearity or change in boundary conditions due to resultant deformations: contact and follower forces.

Contact

Contact conditions are common in most products, and therefore, common in design analysis. Setting up contact conditions in FEA can either be easy or extremely complex and time consuming, depending on preprocessor and solver capabilities. Some codes allow you to specify surface or curve pairs that derive contact stiffness from the material properties of the underlying elements. In others, you are forced to align nodes on adjacent curves and surfaces and connect them with line or gap elements. You must assign a compressive or tensile stiffness to the element that approximates the interface stiffness. A similar method of specifying a contact condition is the slide line. For slide lines to function, they must be set up on a common plane. The nodes used to define the slide lines do not need to be aligned in any particular manner, but the deformation of the system prior to or after contact must not cause the lines to move off a common plane within a given tolerance.

Because contact is typically difficult to set up, it should be used only when necessary. Given that contact is a nonlinear solution, the runs will take longer than a linear solution, and sometimes much longer. Contact conditions allow parts or portions of the same part to touch or lift off each other. This capability may be necessary to model the interactions of certain systems. It is best to use contact judiciously, but do not be afraid to use it. The speed of computers and solving algorithms has reduced the importance of a major objection to its use—long run times. The other major objection, setup complexity, is being improved in some systems. The stiffness of the gap elements can now be adjusted internally, and some preprocessors will automatically align nodes on specified contact entities. However, in most systems, it is still relatively difficult. Up-front planning for the possibility of contact will somewhat relieve setup complexity.

Follower Forces

Follower forces represent loads that are dependent on the orientation of the features to which they are applied. It should be clear that if a feature deflects so much that the orientation of the load becomes of interest, geometric nonlinearity should probably be considered. Nonlinear forces can be defined as displacement or velocity based; they can be defined to follow the orientation of a feature or scale with displacement or velocity magnitudes in a particular degree of freedom. As with contact, follower forces may be straightforward or difficult to set up depending upon the preprocessor.

Other Types of Nonlinearity

In addition to the preceding, more common types of nonlinearity, there are other specialized incarnations of nonlinear FEA that require special training. These less common types typically require even more detailed material data and convergence assistance. A brief description of selected types is included. However, it must be stressed that these nonlinear solutions are much more difficult than the comparatively simple nonlinear material solutions described earlier. Make sure that you have a coach or other support resource prepared to help you through your first few attempts.

Hyperelastic

Fig. 15.10. Rubber and elastomeric parts require a hyperelastic solution for accurate simulation.

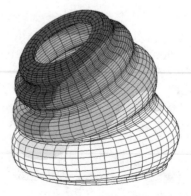

Hyperelastic materials, such as rubber, silicone, and other elastomers, behave differently than standard engineering materials. Their strain displacement relationship is nonlinear even at small strains, and they are nearly incompressible. The Poisson's ratio of hyperelastic materials may exceed 0.50, whereas specifying a Poisson's ratio equal to or greater than 0.50 will result in a failure using any other solution method. Consequently, they must be treated by the solver with special solution techniques that take into account these conditions as well as the large deformations typically associated with hyperelastic response. Most solvers use a strain energy density function formulated for large strain instead of Hooke's law. Consequently, the material properties for a hyperelastic material are much more complex than standard materials.

Material properties are usually entered as Mooney-Rivlin constants or as a set of stress-strain curves. Both sets of data require testing by a credible lab with experience in these types of materials. The Mooney-Rivlin constants consist of distortional deformation constants and volumetric deformation constants. The stress-strain curves for a hyperelastic model are input as force versus stretch for simple tension and compression, shear, and volumetric compression. The mesh used in these types of problems should be constructed such that large deformations do not invalidate elements due to excessive distortion.

Nonlinear Transient

A nonlinear transient analysis is used to calculate the response of a part or a system in which properties or boundary conditions vary with time. A good example is crash or impact analysis. These problems require that both load stepping and time stepping be defined. This can create a convergence nightmare. Some codes are specially designed to handle these types of problems, although the capability exists in a limited format in most general purpose codes.

Nonlinear transient problems are solved with implicit or explicit integration algorithms. An implicit algorithm is the sledgehammer approach where you must specify, and often respecify, the time step size to obtain convergence. Explicit solutions are more time consuming, but also more automatic and may be more accurate for an equal amount of work on your part. Explicit solvers will adjust the time stepping based on the system response. The time step size may be derived from the minimum time constant of the elements in the mesh. In these cases, the mesh must be built with care so that a distorted or unnecessarily small element does not cause the time step to be unreasonably small and extend the length of the analysis. While explicit solutions can be extremely resource intensive, they should always converge to an appropriate result. An implicit solution may require some hand holding.

Creep

Creep is the term used to define time varying, permanent strain due to the long-term application of a constant or near-constant stress level. When a material creeps, it effectively relaxes and the stress levels in the part reduce. Creep analyses require detailed material data that are dependent on duration of load and the magnitude of initial stress and temperature. These data will most likely not be found in any standard text and will need to be determined empirically. Even after creep testing, the time-dependent data must be placed into the format that your particular code requires. For instance, ANSYS provides over 10 different creep models for various conditions, each requiring that the material data be input slightly differently.

Nonlinear Buckling Analysis

A linear buckling analysis simply points out the critical buckling load based on the orientation of an existing load set. The initial buckled shape is also presented by the linear analysis. The linear analysis does not provide data regarding the final shape of the part after the initial collapse. Some structures will collapse to a secondary state of stability, either on their own or due to interaction with other components. To determine the postcollapse response of a structure, a nonlinear buckling analysis is required. These problems are solved using somewhat standard, nonlinear, static solution techniques with special handling for convergence near the point of collapse. A temporary load may be required in the first load case to initiate buckling behavior since an eigenvalue solution is not used. While the linear eigenvalue solution will calculate shapes that are not caused by the applied loading, thus allowing the determination of various collapsed shapes, a nonlinear analysis with a perfectly centered and oriented, compressive load on a thin column will not induce any off-axis deformation. Consequently, the eccentric deformation may need to be manually induced. The shape of the linear buckling solution should help guide you to the orientation of this dummy load. Snap-through events of diaphragms and switchblades are other problems that also fall into the nonlinear buckling category and use the same methodologies.

Bulk Metal Forming

Bulk metal forming analysis, such as sheet metal folding, forging, and cold heading usually require a special software configuration that will automatically clean up elements that become overly distorted by the excessive deformations of the mesh. If a mesh can be defined in a standard, general purpose code that stays valid throughout the deformation, some of these manufacturing type problems can be addressed. However, if analyses of these problems are frequently needed, a code that has made special provisions for bulk metal forming should be considered.

Nonlinear Results Data

The results from a nonlinear analysis will differ from those of a linear analysis in a few ways. First, there will be more results information to deal with because certain data, such as plastic strain, are not calculated in a linear analysis. Second, it is common to output data at intermediate load cases and steps as opposed to final result output only, which is all that is available in a linear analysis. In addition to the quantity of data provided by a nonlinear analysis, more solutions may be required due to the fact that results cannot be linearly superimposed. Unlike linear analyses where results from two or three independent loading scenarios can be combined in an infinite number of scaled conditions, such combinations are not an option because the nonlinearity of a solution is load dependent. Therefore, if you have two or three independent loading scenarios that may be combined, you will have to run each of these combinations independently.

Because nonlinear analyses produce large volumes of data, reviewing study goals and specification of the minimum amount of data required are recommended. Consider requesting only the displacement output after each load case in the first few runs of a large model when several load cases are used, or after each load step otherwise. If convergence becomes an issue, requesting output after each iteration may help provide insight on where things are going awry. However, if these intermediate results are not needed, save yourself some disk space and limit the instances where output is calculated. It is not uncommon for the scratch space requirements of a nonlinear analysis to be two or three times greater than its linear counterpart.

This brings up another important point about load cases and steps. If you know in advance that you require output at particular load levels or when a particular event occurs, it is important that you plan load stepping to ensure that an output set can easily be extracted at that time. Plan your load stepping to produce output for a meaningful animation of results as well. To see the correct timing, distribution, and/or magnitude of deformation and stresses develop in an animation of the results, more output sets might be requested than would be required if the end results were all that mattered.

Nonlinear Solution Methodology

Now that the basic terminology and concepts of nonlinear analysis have been defined, a brief review of the methodology for approaching a nonlinear problem is presented. This methodology is correct for most general cases of problems involving nonlinear behavior. In cases where the nonlinear behavior is special or significant, some deviation from this process may be warranted. Unless you feel extremely comfortable with the system under study and the nonlinear solution techniques employed, do not deviate from this process simply to save time. As mentioned several times earlier, FEA does not provide an environment for someone who craves instant gratification. This is even more true in nonlinear analysis, where patience can easily and often be tried.

Use Test Models to Debug Materials or Contact

Before building a full model, use test models to experiment with nonlinear techniques you are unfamiliar with. If contact is to be used, try out your modeling method on a simple part with similar loading and orientation to the final system. Adjust contact stiffness, mesh requirements, and frictional coefficients (if required) in the test model before building them into the finished system. If the model is large and run time is an issue, use these test models to try several linear approximations of the contact problem. Chapter 8 describes such a test model program to approximate contact in a two-force link.

Use test models to try out nonlinear material data as well. Always be skeptical of data that derive from a test that you did not supervise.

Healthy skepticism is an analyst's best friend. A simple tension or bending model can easily be correlated with a sample plaque of the correct material provided by the material vendor. In addition, put your material model through the paces analytically. Take a test model to the strain levels expected in the final model to help choose load stepping and to ensure that the changes in tangent modulii are not too great at any one point. If the final model is going to require unloading to determine permanent set, take your test model through the unloading process as well. Again, much can be learned about the material model using simple plates and beams. Armed with this information, it will be much easier to debug problems with a larger model.

Building and Running Model as Linear System

Always run your model with a linear solver before submitting the nonlinear problem. This is true for all types of nonlinear problems although the benefits from the linear run may vary from solution to solution. At a minimum, use the linear run to debug boundary conditions and to get a feel for the deformations and stress distribution. Overconstrained models, which result in high Poisson effect stresses, can be identified in the linear run. As mentioned previously, fictitiously high stresses will give a nonlinear material model all manner of headaches. In addition, high stress gradients in unexpected areas may suggest a finer mesh for convergence of the nonlinear solution, over and above the convergence requirements of the mesh for an accurate calculation.

With all this discussion of solution convergence in nonlinear analysis, it must be pointed out that mesh convergence for overall solution accuracy is as important for nonlinear problems as it is for linear problems. However, converging the mesh in a linear model is much more efficient than in a nonlinear one. The model found to be the best for the same loading in a linear solution can usually be used with confidence in the nonlinear solution with a couple of notable exceptions. First, contact problems cannot be completely converged in a linear run due to the redistribution of loads after contact. Second, the mesh in highly deforming models, such as metal forming operations, may not ever see the shapes in a linear solution that would be found in the final run. Essentially, when the load distribution or orientation changes, or the mesh undergoes dramatic changes in the nonlinear run, convergence should be rechecked.

In addition to convergence, the actual need for a nonlinear solution should be double checked in the linear run. In moderately nonlinear situations, you can use the linear analysis to determine if the displacements or stresses are large enough to warrant a nonlinear study. While you may have expected that a nonlinear run was needed, the linear run may show otherwise. Even if the linear run shows that a nonlinear solution is necessary, it will indicate to you the load magnitude that will cause strains crossing the plastic transition point you intend to specify. This knowledge will assist you in setting up load stepping in order for a more efficient solution to be developed. Similarly, in a problem where contact is expected, you may learn from the linear solution that one or more contact regions do not penetrate. If this occurs, the final contact solution will run more quickly with fewer contact regions specified.

Finally, you may learn that the stresses in the linear analysis are so high that it is unreasonable to assume the correction a nonlinear analysis will provide would be sufficient to bring the part into an acceptable range. If you are confident that the part will be unacceptable, make the first few improvements using linear analysis and then fine tune the problem in the nonlinear world. Remember to consider the original goals and failure criteria at this stage. It may have been decided that stresses approaching the yield point or displacements large enough to suggest a nonlinear solution are simply not acceptable from a performance standpoint anyway. Consider the safety factors applicable to each particular situation.

Set Up Nonlinear Solution Parameters

After you have confirmed that a nonlinear analysis is required based on the results of the initial linear studies, the next step is to consider the nonlinear solution parameters. Of primary importance in most problems is the initial estimate of your load case and load step configuration. As described above, use known or expected events such as contact, plastic transition, unloading, or output requirements to determine how to break up your load steps. One important consideration is that the first load step should never cause your material model to cross into the plastic region. Think of your first few load steps as model conditioning steps. Use the results of the linear analysis to guide you.

In most cases, your solver should have defaults that are sufficient for the first pass at a problem and automatic methods for choosing solution algorithms, such as Newton-Raphson versus modified Newton-Raphson. If you choose to change any of the defaults up front, do so based on prior knowledge of similar situations or at the advice of an expert. For subsequent iterations of the problem, review the solution log to determine if manual tweaking of these parameters, including the load stepping, could speed up the run.

Learn how *restarts* work in your particular tool. Requesting a restart before a nonlinear solution is started will save much of the work the solver has to do up front that will not change as you adjust convergence parameters. Upon saving these data for restarts, subsequent convergence iterations will be much faster. Some codes may allow you to save restart data after a completed load case so that you can simply pick up where you left off. Be aware that some codes do not give you much latitude in the adjustments you are allowed to make at the restart. Any changes beyond the adjustments allowed may render the run incomplete and you may end up starting the run from the first load step anyway.

Run Nonlinear Analysis with Large Displacements

Solve the problem with material nonlinearity disabled and large displacement effects turned on. This solution is easier to complete and usually faster to solve. Any debugging of the model that can be made to improve convergence, boundary conditions, or speed should be made here. In addition, the stress stiffening effects may suggest that a nonlinear material model or some of the contact regions are not necessary.

Run Contact Conditions

Enable only the contact regions that previous steps have indicated as necessary. Certain contact regions critical to the behavior of the part or system should have been enabled in the previous step. However, many contact pairs that you may have expected to see here may never require enabling. Moreover, the model may deform in a manner inconsistent with your expectations, thereby causing unexpected areas to contact instead of the ones you had planned.

Of all the steps in this methodology, this one is the most flexible. On smaller problems, it may cost you nothing to simply turn on contact in the first nonlinear run. Use your judgment. Suffice it to say that you should stop and think about the contact conditions before blindly specifying that everything can contact everything else.

Enable Nonlinear Material Model

Once the model has been fine tuned for displacements and boundary conditions, enable the nonlinear material model as necessary. The choices you have made in selecting this model will play a large part in your ability to converge. As mentioned previously, review your load stepping to ensure that at a minimum, the first load step does not cause the solution to use a plastic or nonlinear modulus. If unloading is required, run the solution with the nonlinear material model up to the unloading point first to ensure that it is behaving properly, and then work on the load stepping required to initiate the elastic recovery.

When Convergence Is Not Obtained

While some contact and moderately large displacement runs on well-behaved materials should converge relatively easily, convergence in a nonlinear material model may require several iterations. Do not panic if your model does not converge right away. In most cases, some simple adjustments can bring the solution around. Selected common fixes are listed below.

- Increase number of load increments required.

- Break applied load into multiple load cases.

- Use a more general material definition (i.e., stress-strain curve versus bilinear).

- Look for poorly shaped elements in areas of high stress.

- Redistribute loading to eliminate point or highly localized stresses.

- Examine the most recently completed load step for indications of stress or displacement anomalies.

- Adjust code dependent, advanced solution parameters.

- If the model has multiple nonlinearities, attack one at a time.

- Read the solver's log or message file for clues about specific nodes or elements that cause instabilities in the solution. Repeated messages about specific nodes are indications of problematic areas that require better mesh or repair.

* Introduce initial load steps in the solution to establish equilibrium in the model prior to increasing the nonlinearity.

As a last resort, relax convergence tolerances to obtain a solution and then evaluate the results in search of probable causes. Do not relax tolerances carelessly. If a cause is found, rerun the solution with the tolerances back where they belong. However, if no cause for the convergence problem suggests itself and the model converges with slightly relaxed tolerances, interpret your results in light of the relaxed tolerances. Remember that the accuracy of the solution may be compromised.

Do not hesitate to consult your support resources or FEA coach when experiencing problems with a nonlinear solution. More than any other type of FE problem, nonlinear analyses require experience; there really is a bit of art involved in getting tough models to converge. While there is value to plowing through a problem that is just beyond your current capabilities, you must know when the learning has stopped and only pain is left. This is the point where you call for help.

Finite Element Modeling for Nonlinear Analysis

A few guidelines for meshing and model building are appropriate or more appropriate for nonlinear problems. Building the model correctly in the first place can greatly enhance the speed of the solution and the likelihood that convergence will be obtained. Most of these guidelines center around the concept of keeping the nonlinear model as simple as possible. Due to the length of a nonlinear solution compared to a linear one, every attempt should be made to reduce the model size without compromising solution accuracy. This is where your linear preliminary studies may be useful. Expect that you will run a nonlinear model many times, both to achieve convergence and to adjust the final behavior. Smart modeling can save hours of run time. Keep the following points in mind while building the model for a nonlinear solution.

Use symmetry wherever possible. This is a good idea in a linear model and a great idea in a nonlinear model. The guidelines for using symmetry as described earlier in this book should be used, such as avoiding symmetry in a nonlinear buckling problem or a when a dynamic solution is required.

Use beam, shell, or planar idealizations whenever possible. If a problem is marginal from the standpoint of using an idealization, it may still be worthwhile to consider the simplification as a test model and learn as much as you can from it.

Region your model to use the nonlinear material model only where required. A perfectly valid technique is to restrict the use of nonlinear elements to regions where plasticity is expected. This may require you to mesh the model with multiple properties so that, in reality, two separate materials are used. Chapters 5 and 7 discuss this technique in more detail. By limiting the use of nonlinear elements to the areas that require it, you can speed up the model by forcing the solver to iterate on only a subset of the mesh.

Refine and smooth the mesh in areas of high strain. Nonlinear solutions are sensitive to element distortion, and the discontinuity these elements cause can force the solver to unnecessarily iterate a higher strain, plastic solution when none should have been required. If an explicit nonlinear transient solution is required, distorted elements may skew the time stepping algorithm unnecessarily as well. The mesh at any contact region should be refined to capture the contact stresses that will be developed. As the contact area gets smaller, the need for more refinement increases. You may be able to "unrefine" the mesh where stress is low to reduce the model size as long as the overall model stiffness is not affected.

If large deformations are expected to distort elements such that their accuracy may be called into question, it may be worthwhile to manually distort the elements in the opposite direction somewhat before starting the solution. In this way, the final shape will be closer to the ideal shape. Use your judgment to determine if this is necessary for accuracy or even to prevent the solution from failing due to the presence of highly distorted elements.

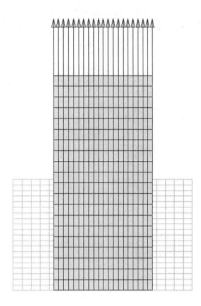

Fig. 15.11. Initially distorting elements can ensure that the final shape will be valid.

Always check your software documentation for element types allowed in a nonlinear solution. Many codes restrict the use of elements in a nonlinear solution to a subset of those available in a linear solution. The use of higher order elements is discussed below. Certain line, rigid, and specialty elements may be restricted for use in linear or dynamic solutions only. Take the time to read your documentation to understand which elements can and cannot be used in a nonlinear problem. Convergence problems can sometimes be resolved simply by changing element types to those better suited for a nonlinear problem.

If you must use a solid model for a nonlinear solution, keep the following additional points in mind while constructing the mesh.

Take another look at small or insignificant features in the model. Due to the speed of most linear solutions, you may have developed the habit of leaving in some fillets or features simply because it took more time to remove them than they added to the run time. While this is acceptable in a linear model, it can cause trouble in a nonlinear run. Consider that in addition to the increased length of a nonlinear solution, many iterations may be required to achieve convergence. While this is not an issue in linear analysis, any run time penalty due to an excessive mesh from unnecessary features will be paid for in each iteration.

Revolve or extrude bricks and wedges wherever possible in the model. Some solvers do not allow the use of higher order elements in a nonlinear solution. These elements can have plastic and elastic regions within the area or volume due to their representation of strain fields. Consequently, some accuracy may be lost in the integration of these elements, especially if a particular portion of an element changes from elastic to plastic and back again in the course of the solution. If only constant strain (linear) tetrahedrons are allowed, it may be worthwhile to develop a simplified version of the part that can be meshed with bricks or wedges, due to the improved accuracy of these elements. This mesh will be much smaller than a mesh with enough linear tests to capture the actual behavior.

Modeling Nonlinear Behavior with Linear Tools

After all of the above discussion of the wonders of nonlinear analysis, the unfortunate fact is that many design analysts do not have access to a nonlinear solver when a situation arises where nonlinear behavior is expected. As mentioned previously, the basic and most inexpensive solver in any FEA code is the linear solver. Some companies do not see the need for the up-front expense of the nonlinear solver and others just do not expect nonlinear behavior to be an issue. Other companies purchase a tool that lacks a nonlinear option. Most of the available p-element based tools have few or no nonlinear capabilities and most of the CAD-embedded systems that run inside a CAD interface lack nonlinear options. Whatever the reason, if analysis is used regularly in product design, it is likely that the need for some nonlinear solution will present itself. Your options are limited in this case. Either you can acquire the nonlinear module, if available, or you can use linear techniques to approximate the nonlinear behavior and interpret your results in light of this approximation. The latter requires additional information discussed in this section.

Moderately Nonlinear Large Displacements

Under an applied force, most large displacement, stress stiffening problems are conservative when run as linear from the standpoint of both stress and displacement. Therefore, if the stress and displacement levels are acceptable when solved as a linear problem, these results will most likely improve when a large displacement solution is used.

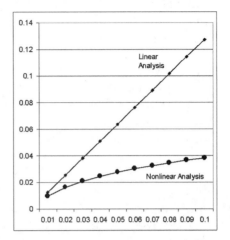

Fig. 15.12. Pressure versus displacement of uniformly loaded flat plate–linear and nonlinear.

Moderately Nonlinear Material Problems

When a material undergoes moderate nonlinearity, the stress results are typically conservative when run as a linear solution under a given load. However, this is not as straightforward as stress stiffening alone. The mechanics of a problem with a nonlinear stress-strain relationship are complex. As the strain increases, any nonlinearity will serve to reduce the modulus or stiffness. Generally, this should cause displacements to rise but stresses to drop. However, if strains are large enough to warrant a nonlinear material model, stress stiffening effects are probably present as well. Therefore, the assumption that stress is conservative should only be made when the linear stress is near yield. For stress levels beyond yield, you should not make assumptions about the nonlinear solution. If you lack the tools to run this problem with a nonlinear solver, contact someone on the outside who can.

Use Linear Methods to Simulate Contact

Some contact problems can be modeled using linear methods to a fair degree of accuracy. These methods include using springs, well-placed constraints, balanced forces, and multipoint constraints. Each has benefits and limitations that must be understood. Use test models to ensure that they are behaving as expected. After this section, it might be worthwhile to review the last example in Chapter 8 on boundary conditions to study the effects of some of these techniques on a part.

Springs

Gap elements, for all practical purposes, are conditional stiff springs. They turn on when displacements between two nodes approach the contact condition specified in the element setup and turn off when the nodes separate. Therefore, one option you can employ in a linear solution *when the contacting surfaces are initially in contact* is to use stiff springs instead of gaps. Mesh the opposing surface or curve pairs that you would expect to contact with aligned nodes as if gap elements were to be used. This technique requires an iterative process because after the first solution, you must identify any of the compression springs that are put in tension by the resultant deformation. These elements should be deleted from the model, or better yet, assigned an extremely low spring constant. By using a nearly zero stiffness spring, the elements are still in the model and can be turned on again by assigning a stiffness if conditions change and the elements show compression. After a few iterations, the contact area should be well defined and provide the same results as if gaps had been used.

As mentioned above, the contact regions must be initially touching because the springs cannot turn on as gaps close. Another limitation of this technique is that sliding should be almost nonexistent. While a gap element can tolerate some sliding, this behavior could cause the spring to go in tension due to lateral movement even if the contact regions are in compression. A workaround available in some codes is to use springs that have stiffness in only one degree of freedom. Using these elements, a coordinate system could be defined that has an axis normal to the contact direction, and only displacement in that direction will have spring stiffness. In this case, a limited amount of sliding may occur without invalidating the approximation.

Constraints

When sliding is the dominant behavior at a contact interface and one of the parts involved is relatively rigid and stationary, constraints normal to the interface could be used in place of an elastic contacting member. To define these constraints, it is best to identify or create a coordinate system with an axis normal to the contact surface. Using this technique, sliding contact on a planar, cylindrical, or spherical surface can be modeled in 3D, or along a line or an arc in a 2D analysis. If the contact geometry is more complex, you must either simplify it to one of these configurations or use an alternate method. Two limitations to this technique follow: (1) the part modeled will never lift off the contact region because it is constrained to stay in contact, and (2) the load transfer between two parts across the contact region is not possible. The latter results from the fact that one of these parts has been replaced with a constraint.

Reaction Forces

Another method for replacing one of the contacting components with a boundary condition is to use an equivalent reaction force. In reality, most applied loads in a model, with the exception of acceleration loads and pressure loads, are simulating contact with an unmodeled component. With this technique, no surface or curve is constrained to a shape but conversely, the contact shape is much more difficult to control. An iterative approach is usually required to develop the area of application and the load distribution to get the response you expect. This technique should be employed when a constraint approximation will add unwanted stiffness to the system.

Multipoint Constraints

A final technique utilizes multipoint constraints or constraint equations to relate the behavior of one part to another. This expands the use of the constraint approximation mentioned earlier by removing the limitation of replacing one part with a boundary condition. A multipoint constraint allows two elastic bodies to interact in a sliding manner with load transfer across the boundary. The limitation that no lift-off is permitted remains. However, by using multiple coordinate systems, some complex interactions can be developed with this method. If the use of multipoint constraints is new to you, refer to Chapter 7 for tips on using these tools.

Summary

For experienced FEA users, a nonlinear analysis represents an excellent chance to push the envelope of simulation by modeling a system that behaves much more like the actual response observed in the lab or the field. Real behavior such as yielding and contact bring the worlds of FEA and testing closer. However, these techniques are more difficult and you should master linear techniques before approaching a nonlinear solution. If boundary conditions and results interpretations are still somewhat confusing in the linear world, do not attempt to move into the nonlinear world. You will just be asking for trouble. However, once you are comfortable enough to give it a try, the opportunities are endless, especially if your design challenges involve plastic parts. The various material models and yield criteria can combine in many different ways to allow you to simulate extremely complex material behavior.

As with any other technique in FEA, proceed cautiously and check yourself at each step along the way. Carefully follow the methodology described in this chapter and amend it based on your own experiences with your company's parts and systems. Again, a notebook of best practices or an on-line collection of tips and tricks is the best place to document material models, as well as load case/load step combinations that enhance the convergence or understanding of your parts. Finally, do not hesitate to look for help when you get started in nonlinear analysis. Discuss your problems with an internal or external coach or your software support resource. Look for opportunities to obtain advanced training, both in your software and in the mechanics of the materials that you will be using. With requisite patience and commitment to learn, nonlinear analysis can greatly enhance your company's ability to design and improve products in the virtual world.

16
Buckling Analysis

As discussed in Chapter 2, buckling involves lack of stability in a structure subject to compressive forces. This instability is independent of material strength and dependent on the structure's shape. Buckling analysis in FEA may be thought of as an extension of static analysis. You should always start by performing a linear static study of the structure under consideration. Once reviewed, the results of this analysis may generally be used by a subsequent buckling analysis, which reports a buckling load factor (BLF). This number is the factor times which the loading applied to the structure will cause it to buckle; that is, the BLF times the magnitude of the applied loading present in the model is equal to the critical load of the structure. Another useful output from a buckling run is the buckling mode shape. This graphical result shows the shape of the instability corresponding to the BLF. Does the shape make sense? Can the structure deform in such a manner? These are questions that must be answered before attaching the BLF to the system.

Always keep in mind that a linear static analysis alone will not predict buckling! To prove this to yourself, try bending a thin-walled tube such as conduit or copper pipe without any internal support. It will buckle very quickly. Repeat this in a linear static FEA and you will not see the tube collapse at any load. It will continue to bend as if it were solid. A nonlinear analysis may, if set up properly, indicate buckling by failing numerically due to structural instability. Learn what your code reports in these situations to ensure that you can differentiate buckling instability from convergence problems.

Of course, it is not necessary to follow up every static analysis with a buckling analysis. This is where engineering judgment enters. Are there high compressive loads in the system? Does the structure appear to be long and slender? Is it thin-walled? When in doubt, conducting a buckling analysis should prove to be a time-efficient way to ensure structural stability.

Possible Scenarios

The following four scenarios may result from a buckling analysis. They are related to the applied loading used in the model and depend on the reported maximum stress (σ_{max}) from the static analysis, the BLF from the buckling run, and the failure strength (S_f) of the structure's material. This last property is equal to the yield strength of ductile materials or the fracture strength of brittle ones.

- $\sigma_{max} < S_f$, $BLF \leq 1$. The study indicates linear instability in the structure as analyzed. Buckling should be considered a likely failure mode and must be addressed by the design team.

- $\sigma_{max} < S_f$, $BLF \geq 1$. The study indicates linear stability in the structure within the safety factor used in the analysis. If this safety factor has been properly chosen, buckling need not be considered a failure mode for the structure as loaded in the model.

- $\sigma_{max} \geq S_f$, $BLF \leq 1$. The study indicates either linear or nonlinear instability in the structure, depending on how close or far these values are from satisfying the equalities. For example, if the BLF is small and σ_{max} is close to S_f, the resulting critical load is likely to give rise to structural stresses below the failure point, and the structure will experience linear buckling. On the other hand, if the BLF is closer to unity and σ_{max} is much larger than S_f, the critical load is likely to place the material in an unsafe stress state, and the buckling phenomenon will be nonlinear. Hence, nonlinear static and buckling analyses may be warranted.

- $\sigma_{max} \geq S_f$, $BLF \geq 1$. Assuming that failure strength stress levels in the material are acceptable, the study would seem to indicate linear and nonlinear stability in the structure within the safety factor used in the analysis. Yet, because of the nonconservative nature of linear buckling analysis beyond the proportional limit of the material (refer to Chapter 2), nonlinear static and buckling analyses should be performed for this structure. If nonlinear analyses are not available to you, a larger safety factor should be employed as well as a substantial test plan.

In many FEA codes, it is possible to conduct a linear buckling analysis without explicitly preceding it by a static analysis. In addition, analysts will often use a unit load to obtain a BLF equal to the critical load of the structure. Although both approaches constitute very efficient ways of quantifying a structure, you must always remember to check the corresponding stresses at the reported critical load to verify that they are indeed below the proportional limit of the material. Otherwise, as mentioned above, this critical load and the structure's quantification will be nonconservative.

Accuracy Issues

Besides ensuring that the results from a linear buckling run correspond to the linear range of the material's properties, there are other accuracy concerns that should be addressed. One is the need to remember that imperfections in the structure will work as instability risers. For example, a soda can may easily withstand your weight if you were to stand on it squarely, yet a single wrinkle on its wall will cause it to promptly buckle under similar conditions. A similar case can be made regarding dimensional tolerances on wall thickness. You must foresee and account for all structural variances whenever possible by either including them in the model or adjusting the safety factor used in the analysis.

Another concern arises from the possible misalignment of the load vector. Standing at the center of the soda can is a lot safer than standing toward its edge. You must always verify that the load vector is at its worst orientation as allowed by system tolerances. "Worst" implies that this orientation will cause the BLF to be the lowest.

Because instability is such a precarious state, you must always ensure that your safety factor is properly selected. Remember that failures due to buckling are usually sudden and disastrous.

Simple Buckling Analyses and Correlation to Theory

Consider standing on an empty soda can that is propped up on level ground. Make the following assumptions.

Fig. 16.1. Simplified soda can with a 200lb person standing squarely on top of it.

- The structure in Fig. 16.1 may be simplified as a cylinder with an outside radius (r_o) of 1.3", a wall thickness (t) of 0.005", a height (L) of 4", and end cap thickness (t_c) of 1/16". The inside radius (r_i) is thus 1.295".

- The structure's material is 6061 aluminum with a modulus (E) of 1e7 psi, a Poisson's ratio (v) of 0.3, and a yield strength (Sy) of 6e4 psi.

- You weigh (F) 200 lbs. and are able to stand directly above the can, on its upper rim, with your c.g. coinciding with its center axis.

- This scenario may be considered to represent a fixed-free condition. The rotations of the bottom of the can are fixed by its geometry and translation can be assumed fixed by the frictional forces. Hence, the length factor (K) is equal to 2 and the effective length of the can is $L_e = KL = 8''$.

First, you must calculate the can's cross-sectional area, moment of inertia, and radius of gyration.

$$Eq.\ 16.1 \qquad A = \pi(r_o^2 - r_i^2) = 0.0408 in^2$$

$$Eq.\ 16.2 \qquad I = \frac{1}{4}\pi(r_o^4 - r_o^4) = 0.0343 in^4$$

Eq. 16.3 $\quad r = \sqrt{I/A} = 0.917in$

With Eq. 2.60, you can determine whether the soda can falls into the Euler ("slender column") category.

Eq. 16.4 $\quad \dfrac{L_e}{r} = 8.72 < \sqrt{\dfrac{\pi^2 E}{S_y}} = 40.6$

Because the first quantity is not greater than the second, the column is nonEuler and its buckling mode will belong to the short to intermediate height regime. As shown below, it is interesting to erroneously continue with a Euler type analysis. According to this type of analysis, the column will buckle under the critical load as seen in Eq. 16.5.

Eq. 16.5 $\quad P_{cr} = \dfrac{\pi^2 EI}{L_e^2} = 52900lbs!(BLF = 265)$

Note that all BLFs reported in this example correspond to the 200lb loading scenario. You probably would not trust the column under anything close to this loading–remember that Euler's hyperbola is being extended into the nonEuler range. At this point, you should calculate the compressive stress caused by the 200lb weight on the can's wall as follows:

Eq. 16.6 $\quad \sigma = F/A = 4910psi$

The calculated stress is well below the yield point of the aluminum; thus, were the can to buckle under your weight, it would be an elastic phenomenon. Calculating the force (F_y) necessary for the can to reach yield is useful, so that a nonlinear buckling analysis can be anticipated.

Eq. 16.7 $\quad F_y = S_y A = 2450lbs(BLF = 12.25)$

It is hard to believe that the can would remain stable under this much weight. If you have ever seen a soda can buckle under a coaxial load, it is likely you have noticed an "accordion-like" crushing effect of the can wall as it collapses. This elastic, buckling crush mode has been quantified theoretically, and the equation describing it appears below.

Eq. 16.8 $\quad P_{cr} = \dfrac{\pi E t^2}{\sqrt{3(1 - v^2)}}\left(2 - \dfrac{t}{r_o}\right) = 949lbs(BLF = 4.75)$

Eq. 16.8 is valid for thin-walled circular tubes under uniform longitudinal compression whose radius to wall ratio is greater than 10 and whose length is several times greater than the quantity $1.72\sqrt{r_o t}$. Note that all conditions are met by the soda can example. In addition, note that the critical load equation is independent of length, which is not entirely intuitive when dealing with buckling phenomena. This crush mode has also been treated empirically, resulting in an approximate equation of the form:

$$Eq.\ 16.9 \qquad P_{cr} = 0.3\pi E t^2 \left(2 - \frac{t}{r_o}\right) = 470 lbs (BLF = 2.35)$$

which assumes the same conditions as its theoretical counterpart. Because it is based on testing data, Eq. 16.9 is the most indicative of the load necessary to crush the can.

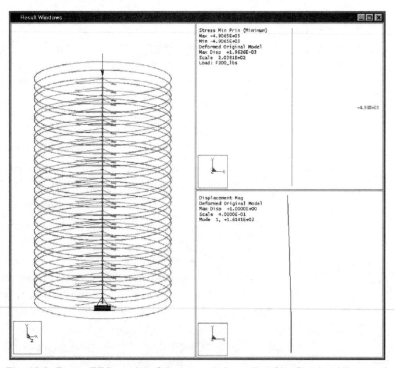

Fig. 16.2. Beam FEA model of the can, and results of its first buckling mode.

Now you are ready to perform FEA on the soda can. The simplest type of FEA you can conduct will make use of beam elements. Keep in mind, however, that because these elements have no real wall, they are not likely to capture the accordion crush mode. Note also that these elements cannot be "capped" (although the constraint at the bottom endpoint will act like a cap). Fig. 16.2 shows a Pro/MECHANICA beam model of the can with corresponding first mode buckling results (BLF = 161). The mode shape is that of a Euler column's first mode, yet the BLF is 60% of the hand calculated value. Note that the BLF is also nearly 70 times greater than that calculated using the empirical crush mode equation. Hence, you should avoid the use of beams for the buckling analysis of nonslender structures. This should not come as a surprise, because beam elements generally should not be used to represent structures that are short relative to respective cross sections.

Next, a Pro/MECHANICA shell model is built. This FEM and corresponding results are shown in Fig. 16.3 (BLF = 5). The mode shape is quite different from that of the beam. It is the accordion shape predicted by the crush mode theory. Note that the BLF is only 7% higher than that predicted by Eq. 16.8, yet over twice the value predicted by Eq. 16.9. As mentioned in the introduction to this chapter, buckling analyses require a robust safety factor and a thorough test correlation plan.

The above numbers mean that the can will not buckle under your perfectly centered weight, yet the comfort zone varies quite a bit depending on the analysis used. In addition, when dealing with such a thin-walled structure, tolerances are a big issue. A single thousandth of an inch difference in wall thickness (t = 0.004") will bring the empirical crush mode BLF down to 1.5–precariously close to unity.

Investigating the change in results caused by increasing the length of the soda can, while keeping all of its other properties and boundary conditions the same is extremely interesting. By reviewing the required conditions stated above, note that as *L* grows, both the theoretically and empirically derived crush mode BLFs become even more valid. Table 16.1 presents the manually calculated and FEA results of three new length scenarios in addition to the original.

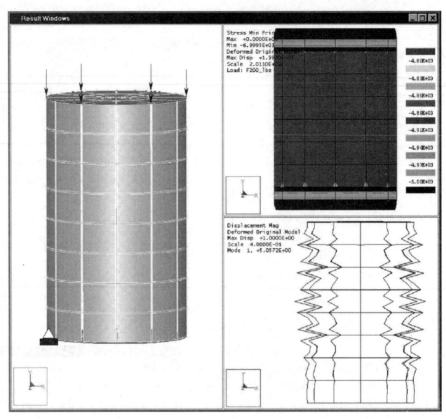

Fig. 16.3. Shell FEA model of the can, and results of its first buckling mode.

Table 16.1. Analytical buckling results of a fixed-free soda can subject to a coaxial 200lb compressive load applied to its free rim: $r_0 = 1.3"$, $t = 0.005"$, $t_c = 1/16"$, $K = 2$, $S_y = 6e4$ psi, $E = 1e7$ psi, $\nu = 0.3$

L	L_e	L_e/r	$\sqrt{\pi^2 E/S_y}$	Euler BLF	Theoretical Crush BLF	Empirical Crush BLF	Beam FEA BLF	Shell FEA BLF
4"	8"	8.72	40.6	265	4.75	2.35	161	5.06
20"	40"	43.6	40.6	10.6	4.75	2.35	10.3	10.3
65"	130"	142	40.6	1.00	4.75	2.35	0.999	1.00
100"	200"	218	40.6	0.425	4.75	2.35	0.425	0.425

Many interesting phenomena are occurring here. First, note once again that the crush mode BLFs are independent of column length. In addition, note that all three longer cans meet the slender, Euler column criterion.

The most difficult analysis, in terms of dealing with its results, turns out to be that of the 20" can, which can be barely considered as a Euler column. (Refer to Fig. 16.4.) Note that both the beam and shell FEA results show an Euler-type buckling mode with identical BLFs, which agree within 3% with the manually calculated Euler BLF. Yet, these values are all higher than their crush mode equivalents. It appears that as the column becomes Euler, FEA shell models no longer predict the accordion mode as the first buckling mode. Alternatively, it may be that to capture this mode, the model requires many more elements (although Pro/MECHANICA reported the results as having converged within 2.5%). Experiment using your own code. Note that all analyses still indicate that the longer can will still handle your perfectly centered weight, but the comfort zone derived from FEA may be dangerously misleading.

Fig. 16.4. 20" soda can, first buckling mode results of beam FEA and shell FEA.

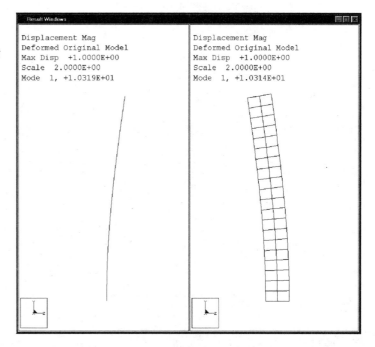

The last two analysis results are easier to reconcile. Both 65" and 100" columns reside overwhelmingly in the Euler regime. Fig. 16.5 shows Euler type, buckling mode shapes from FEA. The corresponding BLFs are identical to the calculated Euler. Note that at the 65" length, these BLFs have reached unity–for the first time the can is no longer stable under your weight. Note as well that these BLFs are smaller than their crush mode equivalents. Had you taken for granted that the first mode shape would always be the accordion type, you would have underestimated the can at these lengths.

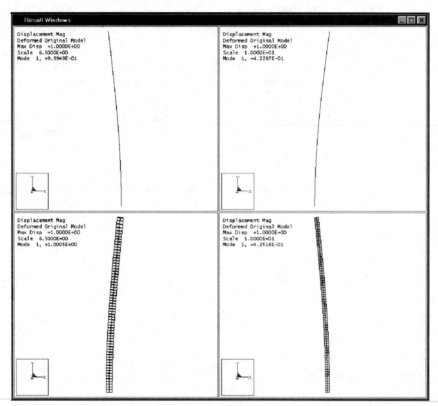

Fig. 16.5. 65" and 100" soda can, first buckling mode results of beam FEA and shell FEA.

Summary

Buckling analysis is a very difficult topic indeed. Any assumptions made regarding surface smoothness, geometric consistency, load vector placement, and orientation variances must be well founded and thoroughly reviewed. Worst case scenarios must always be utilized in the analysis. Manual calculations should be used against the FEA results whenever possible. Always review the buckling mode shape to verify that it makes sense and is not entirely unexpected. Always use large safety factors, especially when dealing with relatively short or nonlinear structures.

Most companies know whether buckling is a commonly encountered phenomenon in their product lines. If it is, it is well known that its results are sudden and spectacularly catastrophic. FEA users in these companies should spend quality time with test models, calculations, and empirical testing before basing a buckling critical design on analysis results. If buckling has not historically been a problem in your company, you should still take to heart the statements made earlier about considering tolerances and all load orientation options.

In general, however, regardless of your company's history, when dealing with high compressive stresses or structures with slender features that are under stress, you should always check for buckling. FEA is a great tool for efficiently accomplishing such checks. Its mode shape results are an excellent way of envisioning this type of failure. Yet, your analysis must always be thorough and you should plan to correlate with testing when stability appears to be an issue. Do not let buckling take you by surprise–it is never a pleasant one.

17
Modal Analysis

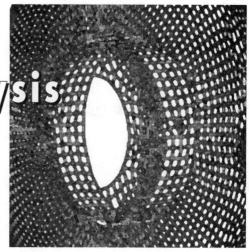

Modal or *natural frequency* analyses are used frequently when parts being designed or verified are subject to vibratory or cyclic loads. This solution type returns the resonant frequencies for a given structure under a specified constraint set. The mode shapes corresponding to those frequencies are also provided. Modal analysis is extremely important for products mounted on an automobile or a motorcycle that experience vibration resulting from the engine's unbalanced forces. Modal analysis is also valuable for products subject to transient (time dependent) loads such as impact or drop forces. The basic steps and terminology of modal analyses should be understood by all FEA users, regardless of the need for further dynamic analysis, due to some of the other uses for this solution presented later in the chapter. However, if vibration due to motor, engine, shaker, wind, or other harmonic excitation is a fact of life in your product challenges, take some time to review the "Dynamic Analysis" section of Chapter 2 before proceeding.

A Simple Modal Analysis

A thin column fixed at one end will vibrate or fluctuate most easily about the fixed point with no "nodes" or additional bends (*inflection points*). This beam is illustrated in Fig. 17.1.

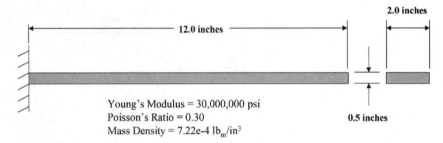

Fig. 17.1. Simple cantilevered beam model.

The natural frequency of the beam is essentially the speed with which it moves from one extreme to the other and back. This speed is defined primarily by three fundamental physical parameters of the beam: (1) weight or mass, M, (2) geometry, and (3) material rigidity or "springback," k. The following equation allows for the calculation of the first natural frequency of this bar:

$$Eq.\ 17.1 \quad f_n = \frac{3.52}{2\pi}\sqrt{\frac{K}{3M}}$$

where fn = first natural frequency, k = spring rate of a cantilevered beam = $3*E*I/L^3$ = 1085 lb_f/in, and M = mass of beam = 0.008664 lb_m. Therefore, f_n for this beam should equal 114.5 Hz.

Fig. 17.2 shows the deformed shape of an FEA model of the cantilevered beam. The first natural frequency was found by MSC/NASTRAN to be 114.0 Hz.

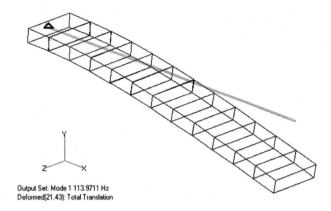

Fig. 17.2. FEA results of simple beam model.

Output Set: Mode 1 113.9711 Hz
Deformed(21.43): Total Translation

Basics of Modal Analysis

In addition to the theoretical basis for modal analysis, it is important to have a practical feel for what physical characteristics of a structure affect its natural frequencies. To understand natural frequencies, it is helpful to first consider the *mode shapes*. In general, simple structures will have a first mode shape corresponding to their most flexible orientation. A cantilevered beam, as shown previously, will bend at its base, and a disc with a pinned circumference will flex in its center. Another way of looking at this is that the first mode shape is the shape with the least potential or *strain energy*. The shapes of the second or third natural frequencies require more energy to generate and have higher internal strain energies.

While many problems involving harmonic input require computation of many higher order natural frequencies, this discussion focuses primarily on the first, or fundamental, natural frequency. This is the most easily predicted and controlled and has many uses beyond dynamic response studies.

As mentioned earlier, each mode shape relates to its corresponding modal frequency through the weight and stiffness of the structure as constrained in the analysis. The contribution of weight is best understood by considering inertia. The more weight, or inertia, the harder it is to change directions when fluctuating. In addition, the mass moment of inertia, which results from the weight distribution, is inversely proportional to the magnitude of the first natural frequency. Thus, increased mass or weight lumped far from the constraint regions such that the mass moment of inertia about these constraints is high, will serve to reduce natural frequency due to inertial effects.

"Spring-back" is the force trying to keep the beam moving. When the beam is bent past its resting position, its elasticity tries to snap it back into place. However, inertial effects do not allow it to stop immediately and cause it to "overshoot" its mark.

To reiterate the description in Chapter 2, the interaction of these two parameters balance out to provide an oscillation speed which is called the "natural frequency." The first mode or natural frequency is the one at which the beam will vibrate after all external excitations are removed. Additional natural frequencies represent the oscillation of the beam in other deformed shapes or "modes." Modal vibration only occurs when the part is being shaken at a frequency which is near a natural frequency.

Without *damping*, these two effects might keep an oscillating body in motion forever. Damping represents inefficiencies of the material due to energy loss at a molecular level or of the system due to component interaction. Higher damping factors cause the oscillation's amplitude to decrease so the beam slowly (or not so slowly) stabilizes. Recall from Chapter 2 that damping has very little effect on natural frequencies at typical structural values. Damping affects the physical, "real" response of the beam when shaken or bumped.

Reasons to Perform Modal Analyses

The preceding discussion is as valid for simple structures as for complex bodies. However, the expected natural frequency is not nearly as easy to calculate. Choosing the equation for the simple example shown earlier required advance knowledge of the mode shape expected. This is not the case with an FEA modal study. As with simulations for stress and displacement, FEA can predict natural frequencies for any general structure.

Knowing the natural frequencies of a design subject to harmonic inputs is important. When a part is excited at a frequency it is "comfortable" vibrating at, the effects of the input are magnified and may cause premature or catastrophic failure. The effect may simply be more noticeable vibration in the system. While this may not result in material failure, it may cause user fatigue in hand tools or perception problems in other consumer goods. For complex structures, FEA is the most efficient (and sometimes the only) means to determine these frequencies.

The goal of a modal study is to ensure that the system does not have a resonant frequency near the operating frequency or in the range of operating frequencies. If the first mode is lower than the operating speed, the product user will notice a "shudder" on start-up as the speed passes that frequency. Most engineers are familiar with the shudder of a grinding wheel as it comes up to speed. This indicates that at least the first mode of the grinder/stand system is below the operating speed. To ensure that resonance effects are avoided within the operating frequency range, certain references suggest that natural frequencies occur only below one-third of the minimum operating frequency and above three times the maximum. In practice, however, this is not always possible. Consequently, it may be easier to push the frequencies down or up, whichever is easier, so that they bracket the operating frequencies.

In some products, it is actually desirable to drive the system at a natural frequency. Ultrasonic welding horns for plastic fastening are designed so that there is a natural frequency at about 20,000 Hz to amplify the input and put more energy into the work. In this case, horn designers are not only interested in the frequency value, but also the mode shape. The excitation is vertical and the ideal horn will have a stretching mode shape at 20 kHz so that energy is not lost in twisting or side-to-side bending. This mode shape is illustrated in Fig. 17.3.

Fig. 17.3. Longitudinal mode shape of a ultrasonic welding horn.

Another powerful use of modal analysis is the assistance it provides in understanding the induced mode shapes themselves. Deformation patterns at an operating frequency may be deemed acceptable if they do not affect an inherently weak section of the system. In addition, if the excitation is constrained to a uniaxial or planar condition, mode shapes that are primarily normal to the input load can be neglected, because they represent deformations which cannot possibly be induced. Take care when exercising this assumption that the input loads in all part, assembly, and "user" tolerances truly do not waver from the expected orientation. If you are uncertain, assume worst case.

Boundary Conditions

A modal analysis does not require boundary conditions but may utilize a constraint case if constraints are present. If the model is not constrained fully in all six DOFs, the first modes will correspond to rigid body motion in each of the unconstrained directions with a frequency of approximately zero. These are called "rigid body modes," and should in theory be of zero magnitude, although numerical round-off may give them a small nonzero value. Some solvers require you to specify that an unconstrained model is unconstrained before initiating the modal study. A linear modal analysis will ignore whichever load cases you have defined. Constraints must be applied with the same care as in a structural analysis. An overconstrained model will behave too stiffly and, therefore, results in an overprediction of the first modes. The stepped cylinder shown in Fig. 17.4 was analyzed with two constraint methods.

Fig. 17.4. Stepped cylinder for modal study.

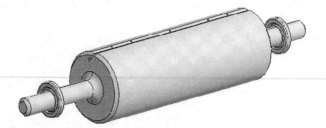

The first method constrained the entire surface of the shaft to be enclosed by the bearings. The second method allowed for rotations at the center of the shaft in line with the bearings. Because most ball bearings will allow some rotation, albeit small, the second method correlated very closely to testing, whereas the first was fictitiously high. This is dangerous because high predictions are nonconservative in a modal analysis. This part could easily have been designed at resonance if the boundary conditions had not been checked.

While it is natural to assume that adding more constraint to a part will increase the natural frequency, the location of the constraint is equally important. The previous cylinder example can be used to illustrate. Testing and a subsequent analysis of the cylinder with no constraints put the first natural frequency at 325 Hz. When the cylinder was analyzed with the correct bearing constraints and tested for correlation, the first natural frequency dropped to 125 Hz. This can be explained by considering how the solver looks at mass concentrations. With no other constraints in the model, the solver will "anchor" the mode shape at the point of maximum mass (inertia) and look to displace away from such point. In the cylinder example, deflection of the short, stiff shafts comprised the bulk of the modal displacement with no other constraints. However, when the shafts were constrained, the cylinder became a cantilevered mass at the end of a rod. Shifting the pivot point forced the modal displacement to the center of the large diameter. Fig. 17.5 illustrates this phenomenon in a schematic representation of the cylinder.

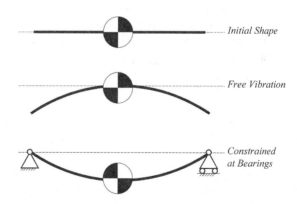

Fig. 17.5. Schematic representation of two vibratory modes of free and constrained cylinder.

Initial Shape

Free Vibration

Constrained at Bearings

Preparing for a Dynamic Analysis

A modal analysis should always be the first step when considering a dynamic analysis. If the modal study indicates that there are no natural frequencies near the operating speed, then the frequency response analysis may not be required. If the period or duration of the transient load is much longer than the period of the first natural frequency, then the input is less likely to excite a resonance. Remember that the period of a sine wave can be calculated as the inverse of the frequency.

Dealing with Resonant Frequencies

A natural frequency near an operating speed does not automatically mean that resonance will be excited and vibration will be a problem. A case in point was a study performed on formed steel bases for cast iron pumps. At the beginning of a cost reduction program, it was agreed that because vibration was never a problem, any new design must prove to be equally impervious to the cyclic pump input. Several new concepts were suggested and the original parts were analyzed to provide a baseline. The baseline analysis showed that most of the current designs had a first natural frequency of nearly exactly the pump operating speed. Because vibration was never noticed, it was clear that any new design that had a different first mode would be even less likely to resonate.

In most cases, however, a first or second mode near an operating frequency will cause noticeable vibration amplification. Adjusting your geometry to move the natural frequencies is somewhat of an art. The natural frequency of a part is related to its weight and its stiffness. However, many techniques for increasing stiffness also add weight. The proper combination of increased stiffness, reduced weight, and redistributed weight is required to fine tune natural frequencies.

Meshing a Part for Modal Analyses

When the end results of a modal analysis are the natural frequencies of a structure, the mesh can be somewhat coarser than would be required for a detailed stress analysis. The density required is similar to that for a gross displacement analysis. Yet, as with a displacement analysis, an overly coarse mesh will result in an overly stiff structure and fictitiously high modes. A modal analysis must be converged as any other structural study.

When a modal analysis is to be used to preprocess a dynamic analysis that requires the actual modal results, the mesh must be developed to capture the behavior desired by the dynamic analysis. If obtaining stresses over a frequency range is the goal of a frequency response analysis, the mesh must be converged for all resultant deformations of interest. If displacement is the only output required, the mesh may be somewhat coarser.

Avoiding symmetry in modal and subsequent dynamic analyses is highly recommended. The danger of using symmetry is that there are many more asymmetric mode shapes for a general structure than symmetric. If symmetry is used, the only mode shapes calculated will correspond to the specified symmetry constraints. If the frequency of interest corresponds to an asymmetric mode shape in a full model, an important result will be missed in a symmetry model. If a frequency response or transient analysis will be using the calculated modes to develop more complex data, this problem is accentuated. A modal-type dynamic solution bases the stress and deformation results on the preexisting modes. If any mode that contributes to the final behavior has been missed due to the use of symmetry, the dynamic results will be meaningless.

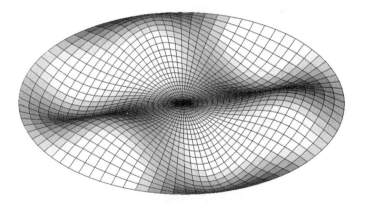

Fig. 17.6. Flat steel disk for symmetry study.

The example below is based on the simple disk shown in Fig. 17.6. A modal analysis was performed on the disk using quarter-symmetry, half-symmetry, and no symmetry. Results are shown in Table 17.1.

Table 17.1. Modal results for a thin disk using various combinations of symmetry. The steel disk diameter is 120 inches, thickness is 0.120 inches, and outer edge is pinned with rotations left free.

Mode	Full disk	Half disk	Quarter disk
1	236.5 Hz	236.5 Hz	236.5 Hz
2	236.5 Hz	396.2 Hz	396.2 Hz
3	396.2 Hz	544.3 Hz	947.3 Hz
4	544.3 Hz	899.7 Hz	1,528.8 Hz
5	544.3 Hz	947.3 Hz	1,667.5 Hz
6	899.7 Hz		
7	899.7 Hz		
8	947.3 Hz		
9	947.3 Hz		
10	1,440.9 Hz		

The actual response of this disk has 10 modes below 1,500 Hz, but the symmetry models capture only about five modes. Note that Mode 8 of the full model, 947.3 Hz, corresponds to only Mode 5 of the half symmetry model and Mode 3 of the quarter symmetry model. Much data would be lost if only a quarter-symmetry model was used.

Experienced analysts can account for symmetry in a modal analysis by using anti-symmetry. Analysts still learning the technology should avoid anti-summetry. This technique could lead to misinterpretation if results are not well documented.

Other Uses for Modal Analyses

Testing for the first mode of a structure can be relatively easy. A simple impact test using a hammer and an accelerometer can pinpoint the magnitude of the first natural frequency. It may not suggest the actual mode shape in a complex part, but simply knowing the frequency can be valuable. A more sophisticated test for the natural frequencies of a part involves mounting it to a shaker table. A table large enough for the part is a prerequisite. The shaker table is then excited with a sweep from low to high frequencies, and the deflection of the part is measured as well as the energy required to shake the table. This technique can pinpoint the modes more accurately.

However determined, the first mode of a part can provide a first pass correlation on geometry and boundary conditions. If the frequency value of the first mode calculated with FEA is significantly different than that measured empirically, constraints should be reevaluated first. If you feel that the constraints accurately represent the test conditions, other properties of the model, such as wall thickness and material stiffness, should be checked.

As described in Chapter 9, a modal analysis will pinpoint an unattached portion of the model or an unintentional mechanism. As stated previously, any rigid body motion will manifest itself in a rigid body mode of zero hertz. If a linear static solution fails due to an insufficiently constrained error, an animation of the first mode results may quickly flag the offending elements. Once the entire model is attached correctly, the first mode should be greater than zero.

Summary

Modal analysis is a fundamental tool for most analysts. Few parts exist in an environment where there is no vibratory or transient loading at some point over the product life cycle. In most cases where vibration is studied, only the modal analysis is required to make design decisions about the system's behavior. Consider both natural frequencies and mode shapes to completely understand the modal response. Push the modes away from operating speeds and try to avoid mode shapes that are similar to the deflection resulting from the load input. Much can be gained by making a thoughtful interpretation of modal results.

18

Dynamic Analysis

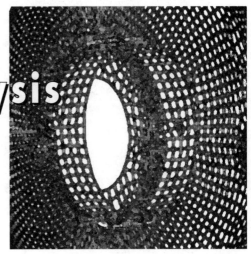

While discussed briefly in previous chapters, dynamic analysis is a broad topic and is covered more thoroughly in this chapter. In brief, a dynamic analysis should be used when the loading is time dependent. Examples of such load inputs include imbalance in rotating machinery, linear oscillatory motion such as that of a piston, or intentional vibration caused by a spinning eccentric. Dynamic analysis is particularly important when the speed or frequency of an applied load corresponds to one of a structure's natural frequencies. Dynamic analyses also provide valuable response data when impact loading sets up a shock wave field in a part, thereby causing amplification of the actual loads. Finally, dynamic analysis may show that the stress or displacement caused by a cyclic load is less than an equivalent magnitude static load due to the fact that the input changes direction before the full response can be developed. In addition to these types of problems, this chapter reviews the importance of damping to each solution type and discusses methods for correlation testing. It is highly recommended that you review the fundamentals of dynamic analysis in Chapter 2 before proceeding with this chapter.

Three primary types of dynamic analysis are required in product design: frequency, transient, and random response solutions. You must select which type of solution is appropriate for the design challenge you are addressing because the inputs to each are quite different. Your analysis code will probably require specification of the solution type during the modeling, and will definitely require the same as you initiate the solution.

Frequency Response

A frequency response analysis requires that all applied loads in a model vary sinusoidally at the same frequency. It will return the steady state response of the structure to the cyclic input. This response may be in terms of stress, displacement, velocity, and/or acceleration. Some important terms used in a discussion of a frequency response analysis are illustrated in Fig. 18.1.

Definitions

Fig. 18.1. Basic components of a dynamic response analysis.

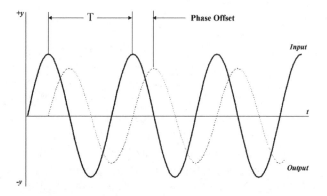

The darker wave of Fig. 18.1 represents a sinusoidal displacement input at the end of a rod. The dashed line represents the steady state frequency response to that input at the opposite end. The time between each peak of the input signal is called its period, *T*, which is the reciprocal of the input frequency, *f*, such that $f = 1/T$. Note that the period/frequency of the response is equal to that of the input. The part of a cycle by which the response lags the input is called the *phase shift*. This quantity as well as the ratio of input to output magnitudes, or amplitude ratio, is constant for a given node at each input frequency, and is a function of the system damping. Theoretically, if there were no damping, the response would track the input exactly. Note that if the *x axis* time scale is scaled by a constant frequency term, ω, the resulting units are those of a cyclic angle term, ωt. Hence, in a steady state frequency response problem, for each individual input frequency there is only one full, repeating cycle of response angle solutions, $y(\omega t)$.

Input

As always, the input to a frequency response analysis is dependent on problem goals. In its most basic form, when a system is excited at only one or a few distinct frequencies, the load is applied as in a static analysis and the solution frequencies are specified. For structures that can experience many excitation frequencies, it is common to input a load versus frequency plot that essentially sweeps each applied load through a range of frequencies scaled according to a defined function. This function is input as distinct frequency/amplitude pairs, as shown in Fig. 18.2.

Fig. 18.2. Typical frequency versus amplitude plot.

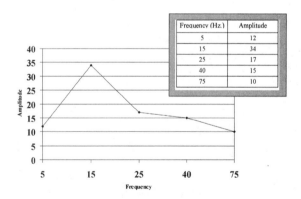

When frequency/amplitude pairs are specified, you may have the option of specifying the number of intermediate frequencies to be included between each major step, as defined by the actual frequencies called out on the tabular input function. This number of frequencies between each step must be large enough to ensure that the peak response is captured. A couple of iterations may be warranted based on the smoothness of your output response profile.

An alternate means of scaling a load based on the solution frequency is required when multiple eccentric masses are driven by a common rotational input. The equation for the centrifugal force, *Fc*, generated by a rotating body follows:

$$Eq.\ 18.1 \quad Fc = MR\omega^2$$

where *M* = the eccentric mass, *R* = the radius of gyration or distance between the mass center and its axis of revolution, and ω = the input speed in radians per second.

The input load, defined by Eq. 18.1, can be applied to the shaft driving the rotating system at the location of the eccentrics. The scale factor in the frequency input table should be set to unity so that the rotating loads are scaled accordingly. Some helpful conversions in problems of this nature relate shaft speed to input frequency as seen in Eq. 18.2.

$$Eq.\ 18.2 \quad 1.0\ RPM = 0.105\ radians/second = 0.0167\ cycles/second$$
$$= 0.0167\ Hz$$

Constraints

A frequency response analysis should be constrained in the same way as a static analysis of the same system, if the constraints are valid. When there is a definite ground in the model, the system will vibrate in accordance with the constraints. However, some systems do not have an obvious ground. Modeling techniques for these free (unconstrained) systems are discussed in Chapter 8. A static analysis requires that ground be specified in some manner. A modal analysis and a frequency response analysis can have one or more unconstrained degrees of freedom and still solve correctly. This capability is important for studying the response of hand-held tools or systems supported by soft mounts with very little stiffness to ground.

Phase Offset

A frequency response analysis oscillates all loads in the load case submitted to the solver at the same frequency. If you absolutely must view the response to multiple frequencies at any given time, you may wish to consider a transient analysis. However, a frequency response study will be faster and easier for steady state solutions and should be used if at all possible. While you do not have much flexibility with multiple frequencies, most solvers allow you to specify a phase offset to different loads in the model. If the true loading of a rotating eccentric is to be modeled, it can be broken down to a vertical load that is phase offset 90 degrees from a horizontal load of equal magnitude. Other uses for this might be cams or other cyclic components of a load that are not perfectly in sync but rotating at the same speed.

Amplitude Offset

In addition to forcing all loads to oscillate at the same frequency, a frequency response analysis cycles the force symmetrically about a mean force of zero. Consequently, if your input has been determined to vary from 5 lbs to 25 lbs at a given frequency, an input to a frequency response analysis of 10 lbs will provide the correct peak-to-peak load, yet excite your system equally in the positive and negative directions. Instead of oscillating between +5 and +25 lbs, it will load your model at -10 to +10 lbs.

To incorporate a nonzero mean force in your model, the dynamic results must be combined with a static analysis with an appropriate offset. For example, results for the 5 lb to 25 lb input mentioned earlier could be developed by combining the 10 lb sinusoidal load with a 15 lb static load as shown in Fig. 18.3.

Fig. 18.3. Combining static and dynamic results to produce a nonzero mean output.

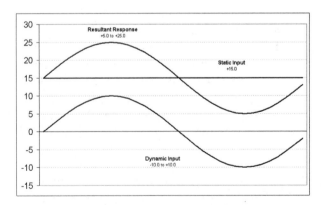

The ability to combine results from two types of solutions, static and dynamic, is not available in all systems. When it is available, it is still somewhat tricky to use. Consequently, a discussion of frequency response results is warranted.

Frequency Response Results

Frequency response results can be calculated in *real-imaginary* format and in *magnitude-phase* format. Magnitude-phase output is the most practical for understanding the response of the system. Magnitude data are available as values of maximum displacement, stress, velocity, or acceleration for each frequency over an entire output cycle. A plot of these data would have frequency values on the x axis and output magnitude on the y axis. Phase data are presented as the value of the phase shift angle for each input frequency.

While these data are excellent for evaluating peak responses and identifying operating frequencies to avoid, they provide little insight on the distribution of the vibrational energy and ways of improving peak responses if these operating frequencies cannot be avoided. To better understand the total response of the system, the magnitude-phase data must be exported to incremental results at a sequence of input frequency angles, ωt, for a given frequency. These output sets can grow in size very quickly if the model is of more than moderate size. The input frequency angle response data at increments of 30 degrees for a single frequency create 12 output sets, which are equivalent in size to 12 static solutions. However, when animated in sequence, it becomes clear how the part or system actually responds to the input. You can examine your system to determine how the energy is being dissipated through it. With this knowledge, it becomes much simpler to implement corrections to better absorb the energy or distribute it to improve performance. Frequency sweep plots at individual nodes or elements cannot provide this insight. Both types of data are important if a part must be studied at multiple frequencies. Input frequency angle response results are more meaningful when the solution is run at only a few frequencies or a few frequencies have been identified as being of interest. An additional output format that might be valuable in a frequency response study is a *node* or *point trace*. The format will show the travel path of a node across the input frequency angle response results. These traces can provide insight to local dynamic response.

Revisiting Amplitude Offsets

With a better understanding of the ways frequency response results are presented, combining dynamic and static results can be revisited. Frequency sweep plots cannot be combined with static data to provide meaningful results. The problem is that you do not know the orientation of the sinusoidal varying load at the reported peak. You may need to add, subtract, or otherwise manipulate the static component. However, the incremental input frequency angle response is, in most respects, compiled in the same format as a static solution. Consequently, the static component can be linearly superimposed on each phase angle result set to provide an accurate picture of the data. Few codes automate the process of combining multiple sets. If this is something you will need to do frequently, a macro (assuming that your postprocessor supports macros) can probably be written to eliminate the repetition.

Transient Response

While a frequency response analysis takes place in the frequency domain, a transient response analysis is a study in the time domain. Forces in a transient analysis are defined to vary with time. Another significant difference between frequency and transient response studies is the boundary conditions used. A frequency response analysis can be solved with no constraints and studied in a free state. Like static analyses, transient analyses must be fully constrained. Results are also more straightforward.

Input

The input to a transient response study is also much like a static analysis except that a time-dependent function can be assigned to a load. Not all loads need a time-dependent definition, and multiple loads can have multiple time functions assigned. These studies are very flexible.

Constraints

As mentioned previously, transient response analyses must be constrained in the same way that their static equivalents would be.

Solution Parameters

One of the critical factors in the accuracy of a transient solution is the *time step* configuration used. Time stepping defines the time increment between calculated steps and output sets. If the time step is too large, peak responses could be missed or truncated. If the time step is too small, the run time may be excessive and the size of the calculated data set might become prohibitively large. Some solutions provide automatic time stepping. If such feature is provided, you should have a good reason not to use it. Otherwise, finding the correct size of time steps will probably require an iterative process. As a guideline, set your time step to one-sixth of the load period or smaller. This procedure will indicate responses of at least half the period in the results. If there is a blip on the output, some response is taking place around that time step. The decision can then be made to improve accuracy by refining the time stepping.

Results Output

Results can be calculated at every time step or at time step increments. If the correct time step is chosen but too few output sets are requested, the results are nearly the same as when too coarse a time step is used. Peak responses may be missed.

The first output to review is a plot of response versus time. The response can be displacement or stress. This will be your best indicator of time stepping accuracy, which is over and above the mesh convergence required for total solution accuracy. If the plot of response versus time suggests that the solution has captured all peaks of interest, the model's results at individual time steps can then be studied. The peaks can be displayed much like static analysis results. In addition, most post-processors will allow you to view a sequence of time step results which will produce a time history animation of the system under the transient loads. Trace paths, as described earlier, can also provide valuable insight in a transient study.

Random Response

Some dynamic excitations cannot be categorized as simply frequency dependent or time dependent. The input fits no simple pattern and is thus considered random. In actuality, while the source of random response loading may be random, the input to the analysis is usually a predefined spectrum of frequencies and amplitudes. The solver does not randomize the data. A random response input might be defined to represent standard road noise for an automotive application, wind load for an aerospace application, or a seismic load for an earthquake study.

Solution Methods

A random response problem may be solved in one of two ways, again depending on the goals of the study and/or the input type or quality. These two methods are known as *time history* and *power spectral density* (PSD). Fig. 18.4 shows the measured response of the 1994 Northridge California earthquake as acceleration versus time (a), and the PSD equivalent of the acceleration history (b). Applying the time history to the system in a transient analysis to determine actual response is feasible. While this can be a very costly analysis due to the duration of the actual event and the number of time steps required to ensure that peak responses are calculated, it does provide certain benefits. For large structures with multiple random inputs or when the input in two or more directions is important, a transient solution to a random input may be the only alternative.

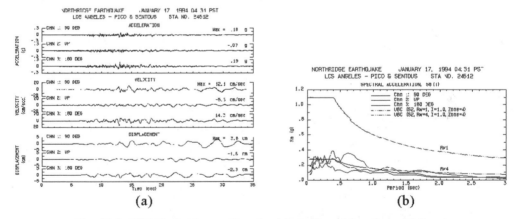

Fig. 18.4. 1994 Northridge earthquake data in time history format (a) and as a power spectral density format overlaid on the Uniform Building Code (UBC) PSD standard for earthquake design (b). (Source: U.S. Geological Survey National Earthquake Information Center.)

It is more common and resource efficient to approximate the input as a generalized spectrum that defines an envelope around the peak values of a given quantity at the various frequencies involved in the event. This can be determined by a Fourier transform on the actual input data. While spectral density plots can be calculated for displacement, velocity, or acceleration, the form most typically required by an FEA solver is that of a power spectral density curve. This curve, as shown in Fig. 18.4(b), is defined with units of acceleration squared over frequency (G^2/Hz) as a function of frequency. A PSD curve is appropriate only for systems with a single input source and when directionality of output is not of primary concern. The results of a random response analysis using a PSD input are always positive; they represent the peak values at the frequencies of interest. These results include a PSD curve of stress (σ^2/Hz) and RMS output.

Constraints

A random response analysis is similar to a frequency response analysis in the way boundary conditions are determined and applied.

Modal versus Direct Solvers

Most codes provide the option of a *modal* or a *direct solution* to a dynamic problem. A modal solution requires that all modes that may contribute to the final response be calculated first. This modal solution space is then used to calculate the actual dynamic response. A modal solution is efficient in many cases because an initial modal analysis should be conducted across the operating frequencies of interest anyway. Once the relatively lengthy modal solution has been completed, the modal dynamic solution is fast. The danger of a modal method approach is that the initial modal solution must include modes outside the operating frequencies. Resonant frequencies occurring above and below the operating range that contribute to the total dynamic response are common. Because the modal method can base a solution only on previously calculated mode shapes, the effects of these extraneous frequencies might be missed. This phenomenon is called *modal truncation*. The effects of modal truncation are impossible to predict. To guard against such effects, some references suggest including all modes of a magnitude up to twice the maximum operating frequency and down to half of the minimum, while others suggest that these numbers be three times and a third, respectively. Another guideline is that the number of modes required to fully predict the dynamic response of a system should increase with the discretization of the applied loads. Hence, whereas for one or more point loads, many modes may be required to fully capture the final response for a uniformly distributed load, such as gravity, only the first few normal modes may be required. Experience with your software and products will enable you to be confident in the decisions on the range of required modes.

For complex structures, the number of modes over the operating frequencies alone may be significant. The modal solution may take hours to complete. Consequently, a *direct method solver* is usually available in most codes with strong dynamics capabilities. A direct solution essentially solves the response formulation. For large models with only a few frequencies of interest, a direct method solution may be much faster than the modal method because the lengthy modal analysis is skipped. You will not need to be concerned about modal truncation with a direct solution either. However, as the number of modes of interest increases, the time savings of a direct solution over a modal solution begins to diminish. Table 18.1 shows suggested guidelines for selecting an appropriate method, as presented in the *MSC/NASTRAN User's Guide for Dynamic Analysis.*

Table 18.1. Modal method versus direct method comparison

Condition	Modal	Direct
Small model		X
Large model	X	
Few excitation frequencies		X
Many excitation frequencies	X	
High frequency excitation		X
Nonmodal damping		X
Higher accuracy		X

A direct solution should also be used if the system being analyzed has heavy damping, or greater than 10% of critical. This is due to the fact that the initial modal solution is calculated without considering damping effects. Because high damping will cause a shift in the natural frequencies, using these in a modal method solution will introduce an error in the reported response.

Damping

The accuracy of any dynamic solution is dependent on the damping assigned to the model. Review the concepts on damping in Chapters 2 and 6. Damping can be applied as discrete damper elements, material specific damping, or overall system damping. The overall system damping is usually defined as a percentage of the critical damping, or the damping value that eliminates overshoot when an impulse load is applied. Table 18.2 builds on the data originally presented in Chapter 2 of some typical system damping values for various types of structures as a proportion of critical.

Table 18.2. Representative damping ratios as percent of critical damping

System	Damping ratio, ζ
Metals (in elastic range)	< 1%
Continuous metal structures	2% - 4%
Metal structures with joints	3% - 7%
Aluminum / steel transmission lines	~ 0.04%
Small diameter piping systems	1% - 2%
Large diameter piping systems	2% - 3%
Auto shock absorbers	~ 30%
Rubber	~ 5%
Large buildings during earthquake	1% – 5%
Prestressed concrete structures	2% - 5%
Reinforced concrete structures	4% - 7%

Your code may include the ability to vary the damping magnitude based on the modal frequency being excited. These values are typically input as a table. The values will not be found in a general reference, but must be determined through testing or from industry or product specific literature.

Note that whenever the results of a dynamic analysis are critical, damping must be determined from testing. However, if the analyses are used for trend studies, consistent damping across the iterations may suffice. Damping is always critical as the input or excitation frequency approaches each of the system's natural frequencies. When a system is driven at any of its natural frequencies, damping effects dominate the calculated response. This is illustrated for a system having a single natural frequency in Fig. 18.5, originally shown in Chapter 2. As mentioned at the beginning of this chapter, a review of these concepts in Chapter 2 will greatly enhance your understanding of the effects of damping on a dynamic analysis.

Fig. 18.5. Response amplification at various damping values as the operating frequency approaches the first natural frequency.

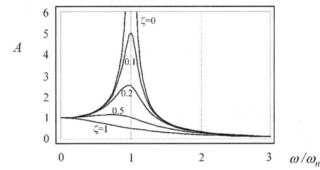

Summary

Dynamic analyses, along with nonlinear analyses, extend the capabilities of FEA beyond simple design verification of stationary metal parts. However, with these added capabilities comes a greater responsibility on the part of the user. All the uncertainty and potential for error is magnified tenfold in a dynamic analysis. The importance of joint stiffness, mass moments of inertia, damping, and the applicability of proper modeling techniques will have a large and immediate effect on the accuracy of a dynamic solution. If the time is not taken to understand the implications of these concepts, it is best to stick with linear static solutions where you must use engineering judgment and correlation testing to bridge the gap between static and dynamic domains. Remember that there is a natural tendency to believe results displayed in Technicolor on your monitor. Spend the extra time reading additional references that cover this topic specifically and use test models to ensure that the results you calculate are of the highest quality.

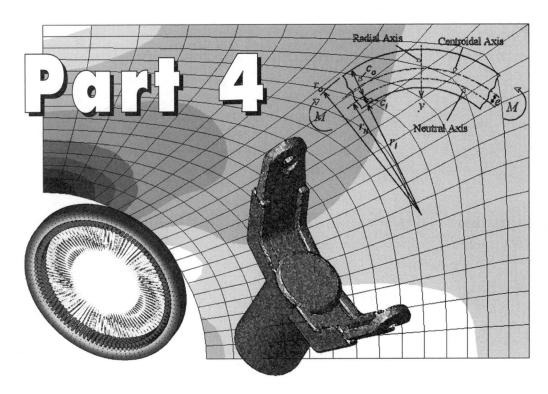

Part 4

Integrating Simulation into Product Design Strategy

19

Overview of Popular Industry Offerings

Armed with all the FEA knowledge presented thus far in this book, you must now be chomping at the bit to get going. If your company has implemented the technology already, you may be questioning whether the chosen code provides the basic, yet crucial, capabilities discussed in previous chapters. If, instead, your company is looking to implement the technology soon, such information is crucial for making the right decision.

This chapter begins by presenting a brief overview of the terminology surrounding the various packages and bundles most commonly offered. The number of choices in the industry today can be overwhelming. However, the information discussed here, as well as that discussed in the following chapter on implementation, should help clear the air somewhat. Of course, the bottom line is that, when used properly, almost any credible FEA package can provide better answers than guessing. However, getting the biggest bang for your buck will require an investment in time to uncover the pros and cons that most directly affect your company's ability to design better products. Following the general overview, the capabilities of some of the more popular and credible tools being used in design engineering will be presented. This presentation is in tabular format to allow you to quickly search each code's content for any of the techniques or capabilities presented throughout this book.

Packaging

Several options are available to the FEA shopper with regards to bundling, integration, capabilities, and ease of use. It is likely that if you are able to sort through the marketing hype and sales smokescreen, the choice for your particular needs will be obvious. It is not easy to see clearly in today's market if you do not have a strong background in the technology. To make matters worse, if you procrastinate for six months, the industry may have changed dramatically! As with CAD tools, the window for making a decision is three to six months. The bright side of this reality is that makers of the stronger codes, such as the ones described later in this chapter, will soon adopt useful features introduced by competitors.

You must make three primary choices in your search for FEA tools: system packaging, the code vendor, and the analysis options you believe that your company needs. Review the previous chapters to identify the analysis options available and their applicability to your company's products. Selected popular vendors will be detailed throughout this chapter. The choice of a vendor should be based on its track record, commitment to service, and ability to provide the tools that you need. Selecting a vendor will require some time for research and collecting references. The one piece of the puzzle that still requires some clarification is the packaging of the technology. There are four basic configurations to choose from, each with specific benefits and drawbacks.

Three Key Components of Any FEA Solution

Prior to discussing the packaging options, a couple of terms should be clarified. The process of FEA requires three key components: preprocessor, solver, and postprocessor. Whatever the final packaging you choose, you will need all of these components. Because a limitation in one component could limit your total ability to solve engineering problems, it is important that you understand the differences. Preprocessors and postprocessors are usually provided as a single package. Most solvers provide a pre and post option, but such option may not be the best for your needs. Resist the temptation to fall into the single vendor convenience trap because few companies do everything the best. The best preprocessor for your needs may be developed by a company other than the provider of the best solver for your needs. Evaluate the entire market.

Preprocessor

The preprocessor provides the interface between you and your solving technology. It allows you to create geometry or input CAD geometry, and provides the tools for meshing the geometry, defining material or element properties, constructing a boundary condition scheme and compiling the data file required by the solver. Most preprocessors are graphical in nature and work much like a CAD system. You should be able to dynamically spin a shaded image of your mesh on the screen and organize your data in groups and layers. More importantly, however, the preprocessor should provide the meshing tools required for you to begin simulating your structures shortly after its acquisition, and provide a growth path for you as your skills develop. While tools should be provided for importing geometry from your CAD system, this is only crucial if your parts are best solved with a 3D solid automesh. As the proportion of parts in your product offering diverge from this mode, the importance of "seamless" CAD integration becomes overshadowed by more efficient meshing and boundary condition tools. Other important considerations when evaluating a preprocessor are availability of qualified support, both from the vendor and outside resources, and speed on your hardware. A screaming tool in a high end UNIX box will be of little use to you if your company is committed to mid-range Windows NT or Windows 98 systems.

Solver

The processor or solver takes the node and element connectivity data provided by the preprocessor and calculates the requested response. Almost every solver in the industry ships with some type of embedded pre and postprocessors. These may simply take the shape of a programming manual and an invitation to fire up a handy word processor, or may be tools that can rival the best pre and postprocessors on the market. The various solvers differ in quality and speed. You should expect a potential vendor to provide a copy of a verification manual that compares the solver's solution to others using classical engineering problems or industry standard benchmarks. The solver should at least have the ability to cover all linear and dynamic needs you expect to have, even if they will not be used in the initial implementation. This growth path is important. As with the preprocessors, it should be fast on your hardware for all the solution types you expect to require. It should have a solid and qualified support network and you should take advantage of it. Do not try to do this alone!

Postprocessor

The postprocessor takes the data from the solver and presents it in a form that you can understand. You should insist on a postprocessor that produces high end color graphics of your results while still providing access to X-Y plots and tabular data. Each type of output has its place. An important capability of a postprocessor is the total user control of the output graphics. You should be able to scale the color legends of your fringe plots and change the colors freely to best represent the goals of the project. Not only should your postprocessor provide animation capabilities, it should provide the ability to output an animation file that can be embedded in a presentation or e-mailed to an internal or external client. This capability, which brings major benefits, is becoming an industry standard. You should also be able to perform error checking on results and manipulate the results data to perform bulk calculations for output quantities that the base tool does not support. Review Chapter 11 for more suggestions on the basic capabilities of a postprocessor.

Open Systems

The traditional packaging of FEA tools is the open system method. A few years ago, companies like The MacNeal-Schwendler Corporation and ANSYS were developing state-of-the-art solving technologies and companies like SDRC and PDA Engineering were providing pre/post tools for most of them. At present, it seems that everyone is providing everything. However, the open system approach is still a viable option and provides the greatest degree of flexibility in an everchanging market. With an open system approach, you choose the one or two pre/post options that work best with your CAD system(s), and the solver that appears to provide the best technology for your needs. As new tools become available, and they will, you can choose to evaluate or incorporate them at your leisure. The disruption to your overall implementation should be minimal because you can phase in the new technologies over time and they should not affect your company's design or CAD practices.

The true, open system pre/post options discussed later in the chapter are FEMAP from Enterprise Software Products and HyperMesh from Altair Computing. Also discussed are pre/posts from companies that provide other pieces of the puzzle, such as I-DEAS from Structural Dynamics and Research Corporation (SDRC), MSC/PATRAN from The MacNeal-Schewndler Corporation (MSC), and, to a lesser extent, Pro/MESH from Parametric Technology Corporation (PTC). SDRC also provides CAD and solver technology, MSC's primary business is solver technology, and PTC's primary business is CAD solid modeling.

Integrated Systems

While most of the players in the FEA market have branched into providing all pieces of the puzzle, it is natural that they would integrate their tools into a system that provides value or benefits that would not be possible if one of their pieces was used separately. Integrated systems might combine CAD and pre/post tools like Pro/MESH, pre/post and solver tools such as MSC/FEA, and all three such as SDRC I-DEAS. One of the most obvious benefits of integrating CAD and pre/post is the availability of specialty tools, such as automatic mid-plane extraction for the development of shell models. Other benefits include continuity of properties such as boundary conditions on surfaces, material properties, and mesh control over multiple iterations of a design improvement process. The need for a linked FEA model to the final CAD model and drawing is still being debated and the pros and cons have been discussed several times throughout the book. However, CAD-pre/post integrated systems may provide this possibility. Choose an integrated system only if the whole is better than the parts for your needs. This usually means that the CAD capabilities of an integrated system are chosen to be the CAD system used for design, or that the specific needs of your chosen solver are best accessed by that vendor's own pre/post.

CAD-embedded Systems

CAD-embedded systems are the FEA industry's response to design engineers who want access to the technology without having to spend time learning the basics. The basic form of a CAD-embedded system is to take a known solver technology and hide it behind a popular CAD tool with a new set of menus that provide access to FEA capabilities. One of the key attractions of this incorporation is that design engineers never have to leave the comfort of their CAD system and brave the challenges of FEA modeling. Most of these tools shield the bulk of the FEA system from the user and limit the available choices to keep the complexity of the technology to a minimum.

The danger of CAD-embedded systems, as they exist today, is that the users are not being educated as to the limitations of the tools. These systems typically restrict users to solid automeshes of single parts or parts that can behave as continuous load bearing assemblies. The best place for these tools is in a supporting role when a company has more complete FEA tools and trained users to oversee the activities of the embedded tools. However, as the popularity of such tools increases, the capabilities are sure to increase as well.

Enter into an implementation of a CAD-embedded system carefully and ensure that you know its limitations. Do not settle for happy references, because they may not know the limitations either. Take some time to discuss the tools with a user of more complete tools whom you trust to ensure that your best interests are considered. The most common CAD-embedded tools on the market are MSC/InCheck from MSC, Design-Space from ANSYS, and various incarnations of COSMOS/M. Each runs in a subset of popular CAD tools. If your company switches CAD systems to one that is not supported by embedded technology, you may find yourself in a much worse position of having to learn a new system that is totally incompatible with the work you had done previously. These tools are the opposite of an open system and you will pay the price of capabilities and flexibility for the convenience.

Proprietary Systems

Proprietary systems typically require you to use a pre/post that is designed specifically for the solver technology. While some CAD interfaces or tools for importing a plain mesh in a standard format might be provided, the bulk of the work done in these systems cannot be shared with other FEA systems. Most p-element technologies are proprietary, including Pro/MECHANICA, which is discussed throughout this book and included in the tables of this chapter. Most manufacturing codes and specialized technology solvers have proprietary interfaces. If these tools are important to your product design process, you may not have a choice. Some degree of openness will make it easier to evaluate up and coming technologies. One danger of implementing a proprietary system is that it will be very difficult to drop it and pick up a tool that shows more potential at a later date. Enter into these implementations only after a thorough review of the options.

Options

Tables 19.1 through 19.3 list important capabilities for building better models more efficiently. The capabilities listings are not exhaustive, and some may not be as important as others. You should reference these tables as you study various parts of the book to determine which of these popular codes provides a technique you have studied and believe would assist you in using FEA. Table 19.1 lists the preprocessor capabilities, which are the most numerous because they have the greatest impact on the accuracy of the solution. Table 19.2 lists solver options, and Table 19.3, postprocessor options. A Y entry in the table indicates that the product has this capability, and N indicates that it does not. An L entry means that the implementation of the capability is limited. In many cases, a note number accompanies an entry. The notes are listed at the end of each table and were provided by both vendors and users of these tools. Blank entries indicate a lack of response for a particular option.

The products reviewed in these tables represent a small slice of the overall FEA market. The respective manufacturers were chosen for their demonstrated resolve to provide a high level of technology to design analysts and a commitment to quality and accuracy. By no means does inclusion in this list represent an endorsement of the codes or confirmation that they are the best options for your company by the authors or publisher. However, they should probably be on your short list as you evaluate the benefits of open or integrated systems.

Another important disclaimer should be made before proceeding. These tables were compiled based on various documents provided by the vendors and discussion with selected users. Some provided more documentation, and others less. Each vendor had the opportunity to review and update the tables before publishing, and their recommendations were taken into account. PTC and SDRC declined to participate in the review, but other vendors were very cooperative. The capabilities listed reflect the most current releases as of September 1998. While much effort was placed into making this overview accurate, version numbers, alternate terminology, and other unpredictable circumstances may have rendered some of the data incorrect. It is best to use this overview as a guide; capabilities you deem as important should be verified live with your parts on your systems.

Table 19.1. Overview of selected industry preprocessors

	HyperMesh	FEMAP	Pro/MESH	SDRC/I-DEAS	MSC/PATRAN	COSMOS/M GeoStar	ANSYS/PrepPost	Pro/MECHANICA
General Tools *Utilities*								
Undo button	L	Y	N		L	N	N	Y
Redo button	N	Y	N		N	N	N	Y
Playback of log file	Y	N			Y	Y	Y	N
Record / play macros	Y	Y			Y	Y	Y	Y
User-programmable macros	Y	Y			Y	Y	Y	Y
Customizable interface	Y	Y			Y	N	Y	L
Icon interface	Y	Y			Y	L	N	N
Menu selection interface	Y	Y			Y	Y	Y	Y
Command line interface	Y	L			Y	Y	Y	N
Units manager	N	N			N	N	N	L[1]
Units conversion utility	N	Y			Y	N	N	N
Self-contained database (single file)	Y	Y	N		Y	N	Y	N

	HyperMesh	FEMAP	Pro/MESH	SDRC/I-DEAS	MSC/PATRAN	COSMOS/M GeoStar	ANSYS/PrepPost	Pro/MECHANICA
Save as new file name	Y	Y	Y		N	Y	Y	Y
Save as previous version of code	Y	Y	N		N	N	N	N
Calculator functions for numeric input	Y	Y	Y		N	Y	N	N
Select by								
Elements/nodes by associated geometry	Y	Y	N		Y	Y	Y	N
Element type	Y	Y	N		Y	Y	Y	Y
Color	N	Y	N		N	N	Y	N
Group / layer	Y	Y	Y		Y	Y[19]	Y	Y
Property	Y	Y	N		Y	Y	Y	N
Check								
Load resultant	Y	Y	Y		Y	Y	Y	Y
Model errors and integrity	Y	N	N		Y	Y	Y	Y
Calculate mass / inertia properties	Y	Y	Y		Y	Y	Y	L
Model summary	Y	Y	Y		Y	Y	Y	Y
Creation of coordinate systems								
Cartesian	Y	Y	Y		Y	Y	Y	Y
Cylindrical	Y	Y	Y		Y	Y	Y	Y
Spherical	Y	Y	Y		Y	Y	Y	Y
Read / write in various mesh formats								
ANSYS	Y	Y			Y	Y	Y	N
MSC/NASTRAN	Y	Y			Y	Y	Y	N
COSMOS/M	L	Y			Y[22]	Y	Y	N
MSC/PATRAN	Y	Y			Y	Y	Y	N

	HyperMesh	FEMAP	Pro/MESH	SDRC/I-DEAS	MSC/PATRAN	COSMOS/M GeoStar	ANSYS/PrepPost	Pro/MECHANICA
SDRC I-DEAS	Y	N			Y^{22}	Y	N	N
ABAQUS	Y	Y			Y	Y	Y	N
Pro/MECHANICA	N	N	L^2	N	N	N	N	Y
Job batching								
Embedded	Y	N			L	N	Y	Y
Operating system enabled	Y	Y	Y	Y	Y	Y	Y	N
Display *Settings*								
Wireframe	Y	Y	Y	Y	Y	Y	Y	Y
Hidden line	Y	Y	Y	Y	Y	Y	Y	Y
Shaded, geometry	Y	Y	Y	Y	Y	Y	Y	Y
Shaded, mesh	Y	Y	Y	Y	Y	Y	Y	Y
Shrunk elements	Y	Y	N	Y	Y	Y	Y	
Manipulation								
Dynamic rotation/translation/ scaling	Y	Y	Y	Y	Y	Y	Y	Y
Simplified dynamic rep for weaker system graphics	Y	Y			Y	Y	Y	Y
Dynamic query of model entities	Y	Y			Y	N	Y	N
Zoom window, rectangle	N	Y			Y	Y	Y	Y
Zoom window, general shape	Y	N			Y	N	N	N
Visibility								
User-defined groups/layers	Y	Y	Y		Y	N	Y	Y
Category visibility, geometry types	Y	Y	N		Y	Y	Y	Y
Category visibility, element types	Y	Y	N		Y	L	Y	Y
General assignment of colors	Y	Y	N		Y	Y	Y	L

	HyperMesh	FEMAP	Pro/MESH	SDRC/I-DEAS	MSC/PATRAN	COSMOS/M GeoStar	ANSYS/PrepPost	Pro/MECHANICA
General hide/display	Y	N	N		Y	Y	Y	Y
Scale load vector display by relative magnitude	Y	Y	N		Y	N	Y	N
Display load magitude	Y	Y	Y		N	N	Y	Y
Dynamic position tracking of cursor	N	Y	N		Y	N	N	Y
Representation								
Beam cross section		L			Y	N	Y	Y
Orientation vector		Y			Y	N	Y	Y
Shell thickness		Y	N		N	N	Y	N
Print								
To printer		Y	Y		Y	Y	Y	L
To file, Print File	Y	Y	Y		Y	Y	Y	Y
To file, BMP	Y	Y			N	Y	Y	N
To file, GIF	N	N			N	N	N	L
To fle, JPEG	N	N			N	N	N	Y
To file, TIF	N	N			N	Y	Y	L
To file, VRML	Y	Y			N	N	Y	N
To file, AVI	Y	Y			N	Y	Y	N
Choose color versus grayscale	Y	Y			Y	L	Y	N
Header/footer	N	Y			Y	L	N	Y
Placement of text anywhere on screen	Y	Y			Y	Y	N	N
Geometry transfer *Import, standard*								
IGES	Y	Y	Y		Y	Y	Y	Y
DXF	Y	Y	N		N	Y	N	Y
STEP	N	N	Y		Y	N	N	N

	HyperMesh	FEMAP	Pro/MESH	SDRC/I-DEAS	MSC/PATRAN	COSMOS/M GeoStar	ANSYS/PrepPost	Pro/MECHANICA
VDA	Y	N	N		N	N	N	L
ACIS	Y	Y	N		Y	N	Y	N
PARASOLIDS	N	Y	N		Y	N	Y	N
STL	Y	Y	N		N	N	N	N
Import, CAD-specific								
Pro/E	N	N	Y	N	Y	Y	Y	Y
UG	Y	N	N	N	Y	N	Y	Y
SolidWorks	N	N	N	N	N	Y	N	N
SDRC/I-DEAS	Y	N	N	Y	N	N	N	N
CATIA	Y	N	N	Y	N	N	Y	Y
AutoCAD	Y	N	N	Y	N	Y	N	N
Cadkey	N	N	N	Y	N	Y	N	N
Export								
IGES	Y	N	Y		Y	Y	Y	Y
DXF	N	N	N		N	Y	N	Y
STEP	N	N	Y		Y	N	N	N
VDA	N	N	N		N	N	N	N
ACIS	N	L	N		N	N	N	N
PARASOLIDS	N	L	N		N	N	N	N
STL	N	Y	Y		N	N	N	N
Geometry Capabilities *Create*								
CAD solids	N	Y	Y	Y	Y	Y	Y	N
Surfaces	Y	Y	Y	Y	Y	Y	Y	Y
Wireframe	Y	Y	Y	Y	Y	Y	Y	Y
Modify								
CAD solids	N	Y	Y	Y	Y	L[20]	Y	N

	HyperMesh	FEMAP	Pro/MESH	SDRC/I-DEAS	MSC/PATRAN	COSMOS/M GeoStar	ANSYS/PrepPost	Pro/MECHANICA
Surfaces	Y	Y	Y	Y	Y	Y	Y	Y
Wireframe	Y	Y	Y	Y	Y	Y	Y	Y
Tools								
Multisurface patch[3]	Y	N	N	Y	Y		Y	N
Surface patch for BC application	Y	Y	Y	Y	Y	L[4]	N	
Volumetric region for property differentiation	Y	L	L	Y	N	N	Y	N
Parametric geometry relations	N	N	Y	Y	Y	Y	L	Y
Surface stitching for volume definition	Y	Y	Y		Y	Y	Y	Y
Midplane extraction	L	N	Y	Y	N	N	N	L[5]
Delete geometry	Y	Y	Y	Y	Y	Y	Y	Y
Translate geometry	Y	Y	Y	Y	Y	Y	Y	Y
Rotate geometry	Y	Y	Y	Y	Y	Y	Y	Y
Scale geometry	Y	Y	Y	Y	Y	Y	Y	Y
Mirror geometry	Y	Y	Y	Y	Y	Y	Y	Y
Model Types								
Beam	Y	Y	L	Y	Y	Y	Y	Y
3D shell	Y	Y	Y	Y	Y	Y	Y	Y
3D solid	Y	Y	Y	Y	Y	Y	Y	Y
Plane stress	Y	Y	N	Y	Y	Y	Y	Y
Plane strain	Y	Y	N	Y	Y	Y	Y	Y
Axisymmetric	Y	Y	N	Y	Y	Y	Y	Y
Cyclic symmetric	Y	N	N	Y	N	Y	Y	N
Planar symmetry[6]	Y	Y	Y	Y	Y	Y	Y	Y
Transitional, mixed-element	Y	Y	L	Y	Y	Y	Y	Y

	HyperMesh	FEMAP	Pro/MESH	SDRC/I-DEAS	MSC/PATRAN	COSMOS/M GeoStar	ANSYS/PrepPost	Pro/MECHANICA
Element Types								
Line								
Beam	Y	Y	Y	Y	Y	Y	Y	Y
Rod	Y	Y	Y	Y	Y	Y	Y	N
Bar	Y	Y	N	Y	Y	Y	Y	N
Rigid	Y	Y	L^7	Y	Y	Y	Y	N
Point-to-point spring	Y	Y	Y	Y	Y	Y	Y	Y
DOF spring	Y	Y	N		Y	Y	Y	N
1D shell, axisymmetric	Y	N	N		N	Y	Y	Y
Solid-to-shell transitional	L^8	L^8	N	L^8	L^8	Y	Y	Y^9
Shell								
Quad	Y	Y	Y	Y	Y	Y	Y	Y
Tri	Y	Y	Y	Y	Y	Y	Y	Y
2D solid, plane stress	Y	Y	N	Y	Y	Y	Y	Y
2D solid, plane strain	Y	Y	N	Y	Y	Y	Y	Y
2D solid, axisymmetric	Y	Y	N	Y	Y	Y	Y	Y
Solid								
Hex	Y	Y	N	Y	Y	Y	Y	Y
Wedge	Y	Y	N	Y	Y	Y	Y	Y
Tet	Y	Y	Y	Y	Y	Y	Y	Y
Contact								
Gap	Y	Y	Y	Y	Y	Y	Y	N
Slideline	Y	Y	N	Y	Y	N	Y	N
Surface/surface	Y	L^{10}	N	Y	L^{10}	Y	Y	Y
Other								
Mass/inertia	Y	Y	Y	Y	Y	Y	Y	Y

	HyperMesh	FEMAP	Pro/MESH	SDRC/I-DEAS	MSC/PATRAN	COSMOS/M GeoStar	ANSYS/PrepPost	Pro/MECHANICA
Element Properties *General*								
Assigned to geometry	N	N	Y	Y	Y	N	Y	Y
Modifiable	Y	Y	Y	Y	Y	Y	Y	Y
Color-coded	Y	Y	N		Y	Y	Y	L[11]
Auto-grouped	Y	Y	N		Y	Y	N	Y
Material								
Read in from CAD	N	N	Y	L[12]	L[23]	L	N	L[14]
Internal library/database	N	Y	Y	Y	Y	Y	Y	Y
Add/remove to database	Y	Y		Y	Y	Y	Y	Y
ASCII input to database	Y	Y			Y	Y	Y	Y
Control of orientation	Y	Y		Y	Y	Y	Y	Y
Orthotropic definition	Y	Y		Y	Y	Y	Y	Y
Anisotropic definition	Y	Y		Y	Y	Y	Y	N
Line								
Offset neutral axis	Y	Y	Y		Y	Y	Y	Y
Offset shear center	Y	Y	N		Y	Y	Y	N
Tapered	Y	Y	N		Y	Y	Y	N
Internal cross-section library/ database	N	Y			Y	Y	L	Y
Add/remove to database	Y	Y			N	Y	Y	Y
Cross-sectional sketcher	Y	N	Y		Y	L	L	N
Proper calculation of "J"	Y	N	N		Y	Y	Y	Y
Control of end release	Y	Y	N	Y	Y	Y	Y	Y
Control of stress recovery point(s)	Y	Y		Y	Y		Y	Y
Control of beam orientation	Y	Y	Y	Y	Y	Y	Y	Y

	HyperMesh	FEMAP	Pro/MESH	SDRC/I-DEAS	MSC/PATRAN	COSMOS/M GeoStar	ANSYS/PrepPost	Pro/MECHANICA
Shell								
Offset neutral surface	Y	Y	N		Y	N	N	L
Variable thickness	Y	Y	N		Y	N	Y	N
Laminate	Y	Y	N		Y	Y	Y	L
Internal thickness library/ database	N	Y			N	N	L	N
Add/remove to database	N	Y			N	Y	L	Y
Control of normals	Y	Y	N		Y	Y	Y	Y
Mesh								
Automesh, general								
Failed mesh ID of problem area	Y	Y	N		L	L	L	Y
Failed mesh suggestions for correction	Y	N	N		Y	N	L	Y
Automesh, solids								
Tet	Y	Y	Y	Y	Y	Y	Y	Y
Tet/wedge	N	N	N	N	N	N	N	Y
Hex	N	N	N	N	N	N	N	N
Hex/wedge	N	N	N	N	N	N	N	N
Automesh, shells								
Tri	Y	Y	Y	Y	Y	Y	Y	Y
Quad/tri	Y	Y	Y	Y	Y	Y	Y	Y
Automesh, lines								
Beams	Y	Y	Y	Y	Y	Y	Y	Y
Automesh control								
Specify element creation shape limits	Y	Y	Y		Y	N	Y	Y

	HyperMesh	FEMAP	Pro/MESH	SDRC/I-DEAS	MSC/PATRAN	COSMOS/M GeoStar	ANSYS/PrepPost	Pro/MECHANICA
Default element size	Y	Y	Y		Y	Y	Y	NA
Local element size, surface	Y	Y	Y		Y	Y	Y	N
Local element size, curve	Y	Y	Y		Y	Y	Y	Y
Local element size, point	Y	Y	Y		Y	Y	Y	Y
Hard mesh curves	Y	Y	Y		Y	Y	Y	Y
Hard mesh points	Y	Y	Y		Y	Y	Y	Y
Manual mesh creation								
Select geometry references for element creation	Y	N	N		Y	Y	N	Y
Extrude curve for shells	Y	Y	N	Y	Y	Y	Y	Y
Revolve curve for shells	Y	Y	N	Y	Y	Y	Y	Y
Extrude surface for hex/wedge solids	Y	Y	N	Y	Y	Y	Y	Y
Revolve surface for hex/wedge solids	Y	Y	N	Y	Y	Y	Y	Y
Element quality checking								
Automatic (on creation)	N	N	Y		Y	N	Y	Y
Manual request	Y	Y	Y		Y	Y	Y	Y
On write to solver	N	N	N		Y	N	Y	Y
Other meshing tools								
Mix auto and manual mesh in same model	Y	Y	N	Y	Y	Y	Y	Y
Mix auto and manual mesh on same geometry	N	N	N		Y	N	Y	Y[13]
Convert linear element to quadratic and back	Y	Y	N		Y	Y	Y	NA
Smooth existing mesh, solids	Y	Y	Y		Y	N	N	N
Smooth existing mesh, shells	Y	Y	Y		Y	N	N	N

	HyperMesh	FEMAP	Pro/MESH	SDRC/I-DEAS	MSC/PATRAN	COSMOS/M GeoStar	ANSYS/PrepPost	Pro/MECHANICA
Refine existing mesh, solids	Y	N	N		Y	Y	Y	N
Refine existing mesh, shells	Y	Y	N		Y	Y	Y	N
Refine existing mesh, beams	Y	Y	N		Y	Y	N	N
Boundary/free edge outline	Y	Y	N		Y	Y		Y
Boundary/free face outline	Y	Y	N		Y	Y		Y
Check continuity at element type transitions	Y		N		Y	Y		Y
Check coincident elements	Y	Y	N		Y	Y	N	Y
Check coincident nodes	Y	Y	N		Y	Y	L	Y
Boundary conditions *General*								
May be applied to geometry	N	L[14]	Y	Y	Y	L[21]	Y	Y
Notification of unsupported BCs[15]	Y	N	N		N	N	Y	N
Load types								
Force	Y	Y	Y		Y	Y	Y	Y
Moment	Y	Y	Y		Y	Y	Y	Y
Pressure	Y	Y	Y		Y	Y	Y	Y
Temperature differential, local versus ambient	Y	Y	Y		Y	Y	Y	Y
Acceleration/gravity	Y	Y	Y		Y	Y	Y	Y
Centrifugal	Y	Y			Y	Y	Y	Y
Bearing[16]	N	N	N		Y	N	N	Y
Total load applied at point	N	N	N	N	N	N	N	Y
Load distribution								
Uniform	Y	Y	Y		Y	Y	Y	Y
Total over all selected entities[17]	N	N			N	N	N	Y

	HyperMesh	FEMAP	Pro/MESH	SDRC/I-DEAS	MSC/PATRAN	COSMOS/M GeoStar	ANSYS/PrepPost	Pro/MECHANICA
Per unit length	Y	Y			Y	Y	Y	Y
Per unit area	Y	Y	Y		Y	Y	Y	Y
Interpolated over selected points	N	N			N	Y	N	Y
Function of coordinates	L[18]	Y	Y		Y	N	Y	L
Load in terms of user-defined coordinate system								
Cartesian	Y	Y	Y		Y	Y	Y	Y
Cylindrical	Y	Y	Y		Y	Y	Y	L
Spherical	Y	Y	Y		Y	Y	Y	L
Constraint types								
Enforced displacement	Y	Y	Y		Y	Y	Y	L
Multipoint constraints	Y	Y	L		Y	Y	Y	N
Constraint equations	Y	Y	N		Y	Y	Y	N
Constraint in terms of user-defined coordinate system								
Cartesian	Y	Y	Y		Y	Y	Y	Y
Cylindrical	Y	Y	Y		Y	Y	Y	Y
Spherical	Y	Y	Y		Y	Y	Y	Y
Solution types								
Linear static	Y	Y	Y		Y	Y	Y	Y
Modal	Y	Y	Y		Y	Y	Y	Y
Buckling	Y	Y	L[18]		Y	Y	Y	Y
Dynamic, transient	Y	Y	L[18]		Y	Y	Y	Y
Dynamic, frequency response	Y	Y	L[18]		Y	Y	Y	Y
Dynamic, random	Y	Y	L[18]		Y	Y	Y	Y
Nonlinear material	Y	Y	L[18]		Y	Y	Y	N
Nonlinear, large deformations	Y	Y	L[18]		Y	Y	Y	N

	HyperMesh	FEMAP	Pro/MESH	SDRC/I-DEAS	MSC/PATRAN	COSMOS/M GeoStar	ANSYS/PrepPost	Pro/MECHANICA
Nonlinear, contact	Y	Y	Y		Y	Y	Y	Y
Additional solution requests								
Track system-defined scalar quantity of interest	Y	Y	N		N	Y	Y	Y
Track user-defined scalar quantity of interest	N	N	N		N	N	Y	Y
Optimization/sensitivity *Design variables*								
Geometric	Y	N	Y		N	Y	L	Y
Material property		N	N		N	Y	Y	Y
Elemental property		Y	N		Y	Y	Y	Y
Solution types								
Local sensitivity			L^{18}		N	Y	Y	Y
Global parameter sweep	Y		N		N	Y	Y	Y
Optimization solvers supported								
ANSYS	Y	N	Y		N	N	Y	N
MSC/NASTRAN	Y	L	Y		Y	N	Y	N
COSMOS/M	N	N	Y		N	Y	N	N
SDRC I-DEAS	N	N	N		N	N	N	N
ABAQUS	Y	N	N		N	N	N	N
Pro/MECHANICA	N	N	N	N	N	N	N	Y

[1] In Pro/ENGINEER integrated mode only.

[2] Many Pro/MESH tools are equivalent in the Pro/ENGINEER interface to Pro/MECHANICA.

[3] Consolidation of small surfaces into large patch.

[4] Requires a shallow solid cut.

[5] In Pro/ENGINEER integrated mode only.

[6] All codes provide this through constraints.

[7] Point-to-point rigid link only.

[8] Only rigid elements supported.

[9] Link element provides continuity across all element types.

[10] Limited support for some solvers.

[11] Only color coded through automatic layering.

[12] Through integrated CAD interface only.

[13] Pro/MECHANICA allows you to start manually meshing a surface or volume and then finish the geometry with an automesh for better mesh control.

[14] Constraints on geometry are limited.

[15] That is, moment loads on solids.

[16] Automatic distribution of bearing pressure or force.

[17] Choose several surfaces and apply a single load that is correctly distributed.

[18] Dependent upon installed solver.

[19] Groups are called "selection sets" in COSMOS/M.

[20] COSMOS/M recommends that imported solids not be changed in the preprocessor.

[21] Geometry must be meshed first.

[22] Use an unsupported applet under Utilities menu.

[23] From a Parasolids based CAD system.

Table 19.2. Overview of selected industry solvers

	MSC/NASTRAN	ANSYS	COSMOS/M	SDRC I-DEAS	Pro/MECHANICA
Hardware and operating system requirements					
Resource usage					
Dynamic allocation of memory resources	L	L	Y	Y	N
Manual allocation of memory resources	Y	Y	N	Y	Y
Node limited	N	N	Y[1]	N	N
Run log or summary	Y	Y	Y	Y	Y

	MSC/NASTRAN	ANSYS	COSMOS/M	SDRC-I-DEAS	Pro/MECHANICA
Element technology support					
H-element support	Y	Y	Y	Y	N
P-element support	L	L	Y	L	Y
Geometric element definition of p-elements	L	N	N	N	Y
Increased p-order of existing h-element mesh	N	Y	Y	Y	N
Solution types					
Linear static	Y	Y	Y	Y	Y
Modal	Y	Y	Y	Y	Y
Buckling	Y	Y	Y	Y	Y
Dynamic					
Transient	Y	Y	Y	Y	Y
Frequency	Y	Y	Y	Y	Y
Random response	Y	Y	Y	Y	Y
Modal method solution	Y	Y	Y	Y	Y
Direct method solution	Y	Y	L	Y	N
Nonlinear					
Materials	Y	Y	Y	Y	Y
Large deformation	Y	Y	Y	Y	Y
Contact	L	Y	Y	Y	L
Buckling	Y	Y	Y	Y	N
Automatic adaptive mesh refinement	L^2	L	Y	L	Y
Superelement support	Y	Y	Y		N
Submodeling support	N	Y	Y		N
Asymmetric loading of axisymmetric models	N	Y	Y		N

	MSC/NASTRAN	ANSYS	COSMOS/M	SDRC I-DEAS	Pro/MECHANICA
Design improvement options					
Geometry optimization	Y	Y	Y		Y
Size optimization of plate or beam properties	Y	Y	Y	Y	Y
Sensitivity studies	Y	Y	Y	Y	Y
Global parameter sweep	N	Y	Y	Y	Y
General					
Local user groups	L	L	L	Y	Y
National user groups	Y	Y	N	Y	Y
Verification manual provided	Y	Y	Y		Y

[1]COSMOS/M provides 64K, 128K, and 256K node/element database structures that are not cross compatible for resource/speed reasons.

[2] Automatic p-element refinement only.

Table 19.3. Overview of Selected Industry Postprocessors

	HyperMesh	FEMAP	Pro/FEM Post	SDRC I-DEAS	MSC/PATRAN	COSMOS/M GeoStar	ANSYS PrepPost	Pro/MECHANICA
Results display options								
Color fringe plots	Y	Y	Y	Y	Y	Y	Y	Y
Color isosurface plots	L[5]	Y	Y	Y	Y	Y	Y	Y
Color isoline plots	Y	Y	Y	Y	Y	Y	Y	Y
Average stress results display	Y	Y	Y	Y	Y	Y	Y	Y

	HyperMesh	FEMAP	Pro/FEM Post	SDRC I-DEAS	MSC/PATRAN	COSMOS/M GeoStar	ANSYS PrepPost	Pro/MECHANICA
Unaveraged stress results display	Y	Y	Y	Y	Y	Y	Y	Y
Continuous tone fringe display	Y	Y	Y	Y	Y	Y	Y	Y
Element centroidal stress plots	L[6]	Y	Y	Y	Y	Y	Y	NA[1]
Color legend magnitude scaling	Y	Y	L	Y	Y	Y	Y	Y
Specify number of colors in legend		Y		Y	Y	Y	Y	L
Display results of specific groups only	L[4]	Y			Y	Y	Y	Y
Vector plots on solid elements	L[4]	Y	N	Y	Y	Y	Y	L[2]
Vector plots on shell elements	L[4]	L	N		Y	Y	Y	Y
Display shell results on "max" or "min" surfaces		L	N		Y	Y	N	Y
Graphing of output data								
Graph results along an edge or line of nodes	Y	Y	Y	Y	Y	Y	Y	Y
Graph across multiple output sets	N	Y	N	Y	Y		Y	N
Deformation and animation								
Animation of deformation	N	Y	Y	Y	Y	Y	Y	Y
Scaling of deformations	N	Y		Y	Y	Y	Y	Y
Animation of contours on deformations	N	Y		Y	Y	Y	Y	Y
Animate results across multiple output sets	N	Y	N	Y	Y	Y	Y	N
Working with multiple output sets								
Combine multiple output sets with linear scaling	N	Y	N	Y	Y	Y	Y	L[3]
Scale output sets linearly	N	Y	N	Y	Y	Y	Y	L[3]

	HyperMesh	FEMAP	Pro/FEM Post	SDRC I-DEAS	MSC/PATRAN	COSMOS/M GeoStar	ANSYS PrepPost	Pro/MECHANICA
Bulk calculations on output data	L	Y	N	Y	Y	N	Y	N
Cross-sectional display								
General creation of cross-sectional results	L	Y	Y	Y	Y	Y	Y	Y
Dynamic dragging of cross-sectional display position	Y	Y			Y	N	N	Y
Miscellaneous options								
Results error estimates	N	Y		Y	Y	Y	Y	Y
Screen-pick query of results display	Y	Y			Y	N	Y	Y
Dynamic query of results by dragging cursor	N	Y			Y	N	N	Y
Tabular output option	Y	Y	Y	Y	L	Y	Y	Y

[1]This display is not applicable to Pro/MECHANICA which supports p-elements only.

[2]Stress vectors (i.e., maximum principal stress vectors) are the only commonly required vectors that are not supported in Pro/MECHANICA.

[3]Pro/MECHANICA allows only output set scaling and combinations if multiple load sets were solved.

[4]Displacements and reaction forces only.

[5]Requires performance graphics mode in HyperMesh.

[6]Solver dependent.

Contact Information for Vendors

Altair Computing, Inc.
1757 Maplelawn Drive
Troy, MI 48084-4603
Tel: 248.614.2400
Fax: 248.614.2411
www.altair.com

ANSYS, Inc.
275 Technology Drive
Canonsburg, PA 15317
Tel: 724.746.3304
Fax: 724.514.9494
www.ansys.com

Enterprise Software Products, Inc.
415 Eagleview Boulevard
P.O. Box 1172
Exton, PA 19431
Tel: 610.458.3660
Fax: 610.458.3665
www.femap.com

The MacNeal-Schwendler Corporation
815 Colorado Boulevard
Los Angeles, CA 90041-1777
Tel: 213.258.9111
Fax: 213.259.3838
www.macsch.com

Parametric Technology Corporation
128 Technology Drive
Waltham, MA 02154
Tel: 617.398.5000
Fax: 617.398.5553
www.ptc.com

Structural Dynamics Research Corporation
2000 Eastman Drive
Milford, OH 45150
Tel: 513.576.2400
Fax: 513.576.2734
www.sdrc.com

Structural Research & Analysis Corporation
12121 Wilshire Boulevard, 7th Floor
Los Angeles, CA 90025
Tel: 310.207.2800
Fax: 310.207.2774
www.cosmosm.com

20

Key Elements of a Successful FEA Implementation

Is push-button analysis FEA for the masses? Has it become so easy that you can simply push a button and get results? Can you really get reliable simulation tools which work within your CAD interface for less than the price of CAD? At this point in the book, you should be skeptical of these claims because, as discussed in previous chapters, there is much more to FEA than CAD solid geometry. However, with these tempting offers, it is no wonder engineers and product development managers are confused about the growing accessibility of performance simulation.

The bottom line is that no product is perfect for all companies, all industries, and all users. In addition, companies are using performance simulation at varying degrees for various reasons with various levels of success. Many factors enter into the success, or lack thereof, of an analysis program. If used correctly, CAD-embedded FEA can produce good results. However, solid geometry transfer is only one part of the picture that includes training, modeling flexibility, test correlation, and documentation, to name a few.

While most companies still consider FEA an afterthought or a "fire extinguisher," some have truly integrated the technology at the initial stages of the design process, or *up front*, before costly tooling or production decisions are made. This is where simulation is most effective and profitable. There are three *key success factors* common to any successful integration of simulation into product design: (1) a constant *evaluation* and reevaluation of the process, (2) a well-conceived *implementation* plan stating the expected benefits and goals, and (3) a documented means to *verify the results* and evaluate the business benefits/measured goals.

Key Success Factor: Evaluation

The integration process must first begin with a comprehensive evaluation. The evaluation process should be comprised of three interrelated tasks: (1) *corporate needs, both economic and technological*, (2) *people and computing resources*, and (3) *current simulation industry offerings*.

Economic Needs

The *economic needs* evaluation should begin with a review of the possible business benefits of simulation. Focus your evaluation on areas of the product design process where you can most quickly achieve the greatest impact. A highly visible product line or component family should be chosen and the business benefit expected should be clarified.

To properly assess the business benefits of a FEA program, a cross-functional team should explore two areas of research: *cost of delay* and *product musts versus product wants*.

A cross-functional team should include representation from engineering, purchasing, manufacturing, marketing, and even accounting. A true understanding of these two concepts can rarely be found in a single group within the company.

Cost of Delay

The first area to research is the cost of delay. The true cost of a delayed project has several facets. Each component of the cost of delay can be used to justify integrating simulation into the product development process. They are also effective measures from which to gauge the value of performing FEA after the implementation.

Product Development Costs

The most tangible costs of delay are the additional direct costs of engineering a product. These include the actual development costs and expenses such as the payroll, and the benefit cost of the engineers involved in the project. Manufacturing down-time, prototyping and testing costs, and expenses for unplanned iterations must be taken into account. The general costs for the multitude of salaried personnel who worry, react, and many times overreact to missed schedules are usually omitted but have a direct impact on the costs associated with schedule delays.

Missed Sales

Another facet of the cost of delay is lost profit from missed sales. Most marketing and sales groups will be able to provide these numbers. If not, a simple approach is to take their unit sales estimates for a year, divide by 365, and multiply by the projected unit profit. Most engineers will benefit from going through this exercise, regardless of any underlying reason for asking, because it helps keep costs and schedule commitments in a tangible perspective. Interestingly, companies seeking to justify simulation may often be able to use this single parameter to make their case. You would be surprised how effective a member of the sales group can be as the champion of simulation when attempting to push through an engineering acquisition. In fact, the *marketing* value of FEA has been taken advantage of by surprisingly few companies. One packaging company executive once remarked that "real" drop-test answers were not nearly as important as answers that looked real when customers came through engineering. Most engineers would not knowingly offer fictitious analysis results, but anyone trying to justify the technology should be aware of their potential allies in other groups within the company.

Lost Market Share

The final aspect of the cost of delay is related to lost profits, which are much harder to quantify. This is the lost market share due to a late product introduction. To better understand the effects of schedule delays on lost market share, a discussion of two different product development strategies is warranted.

There are two types of product development scenarios in successful companies: "market makers" and "fast followers." Market makers are looking to introduce new technologies. They base their profitability calculations, in part, on seizing market attention and must have a product that at least minimally meets claims to gain sufficient momentum in order to ward off the other type of product development organization–the fast follower. A fast follower is geared toward recognizing cutting edge technologies, and in a sense, riding on the coattails of a competitor's successful launch. Fast followers must have processes in place to evaluate new technologies and isolate the strengths and weaknesses of initial offerings. By capitalizing on the competitor's weaknesses, the fast follower can steal market share.

The Sony Walkman and 3M's Post-It Notes are examples of effective market makers because they introduced products that had no rival at the time and sustained their leadership position with quality and marketing. The Japanese auto industry and Microsoft Windows in the early 1980s are classic examples of successful fast followers that either improved on products slow to respond to market needs (Big Three U.S. automakers), or used the best features of an existing product and bundled it differently to overwhelm the originator of the technology (Microsoft Windows versus Apple OS).

To return to the original point, market makers tend to experience little fear of lost market share from a delayed project standpoint. This is due to the fact that there is little or no initial competition. However, the opposite is true if a market maker delivers an inferior product, because it becomes fair game for a fast follower. A fast follower has much to lose by a delayed introduction. The longer the initial offerings are on the market unchallenged, the harder it will be to displace them. Moreover, it is in the fast follower's best interest to be the first fast follower because there may be several others waiting in the wings. It is easier to grab market share when you are only competing with the known and evaluated market maker. Other fast followers mean uncertainty and possible product failure.

Summarizing Cost of Delay

To summarize the cost of delay, all companies should understand the cost associated with project schedule overruns and lost sales. These are very tangible figures that can be used to justify the introduction of simulation. Market makers are less concerned with lost market share than fast followers, who must base their product development process on quickly getting the design, quality, and improvement on the market maker's weaknesses right. Few companies are uniquely either market makers or fast followers. Most companies' product lines consist of a combination of the two. Focusing rapid product development technologies on projects in which the parts within that project have the most to gain is a key part of the evaluation process.

Product Musts versus Product Wants

In this, the second portion of the evaluation, the costs of delay are linked to actual design. Essentially, "product musts" are parts or systems that are likely to cause schedule delays and/or are critical to system performance or safety. These parts must be proven out rigorously using testing, analysis, or a combination of both. Product wants, on the other hand, are parts that historically have been problem free, including parts successfully used on similar products or parts similar to other successful parts. While these parts may not be optimal from a cost or performance standpoint, spending additional development time to optimize them during a crucial project may not be warranted. Such optimization projects are best left for noncritical times.

Product Musts

"Product musts" refer to components or systems within a design that have historically required iterations to fine tune structural performance. Identifying product musts is easiest if your company builds similar products sequentially. If historical data are not available to identify product musts, a design review can identify critical components that are most likely to require extensive prototyping.

From an economic or business benefit standpoint, "product musts" are the components most likely to cause schedule delays and incur related costs. Consequently, the use of FEA can be justified on the basis of reducing or minimizing these delays. Considering the value of costs avoided due to reduced prototype iterations or improved quality instead of attempting to quantify savings is more straightforward, although less tangible. If a prototype iteration typically takes a week and has a predictable cost, the costs avoided by catching a design problem using simulation are the sum of the aforementioned prototype costs and the profit from seven days of additional market presence.

Product Wants

FEA can also be justified with the slightly different "product wants" approach. Consider product wants from the perspective of long-term savings from an optimized design. A 10% cost reduction outcome from an aggressive optimization program is conservative. Savings have been realized in excess of 40 to 60%.

Summarizing the Economic Benefits of FEA

Measures associated with the cost of delay, product musts, and product wants can be quantified to provide the business benefits of integrating a simulation program in the product design process. These same measures should be used throughout the implementation to continually evaluate the benefits of using the technology. It is tempting at times to utilize FEA for the sake of using FEA simply because it is available without stopping to evaluate the effect on product cost and profitability. Yet without business justification this technology will never gain real acceptance within the company and may come to be viewed as a product development team's "computer game." The absence of economic benefit may mean that the simulation program is not implemented correctly, or that FEA technology is not the best approach to reducing costs and improving quality. On the other hand, if the savings attributed to simulation are substantial and well documented, improved hardware, additional training, and expanded software capabilities should be easy to justify.

Ask the following questions: *Is the goal to reduce development time or cost? Will the greatest impact derive from cutting material costs? Will market share increase if safer products are developed or quality is improved?* Identification of the business benefit will keep all other decisions in focus.

Technological Needs

The *technological* portion of the *needs* analysis determines the analysis types required to adequately model your parts or systems. Possible choices at this stage are outlined in Table 20.1. Understanding the requirements and limitations of each type of solution or modeling type is critical when evaluating tools and your ability to use them.

Table 20.1. Information for evaluating required FEA solution types and modeling methods supported by your preprocessor

Product	Modeling type	Solution type	Key simulation challenges
Single bulky, low aspect ratio cast or machined parts	Solids, tet automeshing	Linear	Making efficient simplifications which allow efficient solve times yet do not introduce doubt or error
Single thin-walled or high aspect ratio cast or machined parts	Mid-plane shells, automeshing of surfaces; mapped or extruded solids	Linear	Developing mid-plane models for shells; fighting the urge to automesh tets
Single, injection molded plastic parts	Mid-plane shells, automeshing of surfaces; mapped or extruded solids	Nonlinear material; nonlinear large displacement	Determining the transition point between linear and nonlinear behavior; compiling material properties for nonlinear analyses; developing mid-plane models for shells
Assemblies with closely interacting components	Rigid elements	Nonlinear contact	Modeling accurate interactions; choosing when to isolate parts and when assembly analysis is required
Heavy stampings or welded assemblies	Mid-plane shells; automeshing of surfaces; rigid elements; beams for fasteners	Linear; nonlinear contact	Determining appropriate modeling technique for welds; converting solid CAD data into more accurate surface or shell models
Thin sheet metal stampings	Mid-plane shells; automeshing of surfaces	Linear; nonlinear large displacement; nonlinear contact	Determining the transition point between linear and nonlinear behavior; converting solid CAD data into more accurate surface or shell models

Structural assemblies with beam components	Beam elements; mid-plane shells; rigid elements	Linear; nonlinear contact	Converting solid CAD data into more accurate beam or shell models; modeling accurate interactions
Vibrating or oscillating behavior		Modal analysis, frequency response; transient response; random response	Determining the effects of material damping
Impact or drop-test loads		Modal analysis; transient response	Determining the effects of material damping
Heating or cooling in predictable fluid flows		Thermal stress; general purpose FEA conduction; convection; radiation	Determining convection coefficients; modeling junction or contact losses
Heating or cooling in unpredictable or negligible fluid flow		Computational fluid dynamics (CFD)	Modeling both fluid and solids requires special techniques and much patience

Refer to Chapter 4 for a detailed description of the above modeling techniques and solution types. The key point to remember is that choosing the wrong technology or implementing insufficient technology will make achieving any business benefit difficult. In addition, new users may not be able to properly identify technological needs or acknowledge inefficiencies in a limited implementation. Utilize your software vendors and outside resources to help evaluate current and future needs.

Personnel Resources

Following the needs analysis, the evaluation should focus on the *human resources* required to make the integration successful. First and foremost is the correct choice of users. As described in Chapter 1, a critical trait of a new analyst is simply enthusiasm. The chosen individuals should think the technology is exciting and be willing to devote time (usually personal) to learning theory and practical application techniques. Without this characteristic, growth will be slow or nonexistent and effectiveness minimal. Moreover, the analyst should have a solid background in engineering fundamentals and mechanics of materials with sound engineering judgment so that the problem setup, assumption set, and results interpretation are correct.

Give Users the Best Chance to Succeed

An analysis champion will probably choose himself/herself. The manager's role in this situation is to acknowledge this person's enthusiasm and, if the business benefits have been identified, make time in his/her schedule for growth or adjust other resources to free up some of his/her responsibilities. The best software and hardware in the world will not make up for an unwilling or incapable user. In fact, this is the fatal flaw behind the growing trend of push-button systems. The right person will be willing to learn the technology's requirements. If chosen correctly, this person should quickly identify the limitations of a push-button system and potential for error that this technology presents.

Should Designers Be Doing FEA?

This question has been the topic of much controversy over the last five years. As mentioned in Chapter 19, the entire crop of push-button tools is aimed at a positive conclusion to that debate. Therefore, the subject warrants additional discussion.

Certain terminology must be clarified before proceeding. More precisely, the difference between a designer and a design engineer must be made. While a degree does not necessarily make an engineer, knowledge of materials and failure criteria does. This text will differentiate the two roles by noting that a design engineer should be capable of and responsible for making calculations regarding stress and displacement, at a minimum, with or without the aid of FEA. A designer on the other hand will be tasked with geometry layout and creation. While a designer may be responsible for developing an entire product, s/he should not be responsible for components that have critical structural needs unless given time or resources to evaluate strength and performance through prototype testing. This is an important distinction. All titles aside, if a design team member does not understand the basic principles outlined in Chapter 2, s/he should not be using FEA, period. Of course, this does not imply that such team members cannot learn. However, one must learn to walk before running. The ability to generate an incredible amount of realistic looking data in a single FEA run can overwhelm someone who is not trained or prepared to interpret the results.

An editorial in the July 9, 1998 issue of *MACHINE DESIGN* magazine addressed the subject of designers and FEA. This editorial's premise was essentially, "Why not let designers do FEA without understanding engineering theory?" The discussion was supported by an example of a failed aerospace project conducted by expert analysts that relied a little too heavily on analytical results. The second example described an aerospace project completed by engineers who were admittedly weak on theory and chose to use "make-and-break" methods instead of analysis to great success. Based on these two data points, the author concluded that because expert analysts cannot always get it right, designers without grounding in fundamentals can do no worse. This somewhat irresponsible position is contradictory, assuming that the reader is objective in his/her views on the subject. Taken to its extreme, one might conclude that because successful products have been developed by individuals with no engineering background whatsoever and products developed by highly trained and competent engineers have experienced failures, a basic engineering education should not be a requirement for aircraft development!

An alternate conclusion might be that since experts cannot always get it right with a solid understanding of the underlying theory, less proficient users must strive to understand the workings of the technology as much as possible and validate all assumptions and results to avoid making even bigger mistakes. This must be followed up with the reminder that FEA results, even when completed by experts, are only approximations of a continuous system due to the uncertainty levels described earlier in this book. Consequently, results should always be supported by proper safety factors, vigorous verification testing, and peer review before design decisions are based on them. If an individual does not wish to learn the supporting theory behind FEA as described in Chapter 2, s/he should not shortcut the system by moving to FEA because it is fast and appears easy. Speed without due diligence will not produce better products. It will simply allow bad products to get to market faster.

Good FEA users are not easy to come by. If your company has found someone, either internally or externally, who can and is willing to rise to the challenge after FEA has been justified and proven economically, find a way to keep that person productive and growing. On the other hand, if a suitable candidate or candidates cannot be identified internally, the search should expand externally.

Project Schedule Resources

Another important resource to be evaluated is the availability of time in the project schedule to insert an additional, sometimes lengthy, task. As with CAD in the 1980s, it is fair to expect some delay and redundancy while the integration of simulation ramps up. The trick is to minimize the losses without neglecting the necessary components to ensure success.

It is not uncommon for an analysis implementation to fail simply because the time and place in the project schedule were not properly planned. Chapter 12 addresses emotional commitment and concurrent commitment. These two factors can and will hinder the effectiveness of FEA in a product development cycle. The bottom line is that if the analysis comes too late in the process, the tooling may already be started, the concurrent design of mating parts and assemblies may be too far along, and the crunch of product release activities will take precedence over analysis. The urge to delay analysis until the part is built and tested becomes prevalent. In fact, if the part is far enough along before analysis begins, and comprehensive testing is not cost prohibitive, it may actually be best to wait and see before undertaking the analysis.

The danger of "wait and see" is, of course, the cost of delay if the initial performance estimates or design assumptions were wrong. It is much faster to change the analysis model several times than to make most tooling revisions. It is also more efficient to perform the FEA up front on simplified geometry than to face the task of correlating a failure analysis. These models will require more detail than the initial concept models, and modifications for part improvement will be more involved. In addition, a part that tests successfully may not be the optimal from a cost or performance standpoint. However, cost and inertia may prevent the utilization of simulation to improve it.

Consequently, to get the most out of your investment, resource planning and schedule creation must accommodate up-front simulation on at least the "product musts." Identify and map out time to properly install, train, execute, and verify the simulation process.

Hardware Resources

The next step in the evaluation process is to determine the hardware requirements of the tools under consideration. Discuss the hardware requirements with software vendors and current users. Do not simply ask for the minimum configuration. Ask for a recommended configuration for above average performance. Ask about the performance gains from the major system components. Users may be a better resource for this information. Software vendors have reasons for downplaying the resource requirements of their products. More computing resources raise the total purchase cost of the solution. If one vendor suggests a system configuration which is more costly than a competitor's, less informed buyers might come to the wrong conclusion and buy only based on price. However, because most FEA codes have similar resource needs, hardware pricing on a common platform (Windows NT versus UNIX) should not usually be included in a competitive evaluation.

RAM Requirements

Most codes will benefit from additional random access memory (RAM). Some codes will see a direct relationship between performance and RAM, up to gigabytes of RAM. If your product needs and choices of codes warrant large amounts of RAM, a UNIX workstation may be your only choice due to resource flexibility. On the other hand, some codes will not see an increasing gain with large amounts of RAM; performance may tend to level off between RAM availability of 128 to 256 Mb. This should be discussed with software vendors and confirmed with users.

Hard Disk Requirements

FEA hard disk requirements are significant. There are three primary hard disk requirements: swap space, scratch space, and storage space.

Swap Space

Swap space (virtual memory on Windows NT) is designated by the operating system for rapid, continuous storage. Swap space should be configured to a minimum of twice the available RAM. Swap is used by FEA in much the same way as RAM for volatile storage space. The data placed in swap are never intended to be viewed or saved. Swap space is most effective when it is contiguous and at the front of your hard drive. Defragmentation software can help to improve hard drive efficiency. In addition, swap space can be configured as system or permanent swap, and file system or temporary swap. Permanent swap reserves a portion of your hard drive for swap which cannot be used by any other resource. Temporary swap is space used on an as needed basis. If your disk does not have free space for increased swap usage, temporary swap cannot be created. Typically, if a software requires more swap and cannot find it, it will terminate the application. The software may default to using scratch or free space for storage, but the performance of the system will quickly deteriorate. It is best to identify the maximum swap expected by your particular code and reserve it as swap. Set up swap on your system before loading additional software to ensure that it is contiguous and at the front of the hard drive.

Scratch Space

Scratch space refers to the matrix and database storage requirements of the solution process. Most h-codes provide a disk space usage estimate prior to solving the problem. This is the scratch space required. Scratch space contains data that are deleted automatically or can be deleted when the solution is complete. For instance, the *MSC/NASTRAN Linear Static Analysis User's Guide* for Version 68 provides the following scratch space estimate for a 10,000 node model.

```
10,000 Node Shell Model Scratch = 165 MB
10,000 Node Solid Model Scratch = 2.2 GB
```

While the above estimates are intended to be ballpark estimates, it is interesting to note the significant difference between shell and solid models. The difference is primarily due to the fact that shell models produce sparse matrices compared to dense solid meshes. In a sparse matrix there are many zeros populating the matrix as opposed to significant or nonzero terms. What this means is that a shell mesh of similar size, in terms of nodes, compared to a solid mesh will take better advantage of advances in sparse matrix-solving algorithms despite the fact that a shell node has twice as many degrees of freedom as a solid node. This is another reason to use shells where they are appropriate. It is also important to note that similar mesh size calculations cannot be made reliably for a p-element model. The reason for this is that p-elements use adaptive means to adjust the edge definition in the presence of high stress gradients. Consequently, the final degrees of freedom of a p-element mesh are dependent on the resulting solution. Finally, as mentioned in Chapter 9, it is recommended that you note each model size–in terms of elements for p-element meshes and nodes for h-element meshes–and the corresponding disk usage values in your project notebook in order to make disk space estimates for reviewing similar projects.

Storage Space

Storage space is simply the database and model size of the permanent or final files. The storage space per model is dependent on the number of nodes and elements, the number of result cases for the model, and the number of output quantities for each output set. You can request that the analysis calculate only displacements, which require less space than storing stresses and strains. Storage space is also dependent on solution type. Nonlinear and dynamic runs generate data that are not required for linear static solutions. Model size will also vary from code to code as well as storage format. Most FEA systems will store all geometry, mesh, boundary condition, and results data in a single file. This makes archiving, saving, and sharing data simple. Some codes take the opposite extreme and create a directory structure that must remain intact for project retrieval. This format is more difficult to manage in a multiuser, server based environment because missed files or directories could make it impossible to open results data.

Your storage space requirements will vary based on your company's hard drive policy (local storage versus server), model size, and the number of projects kept active on your system at any given time. Resource estimates should take that last comment into consideration. If you will always be finishing a project and then archiving it immediately, you will need much less disk space than someone who keeps many balls in the air at any given time.

Total hard disk requirements then are the sum of the swap, scratch, and storage space requirements. Because you are performing an estimate, consider increasing the final number by 20 to 30% to ensure sufficient space. Few things in FEA are more frustrating than scrambling for disk space before or during a run.

Graphics Cards

Another resource consideration is the graphics card or device installed in your system. A graphics card that works for most solid CAD systems should be acceptable for FEA. This, again, is software dependent. Remember, however, that hesitant dynamic rotation, lengthy screen repaints, and slow animation are no less annoying and productivity draining in FEA than in CAD.

System Data Bus

The choice of internal system bus comes into play with Windows NT more than UNIX. Most UNIX based engineering workstations are SCSI (*small computer system interface*) by default. This bus is extremely efficient for data transfer to and from the hard drive. Windows NT stations are usually offered with an IDE (*integrated drive electronics*) bus in lieu of a SCSI bus. FEA is very disk intensive so inefficient data transfer becomes a performance issue. While the difference may not be noticeable in CAD, you can expect a 30 to 50% solution speed gain using SCSI versus IDE for the 30 to 50% price difference between the systems.

Hardware Evaluation Summary

The only answer to the question, "What computer should I use for FEA?" is "Biggest, best, fastest...within budget." It is unwise to scrimp on hardware if simulation is going to be a key part of the product design process. A good way to justify additional hardware is to estimate time saved by less down time and simply multiply that by a burdened engineer's hourly cost. There are many other benefits to faster hardware such as increased overall usage and faster design data availability, but these are not as easy to quantify.

When evaluating hardware resources, keep in mind that, while improving rapidly, FEA is still very resource intensive. Tremendous gains in efficiency can be realized by supplying sufficient computing power, that is, processor, RAM, and disk space. If the task of simulation is initially perceived as too lengthy or too difficult to get real problems to run on existing hardware, FEA may never achieve its expected goals. Your computing needs will vary considerably with respect to solution type, software choice, and product needs.

Evaluating FEA Suppliers

Upon reaching an understanding of your needs and resources, the offerings of the industry or FEA marketplace should be evaluated. Inadequate or nonexistent understanding of corporate goals and of the capabilities and limitations of the technology leads many companies to purchase tools based on flashy demos, pretty pictures, or bad advice from vendors. Unlike geometry in CAD, an analysis which looks correct may not necessarily be so. A good example of this is the aforementioned recent proliferation of CAD embedded, tetrahedral meshing solutions from some of the well-known names in the industry. Some engineering managers have been talked into purchasing these systems to analyze sheet metal assemblies because "it looked easy" and "the answers looked right." Unfortunately, these companies may never adequately model their systems with the tools they chose.

Refer to Table 20.1 to review the minimal capabilities based on the product requirements that your chosen tool should possess, and review Chapters 4 through 7 if the reasoning behind this statement is unclear. Next, evaluate suppliers' market share, installed base, and reputation in the market. There is a difference in quality and accuracy between codes and there is a reason why the manufacturing industries with the most experience and the most to lose in terms of human lives due to inadequate analyses (e.g., the automotive and aerospace industries) choose industry leading tools and avoid the highly marketed low-end systems. Chapter 19 provides an introductory overview of some of the more popular offerings in the industry. Use this as a baseline for your evaluation.

Evaluating FEA Systems

Chapter 19 and Table 20.1 should get you started in evaluating software tools. There are many other tool choices out there which, for various reasons, were not included in this book. It is a fair statement to say that a linear static analysis should be reasonably accurate in most publicly marketed codes. These tools will tend to differ from higher end tools in speed as well as in forgiveness for less than perfect elements as well.

CAD Compatibility

The inherent assumption behind equating linear static solutions in the diverse FEA tools is that you can import geometry or create the required analytic geometry with relative ease, and then obtain a good mesh using that geometry. The embedded or integrated systems described in Chapter 19 have made the problem of geometry transfer less of an issue. It is still a major concern with open or proprietary systems. However, do not be fooled into thinking an embedded or integrated system has removed all geometry transfer problems. First, if the best element type for your problem is a shell or beam, an embedded system that supports only solids, regardless of how seamless it seems to bring CAD data across, will not be a viable solution. Moreover, even in the solids world of embedded and integrated systems, a given geometry may still not mesh.

When considering any analysis preprocessor—embedded, integrated, open, or proprietary—ask the vendor to show you what happens when a part does not immediately automesh. All demo and training parts seem to mesh flawlessly. Certain real world, critical path parts will not mesh flawlessly. Do not let the vendor put the responsibility on you to provide a part that will not automesh, because you may not find one. The only exception to this is if the vendor responds with the worst of all scenarios–that there is no recourse after the failed mesh and the offending regions are not identified. Although the meshing in most modern codes is fairly reliable, there would be no reason to assume that this is false. If a graceful and productive recovery from a failed automesh is not provided, your project may grind to a halt.

The current releases of Cosmos/Works, embedded in SolidWorks, and Pro/MESH, integrated with Pro/ENGINEER, both fail a mesh with a useless statement such as, "*Unable To Proceed With Mesh.*" The open system, FEMAP, will complete the mesh on all surfaces it can. It allows the user to manually complete the surface mesh and fill with solid elements. Another option is that the offending surface can be easily identified by examining the areas that were not meshed, and the original geometry can then be adjusted. Pro/MECHANICA, a proprietary system, highlights the curves or surfaces preventing the automesh from completing. While it does not allow the user to hand stitch a few surface elements and complete the solid mesh, geometry tweaking in the Pro/

MECHANICA interface or the original CAD is usually straightforward when you know where to look. If completing your model is dependent on an automesher, you can experience significant delays if the mesher does not provide tools for dealing with less than perfect geometry.

Finite Element Modeling: The Preprocessor

From a finite element modeling standpoint, consider the modeling needs of your product as outlined in Table 20.1. If you expect to frequently use beam elements, most of the beam capabilities described in Chapters 4, 6, and 7 should be available to you. You should be able to automesh a freestanding wireframe model and the postprocessor should provide bending, shear, and moment diagrams in addition to stress and displacement. These modeling requirements hold true for shells as well.

Solution Accuracy

Solution accuracy is extremely important but it is not easy to judge. At a minimum, the vendor should be able to provide a verification manual using known industry benchmarks, such as those provided by NAFEMS. Your vendor should also allow you to repeat a problem that you deem most relevant to your particular product needs. Because these are short problems that solve quickly, there should be no objections to the time involved. If your vendor cannot or will not provide these problems, this should indicate a potential difficulty, at least from a willingness to cooperate standpoint. The aforementioned problems can be obtained directly from NAFEMS at the address listed in the bibliography at the end of the book.

Support and Training

Place a strong emphasis on support and training. Most design engineers do not have the time or patience to pore over manuals and hotline support typically does not provide the application-specific help they need. Software developers you choose to work with must have credible expertise in the training and support of their systems and have developed relationships with experts in the implementation and application of the technology to ensure that your growth will be supported. They must be genuinely interested in partnering with your organization to guarantee success, even if your purchase is small. These are important issues to new users. Ask your suppliers to discuss or demonstrate how they will support you.

Key Success Factor: Implementation

The next success factor to explore is the implementation of the choices made during the initial evaluation. Remember that after the initial evaluation, the evaluation process must continue throughout the life of the implementation as your needs will change, available resources will change, and the simulation industry changes rapidly. If you have properly evaluated your needs, resources, and the market, adjusting your internal processes to take advantage of simulation should be straightforward. Steps taken to implement the technology will play a key role in your downstream flexibility. The three components of a successful implementation are (1) your *plan*, (2) the *training* program, and (3) your *choice of initial and ongoing projects*.

Implementation Plan

Develop a plan. Having a documented plan for evaluation, implementation, verification, and growth is critical. The plan need not be complex, but should contain the following components.

- Goals or business benefits of the technology. Set your expectations because without expectations, your success cannot be evaluated.

- Periodic reevaluation. Do not leave it to chance; schedule a periodic, semi-formal review of your current implementation.

- Process for initiating an analysis or study. Define the optimal interaction between users and others who must work with the data; define priorities and formats for information exchange.

- Develop documentation procedures. Most companies have rigid guidelines for documenting test results, but few have guidelines for analysis documentation.

- Build time for verification into the plan. Failure to correlate analysis to test data and to review methods may lead to inefficient studies, or worse, potentially disastrous inaccuracies.

- Ongoing training and education. Basic software training is not sufficient. Advanced skills training, as well as improved engineering knowledge, such as in structural mechanics, vibration, and material behavior, is important to prevent stagnation and the propagation of bad habits.

The plan should be documented and made available to all that use or interact with the technology. It should be a dynamic plan and open to challenge as you seek for continuous process improvement.

Training

Another key focus of the implementation should be training, or the technical growth of the users. Some companies support periodic brainstorming among FEA users to encourage sharing of tips and techniques. User groups are excellent ways to interact with others and exchange knowledge. Many companies engage experts from outside firms to provide coaching or mentoring for their users. In addition, ongoing software training should be mandatory. Do not just rely on training provided by software vendors. Such training is valuable for learning the interface but may fall short on practical application topics.

Internal Training and Coaching Programs

If your company has a mix of experienced and newer analysts, management should help structure a formal mentor program. For this to be effective, time must be allotted for interaction between mentor and mentored and the reporting structure should accommodate the likelihood of cross-product line or group pairing. In addition, a formal training workbook should be compiled to include examples of projects completed previously and the acknowledged best practices found for modeling methods that are specific to the company's products. A program to teach or refresh the fundamental topics covered in Chapter 2 should precede all other training. Utilizing the Internet or an intranet can allow dynamic data sharing among the various FEA users. Finally, the instructors should be encouraged to provide coaching. It is not easy to be a teacher and it requires great patience. There is also the concern that an experienced person may feel like he or she is giving away some of their value to the company. An incentive program, offering corporate recognition or bonuses, may somewhat alleviate that feeling.

External Training and Coaching

Training admonitions in the previous two sections sound well and good for larger companies, but the fact of the matter is that most companies only have one or two FEA users. Because the one or two users typically have other project responsibilities, rarely is anyone qualified to serve as a mentor. In these cases, it is suggested that an external "coach" be identified and retained. Moreover, the users should be encouraged to attend user groups and other educational forums to expand their awareness of the technology and the chosen code. Many newer users feel better about their challenge after learning that they are not the only ones experiencing these challenges and they gain the benefits of finding peers to talk to.

Choosing a Coach

Finding an outside resource for coaching and project support can be a difficult undertaking. With the cost of entry into FEA dropping, many design and CAD companies are picking up a copy of an FEA code and listing the capability on their business card. Ownership of an FEA license does not in any way imply expertise. In fact, "knowing the code" or being fluent in the terminology cannot guarantee that this organization or individual is reliable and experienced.

As you are learning the technology, remember the following phrase: "Someone who knows a little appears to know everything to someone who knows nothing." Realistically, it will be very difficult for you to judge the competency of a resource if you are not experienced in FEA. A few simple guidelines are listed below.

- Take descriptions of past projects with a grain of salt. Color pictures and glowing descriptions of solved problems may be no more than that: pictures and words. If the showcase projects, however, suggest some glaring errors in modeling or other assumptions, that you, with your limited knowledge of the code, can spot, walk away. The information presented in this book should prepare you to spot some of the problems.

- Remember that a reference client may not know any more than you do and may not be in a position to critique an outside resource's work. They most likely did not correlate the results to test data either. Try to get a feel for the experience level of the reference, although this may be hard to do. Next, remember that the reference and the outside resource in question may not realize their limitations, so do not assume that anyone is presenting you with fraudulent claims. The bottom line is that it is difficult to learn the technology well and due to the wealth of information available, it is hard to know what you do not know.

- Listen carefully to references' descriptions of the resource's timeliness, cooperativeness, and willingness to take the problem to completion. Any good engineer can spot a resource who cannot or is not willing to solve a problem beyond the FEA portion of the job.

- Years of software usage and number of college degrees are dubious indicators of competence. Fourteen years of doing things the wrong way is not a great credential. Similarly, with all due respect, even though many of the brightest and most capable people involved in FEA have a Ph.D., an advanced degree only proves that someone had the money, patience, and persistence to earn it. It is what they have done with such additional knowledge that is important.

Look for a resource that has shown a commitment to education. Active membership in user groups with a leadership role is a good indicator of an individual's commitment to the technology. Presentation of open seminars and publication of articles puts this person's knowledge of FEA "out there" for public criticism. If a person is a respected speaker at user groups and educational seminars, others at her/his level and above will be in a position to evaluate his/her knowledge. At a recent software user group meeting, an FEA consultant was presenting a vibration analysis he performed on a highly flexible plastic tray. He used quarter symmetry. When questioned about the use of symmetry in a vibration analysis and the failure to account for nonlinearity, this speaker made it clear by his lack of response that he was not even aware of these concerns. The entire group questioned his credibility because he had billed himself as a consultant and an expert, but had "missed it" on such basic issues. As a consequence, this FEA consultant has not been seen at this particular user group since.

Do not let the previous account scare you away from presenting a project at a user group or conference. Regardless of your level of expertise, you should approach a presentation as an opportunity to learn from your peers. Remember the statement made earlier, "One must separate what one knows from what one assumes." If you position your presentation as something you know and that you have "all the answers," be prepared to be challenged. If you position your presentation as a set of assumptions that may or may not be the best ones, you can expect to receive valuable advice and possibly some creative input from peers. Worthwhile resources should not be afraid to share their knowledge publicly and should expect to continue learning. These resources can be counted on to share their assumptions with you and open themselves up for critique. This will ensure that the final results are the best combination of your product knowledge and their FEA knowledge.

After All That. . .

After presenting these guidelines for choosing a coach and an outside resource, it is important to note that it may still come down to hiring a resource and evaluating him/her on an actual project. Use the knowledge gained from this book to evaluate the results of the individual's efforts. Another option, which is not widely used, is to get a second opinion. No one thinks twice about a second opinion in terms of a health or car problem that is not clearly understood, yet companies hesitate to question their analysis support. Asking a second resource to critique the results of the first maximizes the chances that the data presented are of the best quality. If there are discrepancies, ask the first to resolve these issues. When a resource has been found that produces reliable information and is genuinely interested in helping your company understand the technology, hang on to him or her.

As mentioned previously, "Good users will readily admit they need to know more. Those who feel satisfied are probably in trouble." Surprisingly, few design analysts model, analyze, and interpret results efficiently or even accurately. However, most do not realize it unless a major mistake costs their company money. Do not wait—if the value of introducing the technology existed in the first place, the value of ongoing training should be self-evident.

Initial Project

The choice of initial projects can go a long way toward ensuring the success of the FEA implementation. The initial projects should be chosen to allow for easy verification of results. Such projects should consist of one or two parts with well-known and well-behaved performance. This choice serves two purposes. First, it builds confidence in the modeling techniques and the technology in general. Second, it provides an opportunity to show the rest of the company the potential of simulation. If the part behaves as all who know the product expect it to, then buy-in from other members of the design team can be had more quickly.

Management Buy-in

Management and/or corporate buy-in is crucial to growth of the program. If the initial project was chosen correctly and the results were deemed successful, advertise these successes internally; in other words, promote the achievements. However, if the initial projects are judged unsuccessful due to insufficient solution power, modeling capabilities, or results interpretation, a widespread simulation program may never gain corporate acceptance. Lack of success can usually be attributed to insufficient training and the choice of an initial project that is beyond the capabilities of a new user. The correct choice of initial projects is key.

Key Success Factor: Verification

Lack of verification and correlation causes simulation to remain a small part of the overall design process at most companies. The verification process should address both the economic and technological goals set in the initial needs evaluation.

Economic Verification

Key to the economic verification process is the confirmation of the business benefits expected, as outlined earlier in this chapter. When these expectations are met or exceeded, growth of the program is easily justified. A solid understanding of the business benefit is also important when verifying your return on investment. The price tag of analysis resources is typically higher than CAD and will require more justification.

Technical Verification

Verifying the technical accuracy of the results should be more straight-forward. Correlation is a goal all analysts strive for. However, dead-on accuracy and correlation may not be as important as qualifying the results. "Qualifying results" basically means quantifying the error and uncertainty introduced as a result of modeling and boundary condition assumptions in terms of the study goal. In many situations, loads, geometry, and properties are not well understood, and the stress or strain in an actual test unit cannot be measured. In these cases, accuracy becomes rather academic and the quality, or applicability, of results becomes the important measure of success.

A Practical Verification Process

The needs of the design analyst are different than those of the traditional analyst. Design analysts are frequently called on to quickly make product decisions based on incomplete data to keep project cycles on track. Consequently, sound engineering judgment plays a key role in interpreting and qualifying test, field, inspection, and QA data. Simulation data fall into the same category. To be able to qualify analysis data accurately and quickly, follow the four steps below to be used before, during, and after the simulation.

- *Develop an understanding of the full system, including design tolerances, manufacturing tolerances, and assembly methods.* Use manual calculations to quantify rough expectations of stress and displacement. If historical product data or previous analyses are available, review them and form expectations of your results based on what you learn. Conduct a design review with all players in the design and manufacturing process to debug your assumptions. It does not take an analyst to suggest that there may be variability in the properties or assembly methods. While this step exposes the analyst to criticism, it is the best way to grow and the only way to ensure that all possible scenarios are accounted for in the simulation. This also helps promote that crucial corporate buy in.

- *Perform a sanity check on methods and assumptions after the analysis.* If possible, reconvene the design review team mentioned previously to discuss the results in terms of the assumptions and expectations. Use common sense. If the results do not match expectations, discipline yourself to resolve these discrepancies. You may learn that long-held assumptions about system behavior were not correct and that the simulation enlightened everyone as to the actual situation. At this stage, the experience of an expert—internal if available, external if needed—should be engaged to bulletproof the methods used and results calculated.

* *Assemble a testing program for assumption verification and data correlation which should be performed by or observed by the analyst.* Keeping the analysis and testing functions linked minimizes errors in communication and unnecessary assumptions. Isolate assumptions or variables using test or prove-out models. These should be simple configurations that are easy to reproduce in testing. *Test what you analyze; analyze what you test.* Always resolve any differences between test results, analysis results, and field test data.

* *Document your assumptions, methods, results, and verification steps.* Develop your documentation so that a project outsider can quickly qualify the assumptions and results of your simulation and determine the applicability to his or her needs. In a particular company's experience, the only person who performed analysis left abruptly. When those left behind tried to pick up the pieces, they found that there was little or no documentation and little reason to believe that any work done previously was based on sound assumptions or had any bearing on actual system behavior. Consequently, management at this particular company had serious doubts about continuing a simulation program because they could not verify that they ever achieved any business benefits. Build documentation into your implementation plan, standardize it, and utilize it to summarize your data and reinforce the verification process.

Summary

Simulation does work. The technology has been proven to save time and money many times over. However, FEA success means more than a just creating a mesh! Proper integration of simulation into the design process takes planning and will require a change in the way projects are structured. Set goals in terms of the desired business benefits. Keep these in mind throughout the evaluation, implementation, and verification stages and let them guide your decisions. Take the time to understand your needs and the limitations of available tools. Develop a plan and stick to it. Finally, verify both the business benefits of simulation and the quality of your results. Adhering to these general guidelines and developing a strong implementation program will minimize your chances of error and maximize opportunity. As clichéd as it sounds, do not try to cheat on the technology because, in the end, you are only cheating yourself.

One of the most important points to learn from this discussion is that a full engineering understanding of the system being studied is critical to the success of simulation or analysis. Choose to be skeptical, even cynical, about your results and your capabilities. Use simple models to prove out your assumptions and qualify the data in terms of your eventual goals. Only after initial models and a thorough review of the overall process should you attempt to model a complex assembly or behavior. Do not hesitate to ask for help and always be prepared to justify your assumptions and methods.

Popular trends in the industry have served to downplay the importance of thoughtful modeling, multiple simulation options, and careful results interpretation. Companies entering into design simulation are best served by researching the technology and the use of it at cutting-edge corporations. Long-time analysts in major industries, such as automotive, aerospace, and defense, who equate accuracy with consumer safety, do not pay a premium for full-featured interfaces and industry leading software because they have money to burn. There is a difference in accuracy for advanced solutions and modeling techniques between the various software packages. In most cases, software which seems to require little or no training will be so limiting that good answers are truly hard to come by. Choosing a full-featured interface and finding a trusted resource to guide you is the best way to get started. Above all else, be patient and use common sense. Design engineers are trained to find solutions to tough problems and the growing role of the design analyst is the logical progression of these skills. Use it with the same checks and balances with which other tools are employed and you will be successful.

21

Trends and Predictions for the Future of FEA

This final chapter makes a few predictions on the future of FEA by reviewing the current trends in simulation technology–a technology moving at such a fast rate that it has become increasingly difficult to even keep track of the advances made from day to day. This pace can be both exciting and intimidating to the user, regardless of his or her degree of experience. Therefore, by highlighting the major directions that the technology is taking, this chapter's intent is to provide information that will allow you make some sense of all these changes.

Faster Computing Speed and Algorithms

The trend that should be most apparent to the user is the seemingly exponential rate of increased computational speed, which is often provided at a declining cost. The current state of computer speed and affordability has given a much larger audience access to FEA–a technology formerly reserved for high end users tied to high end workstations.

The speed of today's desktop computers rivals the supercomputers of yesteryear. In addition, the increased efficiency of today's FEA algorithms is a big contributor to this cause. You should expect to see more of the same in the future; laptop analysis has already become a common sight. The fact that design analysts always seem to keep bumping into the limits of available speed is perhaps the most surprising element in this scenario. It appears that no matter how much more powerful computers/solvers become, the desired analyses simply get more complex accordingly, and the user keeps asking for more speed. Will this cycle ever stop? Not likely–at least not in the near future. Hence, what you should really expect is to be able to complete today's analyses much more quickly in the future, but future analyses (read more complex) will run in about the same time as today's.

Self-adapting, Self-converging Technologies

Another very time-consuming task in the FEA process is the need to refine or otherwise modify a mesh, as made necessary by the review of results, in order to improve accuracy through manual iteration. Many codes have been perfecting the automation of this process through self-adapting, self-converging technologies, whereby the mesh may locally adapt itself in regions of high strain in order to converge to a solution of an acceptable error estimate. P-element codes achieve this by being able to increase the resolution of each element through an increase in the polynomial definition of the element's edges. Other codes, however, are only starting to provide these tools. Notably, there are now h-element codes that, based on user-defined error estimates, adjust mesh density calculations and initiate a remesh. Although in their current state these h-element technologies are not always successful, it is only a matter of time before they are perfected and gain the trust that is now bestowed on their p-element counterparts.

Multiphysics, Multisolution Technologies

Beyond FEA programs that are available strictly for structural analysis, there are other codes that allow the simulation of different phenomena, such as complex thermal states, electromagnetic conditions, fluid flow, acoustics, kinematic/kinetic motion, and so on. In their present state, these codes generally are exclusive as well, that is, they tackle only one of these phenomena and provide a single solution. Yet companies such as ANSYS are beginning to provide multiphysics, multisolution simulation technologies. This trend is a logical extension of basic FEA. Because most structural components or systems of components are in reality subject to many physical phenomena, the more of these you can include, the fewer assumptions you need to make, the more accurate the solution should be. Of course, computational speed will have to support these analyses, which will most likely be very resource intensive. Yet, as discussed above, this is a definite possibility in the near future.

CAD-embedded Systems

This trend is probably the most aggressive at the present time. Affordable supercomputing tied to efficient solvers is opening the FEA doors wide to a great number of design analysts. Hence, the requirement that these tools be extremely user friendly logically leads to their complete transparency. What better place to include these tools for the designer-turned-analyst than within the CAD program itself? There is quite a bit of controversy within the analysis community with regards to the safety of allowing inexperienced users access to such hidden power. However, because of high demand, this trend will be hard to stop. Hence, the hope is that a sophisticated system of checks is developed in the future and embedded in the system to prevent misuse.

Other Advancing Technologies

In addition to the main trends outlined above, others are just emerging, but are expected to take off quickly and soon.

Internet Based Support, Training, and Solvers

With the Internet becoming a mainstay in today's society, look for its possible and powerful impact on simulation technologies. A vast amount of information is already available on the Web regarding the capabilities of various codes. Virtually every simulation technology company has a very detailed Web site that provides easy access to this information and allows you to get in contact with a local representative. User groups, some of which are not code specific, have formed extensive e-mail connections to provide forums for questions, discussion, and sharing of techniques. Many companies have made their technical support available through the Internet, and some have even started remote technical training using this medium. There are signs that companies might provide solver services in the near future, whereby models are submitted across the Internet to a remote supercomputer, and the results electronically sent back at the conclusion of the run. This type of service would make the technology even more accessible by making the initial investment in powerful hardware a nonissue.

Windows NT/UNIX Convergence

Access to FEA technology required a high end UNIX workstation in the recent past. Because of its long association with sophisticated, resource intensive, scientific work, the UNIX operating system has become very stable. However, in the last few years, the advent of affordable super-computing was mainly spearheaded by the robust personal computer market running on a Windows NT operating system. Windows NT is relatively new and, hence, not as stable. The competition has thus been set along the lines of stability versus affordability. Thus far, it seems that affordability has had the upper hand, with a recent explosion of Windows NT machines flooding the marketplace. Look for this to change in the future. It is expected that as UNIX workstations become more competitively priced and Windows NT becomes more stable, crossbreed machines will arise that will be able to run both systems, thereby giving the user the choice of the most efficient solution.

Automation Technologies for Results Verification

The growth of rapid prototyping is curiously tied in a historical timeline to the growth of simulation technology. The past few years have seen the explosion of both industries, as they have virtually become required and expected players in the rapid product development process. Currently, many designs are first analyzed for structural integrity and then rapid prototyped to confirm form and fit. A logical extension of this process is to actually use the rapid prototyping to produce analysis derived, deformed versions of these components to further refine the form and fit verification process. Other automation technologies may actually be used for analysis results correlation. An example of such a technology is known as laser scanning. This is commonly used to verify that a produced part matches the intended geometry by overlaying a 3D laser scan of the physical component on its CAD solid model and investigating discrepancies that are often displayed in an accuracy fringe plot. This same process could possibly be used to correlate an FEA by scanning a component in its deformed shape under its real life application, and overlaying this on its analysis derived, deformed shape equivalent. Expect this and other automation technologies to enter the simulation technology world.

More Accessible Use of Manufacturing Technologies

Some manufacturing methods are still viewed as rather magical. Castings, forgings, stampings, and other techniques have historically been based on empirically derived formulas and tables, backed by almost artistically regarded, trial-and-error processes used to obtain the desired results. Simulation technologies are starting to revolutionize these fields. For example, mold flow analysis packages are the hot ticket item these days. Although most of these codes are being developed as proprietary solutions, expect them to join the mainstream of complete design solutions in the future.

Sophisticated Bulk Calculations on Results

As mentioned in Chapter 11, some FEA packages currently allow the manipulation of results data files to conduct specific calculations for evaluating such extremely relevant quantities as the fatigue life of the structure. Expect these types of calculations to become more automated in the future. Current trends show that affordable software for carrying out fatigue life estimates should be just around the corner.

Improved Optimization Algorithms

The ultimate strength of FEA lies in its ability to provide results essential to the optimization of a component. Some codes have already automated this optimization process as an extension of their overall capabilities. You should expect to see more codes following this path and, in turn, expect improved optimization algorithms that will make this type of study even more efficient.

Summary

Simulation technology has come a very long way in the past few years. FEA was considered a job that only a specialist could conduct not long ago, but is now being placed at the fingertips of engineers more commonly associated with the design process. Hence, the birth of the *design analyst*. This transformation of the technology may be attributed to many factors. Faster solution algorithms and more user friendly interfaces have surely conspired to make this transition possible. However, had affordable power not been made available by the quantum leaps in PC technology, these code enhancements would probably not have been enough. The amazing fact is that all of this is still going strong. Computers are getting faster and cheaper at an incredible rate. UNIX and Windows NT systems look to be converging into very robust systems in the near future. Virtually every software release of a simulation code increases solution speed, adds functionality, and becomes more intuitive. This is expected to continue and, although this is exciting news for everyone involved in the engineering design process, due care must be taken that the technology is not abused. As with any other tool, the solution can only be as good as the quality of the input.

To further automate the FEA process, self-adaptive, self-converging algorithms will also become more pervasive. These will remove some of the error caused by a lack of diligence in the analysis process, but are also likely to take away the artistry and intuition acquired through having to become intimately familiar with the mesh. Speaking of meshes, you should also expect to see more and more systems that are CAD-embedded and may actually hide the mesh completely from the user. Be very careful when utilizing this type of code; it is very rare that the CAD model will be appropriate for analysis purposes and may thus require modification. Hopefully, a system of checks will also be written into these CAD-embedded codes in the near future to catch some of the more basic modeling and setup mistakes.

For those who are weary of making assumptions, multiphysics, multisolution technologies should eventually become available to simultaneously analyze everything under the sun that may be affecting your product. Of course, you will probably not want to model the sun itself, so a limited set of assumptions will always be required. The Internet is likely to become a bigger part of the FEA picture as well. Look for more user participation in electronic forums, increased presence of technical support and training, and even the availability of solvers across the wire for you to kick your analysis runs off without tying up your own resources. More accessible manufacturing simulation technologies loom on the horizon. Automation techniques appear to be promising correlation tools for analysis. In addition, to bring simulation technology to its ultimate expression, improved optimization algorithms will likely become staples of all codes.

It is an exciting time to be in the field of simulation technologies. The pace is moving at an astonishing rate, and the possibilities seem endless. As with any other powerful tool of this magnitude, you must be extremely careful with its use. You should become very familiar with the concepts presented in this book. Review the fundamentals of structural analysis. Consult your software documentation often when you have questions regarding usage of a new element or technique. Build simple test models to help you understand the implications of each of your assumptions. Use the technical support provided by your software vendor and for which you are probably paying through a maintenance contract. Become involved by joining local user groups and attending conferences in your area. Subscribe to technical publications that cater to the simulation technology's audience. If you become active in this field, your success will only be limited by your creativity. Remember to have fun!

Bibliography

Adams, V. "The Fundamentals of Structural Analysis." Schaumburg, Illinois: WyzeTek, Inc., 1997.

_____. "MSC/NASTRAN for Windows Introductory Training Guide v3.0." Schaumburg, Illinois: WyzeTek, Inc., 1998.

_____. "MSC/NASTRAN for Windows Introduction to Nonlinear Analysis v3.0." Schaumburg, Illinois: WyzeTek, Inc., 1997.

_____. "MSC/NASTRAN for Windows Introduction to Dynamic Analysis v3.0." Schaumburg, Illinois: WyzeTek, Inc., 1998.

_____. "Using Boundary Conditions in Finite Element Analysis." Schaumburg, Illinois: WyzeTek, Inc., 1997.

Baguley, D. and D.R. Hose. *NAFEMS: How to Understand Finite Element Jargon.* Glasgow, Scotland: NAFEMS, 1994. (For information on NAFEMS, contact NAFEMS, Scottish Enterprise Technology Park, Whitworth Building, East Kilbride, Scotland G75 0QD, www.nafems.org.)

Blakely, K. *MSC/NASTRAN Basic Dynamic Analysis User's Guide, v68.* Los Angeles, California: The MacNeal-Schwendler Corporation, 1993.

Boyle, J.T. "Accurate Composites Analysis." *Computer-Aided Engineering* (June 1996), 11.

Brauer, J.R. *What Every Engineer Should Know About Finite Element Analysis.* New York: Marcel Dekker, Inc., 1988.

Buchanan, G.R. *Finite Element Analysis.* New York: McGraw-Hill, Inc., 1995.

Burton, T.D. *Introduction to Dynamic Systems Analysis.* New York: McGraw-Hill, Inc., 1994.

Caffrey, J.P. and J.M. Lee. *MSC/NASTRAN Linear Static Analysis User's Guide, v68.* Los Angeles, California: The MacNeal-Schwendler Corporation, 1994.

Collins, J.A. *Failure of Materials in Mechanical Design.* New York: John Wiley & Sons Inc., 1993.

Deutschman, A.D., W.J. Michels, and C.E. Wilson. *Machine Design.* New York: Macmillan Publishing Co., Inc., 1975.

Dvorak, P. "A Finite-Element Analyst's Guide to Instructing Designers." *MACHINE DESIGN* (March 10, 1998), 224-26.

_____. "Test Your Foundation in Finite Elements." *MACHINE DESIGN* (February 5, 1998), 64-69.

_____. "What to Look For In FEA Programs For Designers," *MACHINE DESIGN* (May 7, 1998), 62-64.

_____. "What You Should Know About FEA." *MACHINE DESIGN* (November 6, 1997), 55-58.

Florman, Samuel C. *The Civilized Engineer.* New York: St. Martin's Press, 1987.

Hinton, E. *NAFEMS Introduction to Nonlinear Finite Element Analysis.* Glasgow, Scotland: NAFEMS, 1992. (For information on NAFEMS, contact NAFEMS, Scottish Enterprise Technology Park, Whitworth Building, East Kilbride, Scotland G75 0QD, www.nafems.org.)

Horton, H.L., F.D. Jones, E. Oberg, and H.H. Ryffel. *Machinery's Handbook.* New York: Industrial Press, Inc., 1988.

Jackson, R. *Providing Competitive Advantage.* Atlanta, Georgia: The MacNeal-Schwendler Corporation, 1992.

Khol, R. "Why You Don't Have to Understand (Much) Engineering Theory." *MACHINE DESIGN* (July 9, 1998), 6.

Lee, J.M. "Common Questions and Answers." New York: The MacNeal-Schwendler Corporation, 1992.

Lee, S.H. "Handbook For Nonlinear Analysis, v67." Los Angeles, California: The MacNeal-Schwendler Corporation, 1992.

The MacNeal-Schwendler Corporation. "Introduction to MSC/NASTRAN." Los Angeles: The MacNeal-Schwendler Corporation, 1994.

_____. *MSC/NASTRAN Encyclopedia Versions 68 and 69* (CD-ROM). Los Angeles, California: The MacNeal-Schwendler Corporation, 1993-1998.

Meriam, J.L. and L.G. Kraige. *Engineering Mechanics Volume 1: Statics.* 2nd ed. John Wiley & Sons, Inc., 1986.

_____. *Engineering Mechanics Volume 2: Dynamics.* 2nd ed. John Wiley & Sons, Inc., 1987.

NAFEMS. *A Finite Element Primer.* Glasgow, Scotland: NAFEMS, 1992. (For information on NAFEMS, contact NAFEMS; Scottish Enterprise Technology Park; Whitworth Building; East Kilbride, Scotland G75 0QD; www.nafems.org.)

Nimmer, R. and G. Trantina. *Structural Analysis of Thermoplastic Components.* New York: McGraw-Hill, Inc., 1994.

Parametric Technology Corporation. *Pro/MECHANICA Release 18 Documentation Set.* Waltham, Massachusetts: Parametric Technology Corporation, 1997.

Penton Publishing. *Machine Design: Basics of Finite Element Analysis.* Cleveland, Ohio: Penton Publishing, 1993.

Popov, E.P. *Introduction to Mechanics of Solids.* Englewood Cliffs, New Jersey: Prentice Hall, Inc., 1968.

Raftoyiannis, J. and C.C. Spyrakos. *Finite Element Analysis.* Pittsburgh, Pennsylvania: Algor, Inc., 1997.

Russell, R. "Don't Trust the Pretty Pictures." *MACHINE DESIGN* (May 23, 1996), 68-84.

Shigley, J.E. and C.R. Mischke. *Mechanical Engineering Design.* 5th ed. McGraw-Hill, Inc., 1989.

Steidel, R.F. *An Introduction to Mechanical Vibrations.* 3rd ed. John Wiley & Sons, Inc., 1989.

Timoshenko, S.P. *History of Strength of Materials.* New York: Dover Publications, Inc, 1983.

Young, W.C. *Roark's Formulas for Stress and Strain.* New York: McGraw-Hill, Inc., 1989.

Index

More OnWord Press Titles

INSIDE CATIA
$89.95   Includes CD-ROM

CATIA Reference Guide
$52.95

INSIDE the New Pro/ENGINEER Solutions
$54.95   Includes Disk

Pro/ENGINEER Tips and Techniques
$59.95

INSIDE Pro/SURFACE
$90.00

Automating Design in Pro/ENGINEER with Pro/PROGRAM
$64.95   Includes CD-ROM

INSIDE SolidWorks
$56.95   Includes CD-ROM

SolidWorks for AutoCAD Users
$59.95

INSIDE TriSpectives Technical
$52.95   Includes CD-ROM

OnWord Press

OnWord Press books are available worldwide from OnWord Press and your local bookseller. For order information, terms, or listings of local booksellers carrying OnWord Press books, call toll-free 1-800-4-ONWORD (1-800-466-9673) or 505-474-5130; fax 505-474-5030; write to OnWord Press, 2530 Camino Entrada, Santa Fe, New Mexico 87505-4835, USA, or e-mail orders@hmp.com. OnWord Press is a division of High Mountain Press.

Comments and Corrections

Your comments can help us make better products. If you find an error, or have a comment or a query for the authors, please write to us at the address below, send e-mail to cleyba@hmp.com, or call us at 1-800-223-6397.

OnWord Press, 2530 Camino Entrada, Santa Fe, NM 87505-4835 USA

Visit us on the Web at http://www.onwordpress.com